Isoquinoline Alkaloids Research

1972-1977

W0235285

Isoquinoline Alkaloids Research

1972–1977

Maurice Shamma
and Jerome L. Moniot

The Pennsylvania State University
University Park, Pennsylvania

Plenum Press · New York and London

Library of Congress Cataloging in Publication Data

Shamma, Maurice, 1926-
  Isoquinoline alkaloids research, 1972-1977.

  Includes bibliographical references and index.
  1. Alkaloids. 2. Isoquinoline. I. Moniot, Jerome L., joint author. II. Title.
QD421.S543   547 .72   77-26929

ISBN-13: 978-1-4615-8821-4   e-ISBN-13: 978-1-4615-8819-1
DOI: 10.1007/978-1-4615-8819-1

©1978 Plenum Press, New York

Softcover reprint of the hardcover 1st edition 1978

A Division of Plenum Publishing Corporation
227 West 17th Street, New York, N.Y. 10011

The authors who set out to say something that no one has said before are to be regarded with mistrust.

Max J. Friedländer

PREFACE

Substantial advances in the realm of isoquinoline alkaloids have occurred since *The Isoquinoline Alkaloids, Chemistry and Pharmacology*, was published in 1972. The present volume represents an effort to describe important developments since that time.

The organization of the present book is essentially the same as in *The Isoquinoline Alkaloids*. Each chapter begins with a discussion of structural elucidation and synthesis, a description of typical reactions then follows, and the chapter ends with coverage of biogenesis, pharmacology, and spectroscopy. New chapters have had to be added to describe the completely new alkaloidal types discovered since 1972. These include baluchistanamine (an isoquinolone–benzylisoquinoline dimer), the aporphine–pavine dimers, the 4,5-dioxoaporphines, the secoberbines, the 3-arylisoquinolines, eupolauridine, and very recently imerubrine. Another new chapter discusses the chemistry of the aristolochic acids and aristolactams, a group of substituted phenanthrenes, obviously of isoquinoline derivation in spite of the fact that they do not incorporate a basic nitrogen function. The aristolochic acids and aristolactams were not included in *The Isoquinoline Alkaloids* even though they were known at the time that book was written.

On the other hand, one group of alkaloids which was included in *The Isoquinoline Alkaloids* and nevertheless was deemed not to belong properly in the present work is the naphthalenoisoquinolines, which include ancistrocladine and its relatives. These bases do not originate biogenetically from tyrosine, and beside incorporating tetrahydroisoquinoline moieties show no clear structural relationship to the more orthodox isoquinoline alkaloids.

As with *The Isoquinoline Alkaloids*, not all known or new isoquinoline alkaloids have been presented. Rather, selected examples have been discussed to illustrate specific principles or methods. For a complete listing of the isoquinoline alkaloids, the reader is referred to T. Kametani's *The Chemistry of the Isoquinoline Alkaloids*, Vol. 2,* and also to the annual reviews on isoquinoline alkaloids and the aporphinoids included in *Specialist Periodical Reports, The Alkaloids.*†

* The Sendai Institute of Heterocyclic Chemistry, Sendai, Japan (1974).
† The Chemical Society, London.

Percentage reaction yields are given in parentheses under the proper structures. Infrared absorptions are quoted both in terms of wavelength and frequency. Ultraviolet log ϵ values are given in brackets following the λ_{max} values. The literature up to about mid-1977 has been covered. All nuclear magnetic resonance data are given in δ values, and were obtained in deuteriochloroform solution unless stated otherwise. A special effort was made to include C-13 nuclear magnetic resonance values where available. The term "tetrahydroisoquinoline" refers to the 1,2,3,4-tetrahydro system, and the prefix "nor" relates solely to the N-nor series.

The authors would like to thank Drs. A. Patra and A. S. Rothenberg, as well as Mr. P. Chinnasamy, for commenting on the completed manuscript. A special debt is due senior editor, Mr. Ellis Rosenberg, and assistant managing editor, Ms. Betty Bruhns, for a dedicated effort in facilitating our work.

Maurice Shamma
Jerome L. Moniot

CONTENTS

THE SIMPLE ISOQUINOLINES

1

1.1. Introduction

The chemistry, synthesis, and biogenesis of peyote, *Lophophora williamsii* (Lemaire) Coult., and other cactus alkaloids have been thoroughly reviewed.[1-5]

New natural sources of simple isoquinoline alkaloids are the Euphorbiaceae and the Rhamnaceae. Some recently isolated alkaloids of interest are shown in Scheme 1.1.*

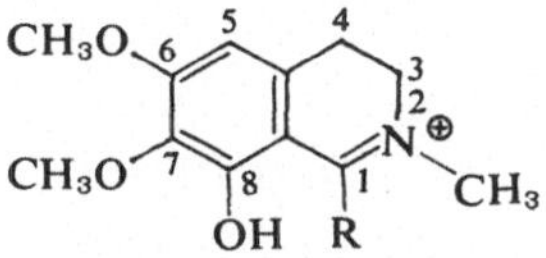

Anhalotine,[6] R = H
Peyotine,[6] R = CH$_3$
Lophophora williamsii
var. *caespitosa*.

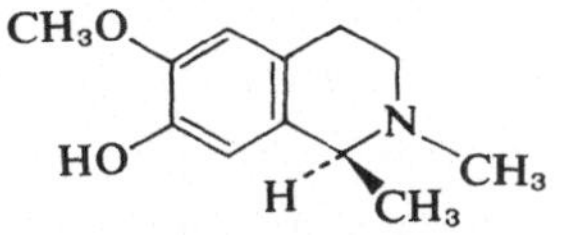

(+)-1-Methylcorypalline[7]

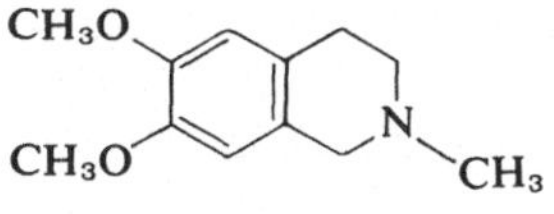

O-Methylcorypalline[8]

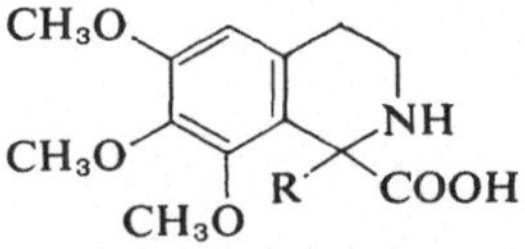

O-Methylpeyoxylic acid,[9]
R = H
O-Methylpeyoruvic acid,[9]
R = CH$_3$

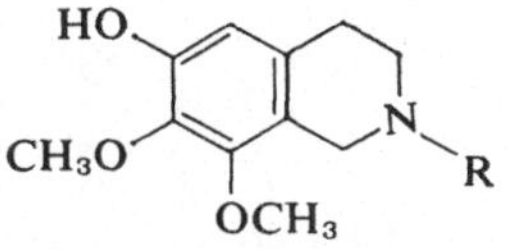

Isoanhalamine,[10] R = H
Isoanhalidine,[10] R = CH$_3$

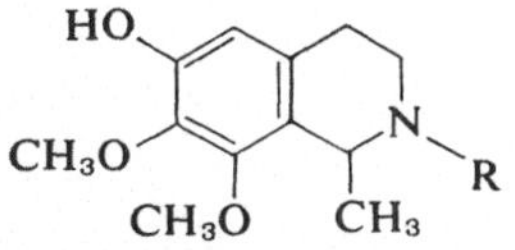

Isoanhalonidine,[10] R = H
Isopellotine,[10] R = CH$_3$

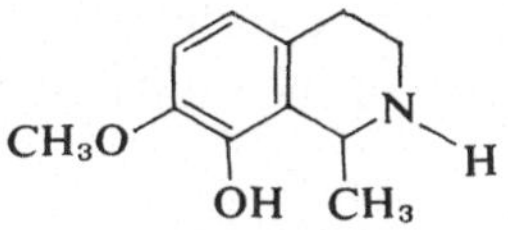

Arizonine[11]

SCHEME 1.1a

* The names anhalotine and peyotine had previously been given to the *N*-methyl salts of anhalidine and pellotine, respectively, also found in *L. williamsii*; see G. J. Kapadia, N. J. Shah, and T. B. Zalucky, *J. Pharm. Sci.*, **57**, 254 (1968).

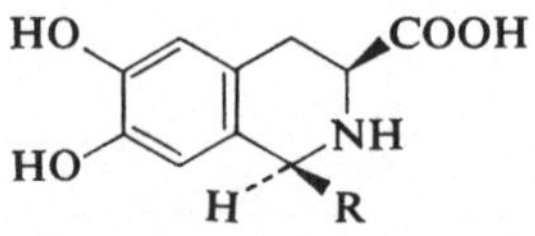

L-3-Carboxy-6,7-dihydroxy-1,2,3,4-
tetrahydroisoquinoline,[12] R = H; and
the C-1 methyl homolog,[13,14] R = CH$_3$

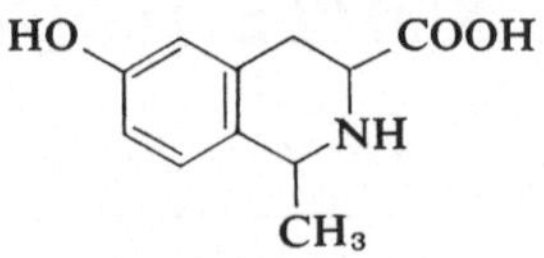

1-Methyl-3-carboxy-6-hydroxy-1,2,3,4-
tetrahydroisoquinoline
Euphorbia myrsinites L. (Euphorbiaceae)[15]

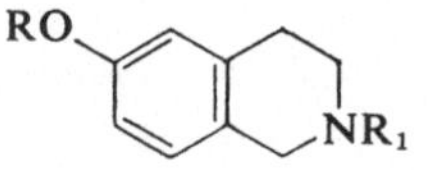

Longimammatine,[16] R = CH$_3$
                R$_1$ = H
Longimammosine,[16] R = H
                R$_1$ = CH$_3$

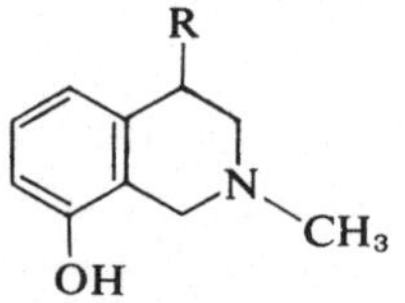

Longimammidine,[16] R = H
(−)-Longimammamine,[16] R = OH

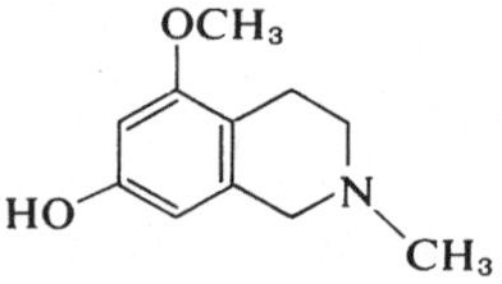

Uberine[16a]

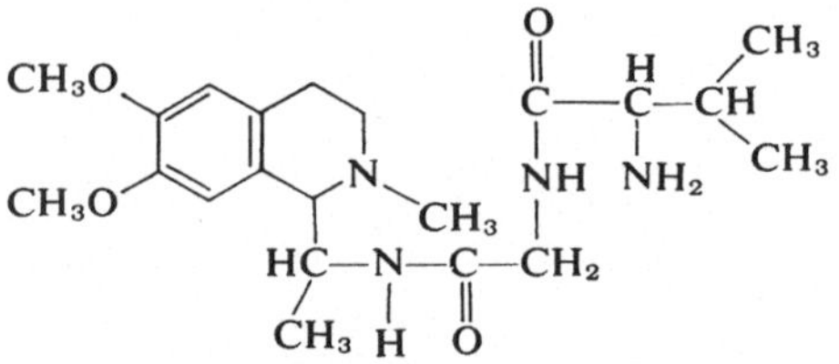

(−)-Amphibine-I *Ziziphus amphibia*
A. Cheval. (Rhamnaceae)[17]

SCHEME 1.1b

Combined GLC–mass spectrometry has proved a useful tool in the rapid identification of minute amounts of the simple isoquinoline alkaloids.[3,18]

1.2. Synthesis[19]

1.2.1. Pictet–Spengler Cyclization

A variety of optically active simple tetrahydroisoquinolines (including the salsolines and salsolinols) have been prepared, and some of the more interesting examples are shown in Scheme 1.2.[20-22]

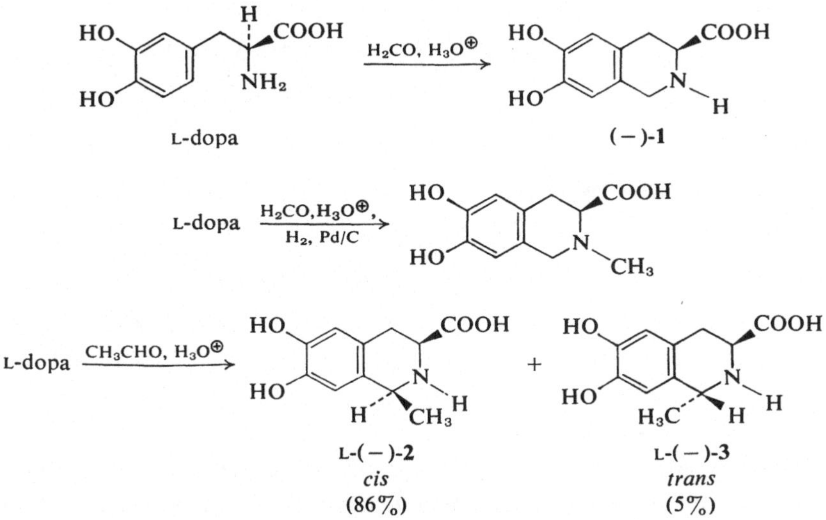

SCHEME 1.2

To ascertain the stereochemistry of the two isomers obtained from condensation of L-dopa with acetaldehyde, the cis acid **2** was converted to the ethyl ester hydrochloride and subjected to X-ray analysis.[22] By effecting the cyclization at pH 7 instead of in dilute mineral acid, the yield of the trans acid **3** was increased from 5 to 25%.[23] Amino acids **1** and **2** are natural products, the former being found in the seed embryos of *Mucuna mustisiana*,[12] and the latter in the velvet bean.[13,14] It is worth noting that base-catalyzed epimerization of derivatives of **2** and **3**, where all the active hydrogens were protected, led mostly to the 1,3-*trans* products.[24,25]

Alternative routes to optically active tetrahydroisoquinolines employ reductive decyanation of chiral α-amino nitriles using sodium in liquid ammonia[26] or NaBH₄,[27] or the reduction of imines or immonium salts in an asymmetric environment (see Sec. 1.2.2, below).

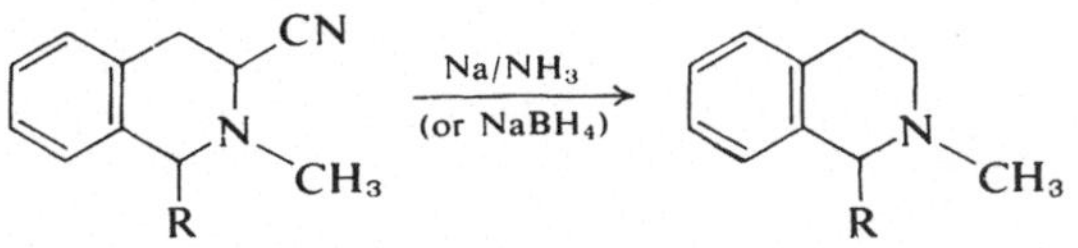

The Pictet–Spengler cyclization has also been used to prepare the alkaloids *O*-methylpeyoxylic and *O*-methylpeyoruvic acids from mescaline hydrochloride.[9]

CH$_3$O, CH$_3$O, CH$_3$O NH$_2$·HCl

1. H—C—C—O-n-Bu, H$_2$O, Δ
2. OH$^{\ominus}$

CH$_3$O, CH$_3$O, CH$_3$O COOH N—H

O-Methylpeyoxylic acid
(45%)

CH$_3$—C—C—O—CH$_3$, H$_2$O, Δ

CH$_3$O, CH$_3$O, CH$_3$O CH$_3$ COOH N—H

O-Methylpeyoruvic acid
(51%)

A systematic study of the synthesis of simple tetrahydroisoquinolines via the cyclization of Schiff bases under neutral or weakly acidic conditions has been carried out. As expected, a phenolic group ortho or para to the cyclization site greatly facilitates the reaction.[28,28a] Some Pictet–Spengler reactions have been run in benzene without the addition of acid.[29]

1.2.2. Bischler–Napieralski Cyclization

Salsolidines with optical purities ranging from 15 to 44% can be prepared by the Bischler–Napieralski route, using optically active immonium salts.[30]

Alternatively, the optically inactive imine from a Bischler–Napieralski cyclization may be reduced asymmetrically using an optically active rhodium complex with (+)-diop as ligand.[31]

CH$_3$O, CH$_3$O CH$_2$—CHO

+ (*S*)-(—)-Ph—C—NH$_2$ $\xrightarrow{\text{benzene}}$

H, CH$_3$

CH$_3$O, CH$_3$O N—C—Ph, H, CH$_3$

1. NaBH$_4$
2. Ac$_2$O, py.
3. POCl$_3$, toluene
4. NaI

CH$_3$O, CH$_3$O N$^{\oplus}$—C—Ph, CH$_3$ CH$_3$ I$^{\ominus}$

1. NaBH$_4$
2. H$_2$, Pd/C

CH$_3$O, CH$_3$O N—H, H CH$_3$

(—)-Salsolidine
(36–44% opt. pure)

A new variation of the Bischler–Napieralski approach uses an ω-phenyl-isonitrosopropiophenone which is reduced to an amino alcohol. This material is then *N*-formylated and cyclized. The product is a 3-benzylisoquinoline derivative.[32]

An ω-phenylisonitroso-propiophenone

It is relevant to point out that Fodor and co-workers have shown that nitrilium cations are intermediates in the Bischler–Napieralski cyclization whenever secondary amides are used. A nitrilium ion was trapped as its crystalline hexafluoroantimonate salt.[33]

A nitrilium ion

A new acid catalyst of potential use in the Bischler–Napieralski cyclization is P_2O_5 in methanesulfonic acid.[34] The phenethylamines used in isoquinoline syntheses are usually prepared from the reduction of the corresponding β-nitrostyrenes. A more versatile procedure starts with a substituted benzyl chloride which is converted to the nitrile using sodium cyanide in DMSO (dimethylsulfoxide). Reduction of the nitrile with $LiAlH_4$ in the presence of $AlCl_3$ gives the desired amine in excellent yield.[35] Benzylamines or their quaternary salts may also be utilized in appropriate solvents in place of benzylic chlorides, so that they too may act as nitrile precursors.[36]

1.2.3. Pomeranz–Fritsch Cyclization

The Bobbitt variation of the Pomeranz–Fritsch cyclization to obtain 4-hydroxytetrahydroisoquinolines continues to be a useful approach.[19,37–39] Jackson and co-workers have worked out a superior method for the preparation of substituted isoquinolines which involves cyclization, under mildly acidic conditions, of acetal sulfonamides.[40]

An acetal
sulfonamide

(64% overall yield
from veratraldehyde)

Super polyphosphoric acid (PPA + P_2O_5) is an effective reagent in the Pomeranz-Fritsch cyclization, replacing the usual $6N$ HCl. Thus when imine acetal **4**, prepared from o-methylacetophenone and aminoacetaldehyde dimethyl acetal, was heated briefly with super PPA, a 30% yield of 1,8-dimethylisoquinoline was obtained. The product was then oxidized and decarbonylated to supply 8-methylisoquinoline.[41]

4

8-Methyliso-
quinoline

Imine acetal intermediates can also be efficiently cyclized using the BF_3–HOAc complex in trifluoroacetic anhydride (TFAA).[42]

(60–82%)

If the imine acetal is cyclized using chlorosulfonic acid without heating, the main product is a 3-chloroisoquinoline derivative.[43]

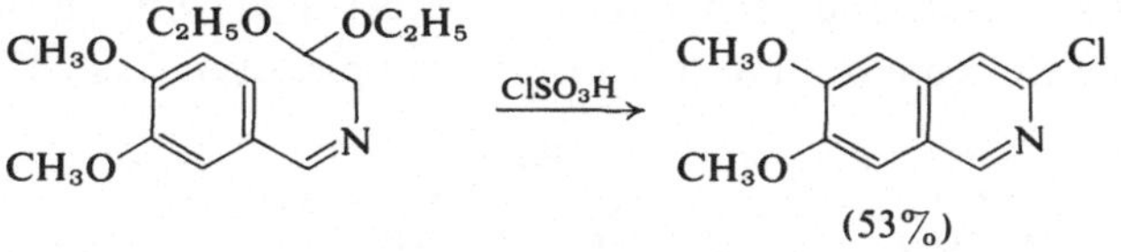

(53%)

1.2.4. Photocyclization of N-Chloroacetylbenzylamines

A novel route to isoquinoline derivatives involves irradiation of an N-chloroacetylbenzylamine to afford an isoquinoline lactam.[44]

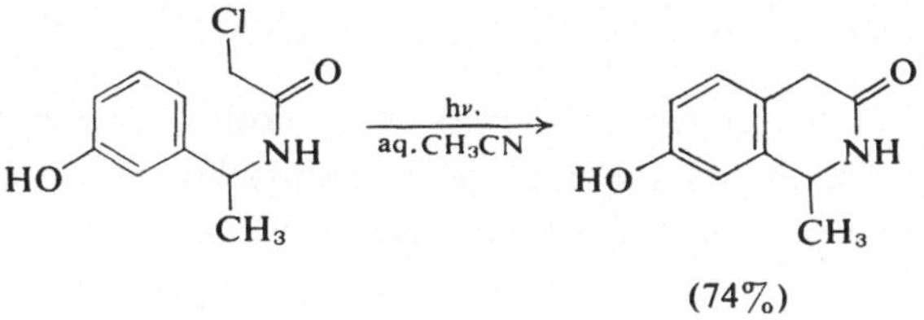

(74%)

1.2.5. Amination of Benzopyrylium Salts

Acylation of 3,4-dimethoxyphenylpyruvic methyl ester with acetic anhydride followed by acid-catalyzed cyclization provides a benzopyrylium salt. Further treatment with ammonia affords an isoquinoline derivative in good yield.[45]

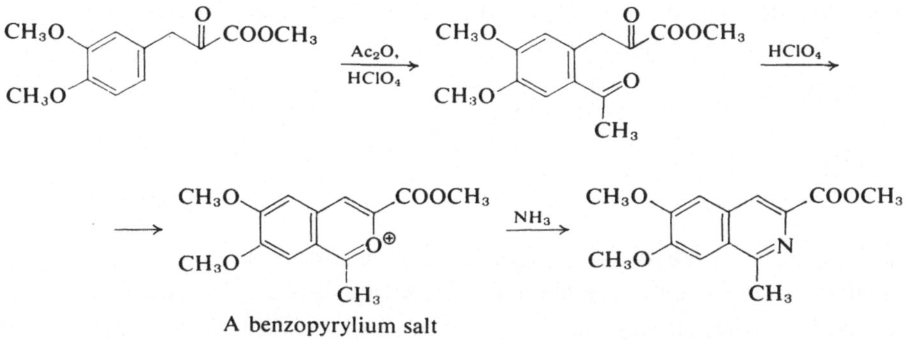

A benzopyrylium salt

1.2.6. A New Isoquinoline Synthesis via Ortho-Substituted Benzylamines

A synthesis of isoquinolines has been developed taking advantage of the fact that the o-tolunitrile carbanion can be acylated with an aromatic ester. The overall yield from o-tolunitrile in the example given was 24%.[46]

1.2.7. Cyclization of Aralkenyl-Substituted Quaternary Ammonium Salts

Quaternary ammonium salts possessing both a β-alkenyl and a benzyl substituent cyclize in the presence of polyphosphoric acid to furnish tetra-hydroisoquinoline salts in respectable yields.[46a]

1.3. Some Reactions of Simple Isoquinolines

1.3.1. Oxidation

A new mild and stepwise dehydrogenation of tetrahydroisoquinolines involves the use of Fremy's salt (potassium nitrosodisulfonate) at room temperature. Best results were obtained when no substituent or only small alkyl substituents were present at C-1.[47]

Tertiary tetrahydroisoquinolines may be oxidized to their 3,4-dihydro analogs using mercuric acetate and EDTA (ethylenediaminetetraacetic acid)[48] or NBS.[39] Alternatively, mercuric acetate can be used in conjunction with acetic and dilute sulfuric acids.[49] 3,4-Dihydroisoquinolines can be further oxidized to the fully aromatized isoquinolines by photochemical oxidation in acid solution[50] or better through dehydrogenation with diphenyl disulfide.[51]

Oxidation of 3,4-dihydroisoquinolines with perbenzoic acid yields nitrones as well as oxaziranes, with nitrone formation being favored in aprotic media.[52]

R = H or CH₃   An oxazirane   A nitrone

Hydrogen peroxide oxidation of the amino ester **5** formed from Michael condensation of tetrahydroisoquinoline with ethyl acrylate produces an N-oxide which eliminates acrylic acid in the presence of hydroxylic base. This sequence provides a convenient route to N-hydroxylated tetrahydroisoquinolines.[53]

The lead tetraacetate oxidation of 7-hydroxytetrahydroisoquinolines has been extended to the 6-hydroxy analogs. Thus oxidation of **6** gave the 4-acetoxy derivative **7**.[54]

p-Quinol acetates can also be used in halogenation, as in the C-8 bromination of corypalline through treatment with lead tetraacetate followed by 48% HBr.[55] Tetrahydrobenzylisoquinolines possessing a phenol at C-7 can also be chlorinated by a similar procedure using concentrated HCl in lieu of HBr.[55]

Corypalline

(75%)

It is sometimes required to oxidize the C-1 methyl group of a substituted isoquinoline to an aldehyde function. Such a transformation is usually carried out using selenium dioxide, in which case anhydrous conditions using dioxane as solvent often give superior yields.[51] An alternate route, however, involves initial formation of the isoquinoline N-oxide followed by acetylation as indicated below. The final step is a manganese dioxide oxidation of a benzylic alcohol.[51]

1.3.2. Alkylation and Acylation

7-Hydroxytetrahydroisoquinolines are methylated at C-8 by thiomethylation followed by Raney nickel desulfurization. 8-Methylcorypalline was thus obtained from corypalline in high yield.[56]

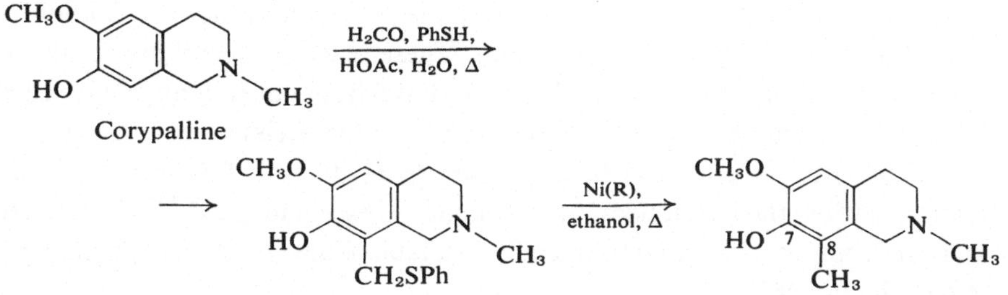

Corypalline

3,4-Dihydroisoquinolines are readily alkylated at C-1 using the nucleophiles indicated in Scheme 1.3.[57–59]

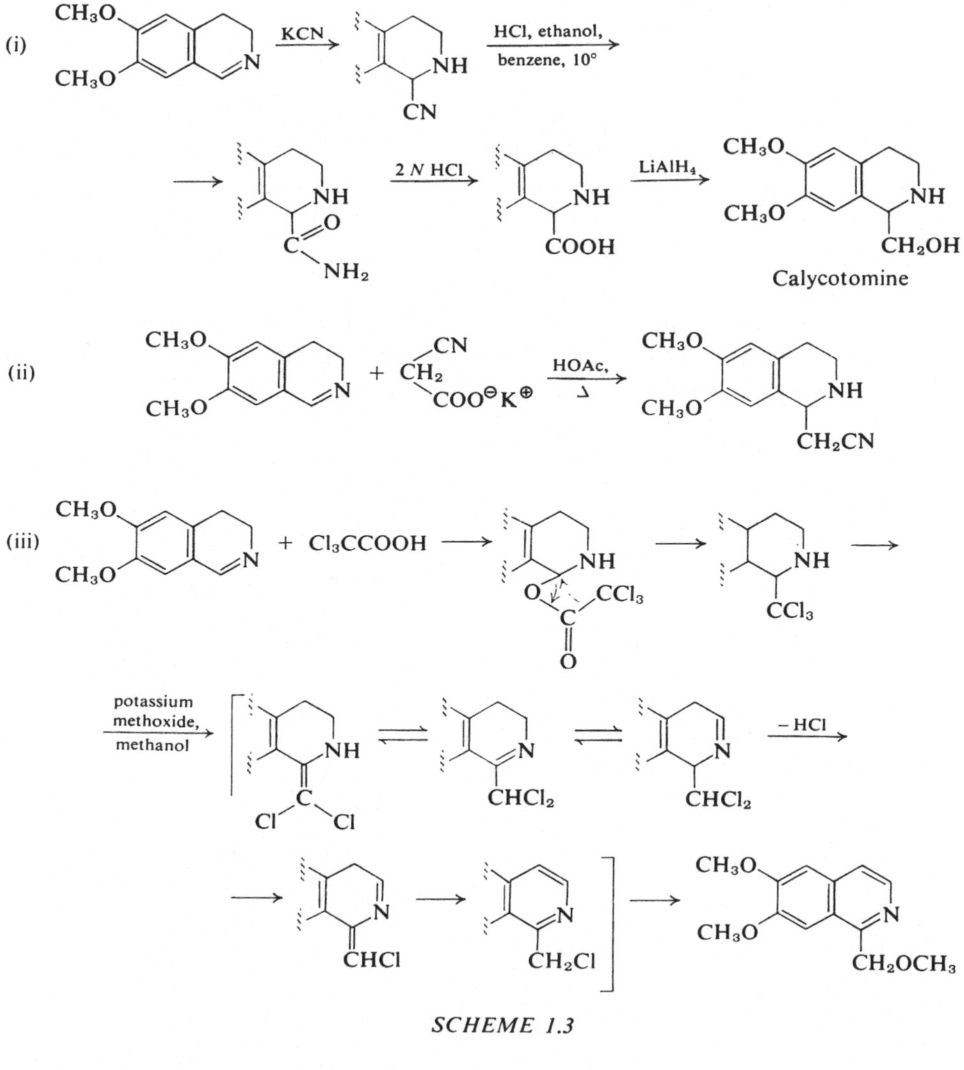

(i)

Calycotomine

(ii)

(iii)

SCHEME 1.3

Still a method of choice for the alkylation or acylation of isoquinolines at C-1 is the use of Reissert compounds. These reactions, carefully studied over a number of years by Popp and others, have been thoroughly reviewed.[60-63] To cite an example, reaction of the anion of the Reissert compound **8** with formaldehyde furnished the benzoyl ester **9**. Hydrolysis followed by catalytic hydrogenation then supplied calycotomine in excellent yield.[64] (A number of Reissert compounds are commercially available from the Parish Chemical Co., Provo, Utah 84601.)

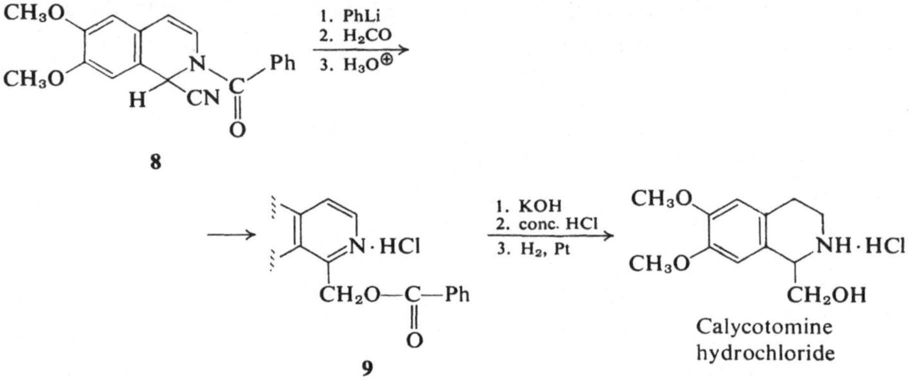

8

9

Isoquinolines that are methylated at C-1 have been alkylated at C-4 in the presence of NaH and DMSO. The alkylating agents used were 2-methylcyclohexenone and phenethyl bromide.[65]

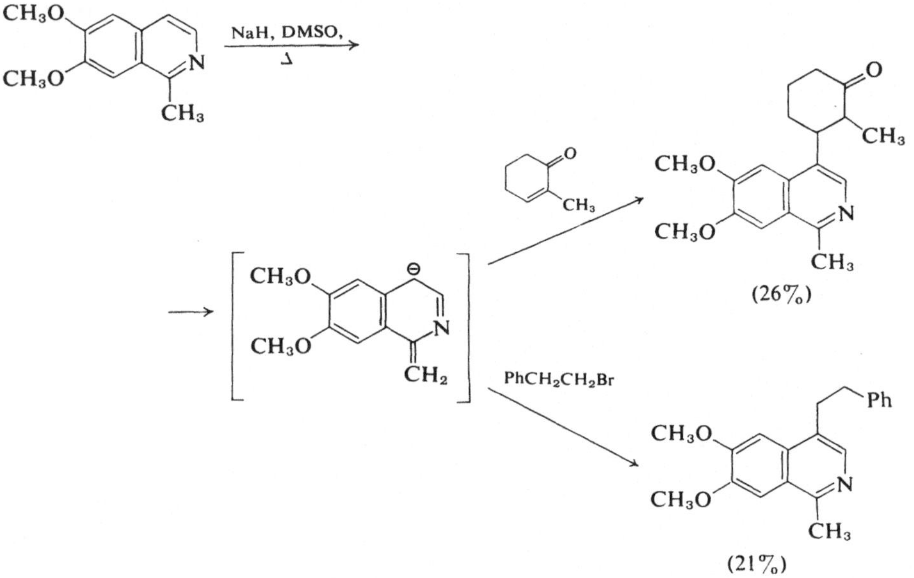

(26%)

(21%)

If a methyl group is not present at C-1, enamine alkylation or acylation at C-4 can still be achieved using the acetone pseudobase to block the C-1 site of an isoquinolinium salt.[66]

A new method for acylation at C-4 employs reduction of an isoquinoline with NaH in hexamethylphosphoric triamide, *N*-acylation, and subsequent *C*-acylation of the enamide.[67]

N-Methyl-1,2-dihydroisoquinolines can also be formylated at C-4 with POCl$_3$–DMF.[39] Alkylation of 1,2-dihydro-2-methylisoquinoline with methyl iodide favors nitrogen over C-4.[68,69]

When isoquinoline is refluxed with acetic anhydride, *N*-acetyl-1-carboxymethyl-1,2-dihydroisoquinoline is obtained in analogy with the Reissert reaction. If acetone is also present during the refluxing, *N*-acetyl-1-acetonyl-1,2-dihydroisoquinoline is formed in addition.[69a]

N-Acetyl-1-carboxymethyl-
1,2-dihydroisoquinoline

N-Acetyl-1-acetonyl-
1,2-dihydroisoquinoline

1.3.3. Fluorophore-Forming Reactions

When dopamine, noradrenaline, or dopa react with glyoxylic acid in polar solvents, highly fluorescent compounds are formed through an intermediate tetrahydroisoquinoline. This transformation is the basis of a histochemical test for catecholamines. In the case of dopamine, the fluorescent tautomeric compound **10** is formed in 90% yield.[70]

10

1.3.4. Electrolytic Oxidation

Corypalline has been dimerized electrochemically in overall yields of 44–85%, depending upon experimental conditions. In general, the product was mainly carbon–carbon dimer with a small amount of the carbon–oxygen dimer.

Corypalline

Major product

Minor product

If a steric factor R was introduced at C-1 in the form of a methyl or ethyl group, and the oxidation carried out on a platinum anode in an aqueous system, the product distribution drifted toward the carbon–oxygen dimer. On the other hand, when the oxidation was carried out on the sodium salt of a phenol in wet acetonitrile, the product was mainly the carbon–carbon dimer, even if R was methyl.

When the isomers **11** and **12** were oxidized in acetonitrile in the presence of base, contrary to expectations, carbon–oxygen dimers were formed because of nitrogen participation as indicated in Scheme 1.4. However, when the nitrogen is protonated (acid medium), or acylated, the products are all carbon–carbon dimers.[71-73]

Electrooxidation of the sodium salt of (±)-1-methylcorypalline in wet acetonitrile exhibits remarkable stereoselectivity to give in good yield only the carbon–carbon dimer **13**, which has restricted rotation around the biphenyl bond. A surface mechanism in which the isoquinoline rings are adsorbed in a planar fashion on the electrode with the C-1 methyl pointing away from the surface must operate here. Therefore, only isomers having the same configuration at C-1 can come close enough to couple at C-8.[74,75]

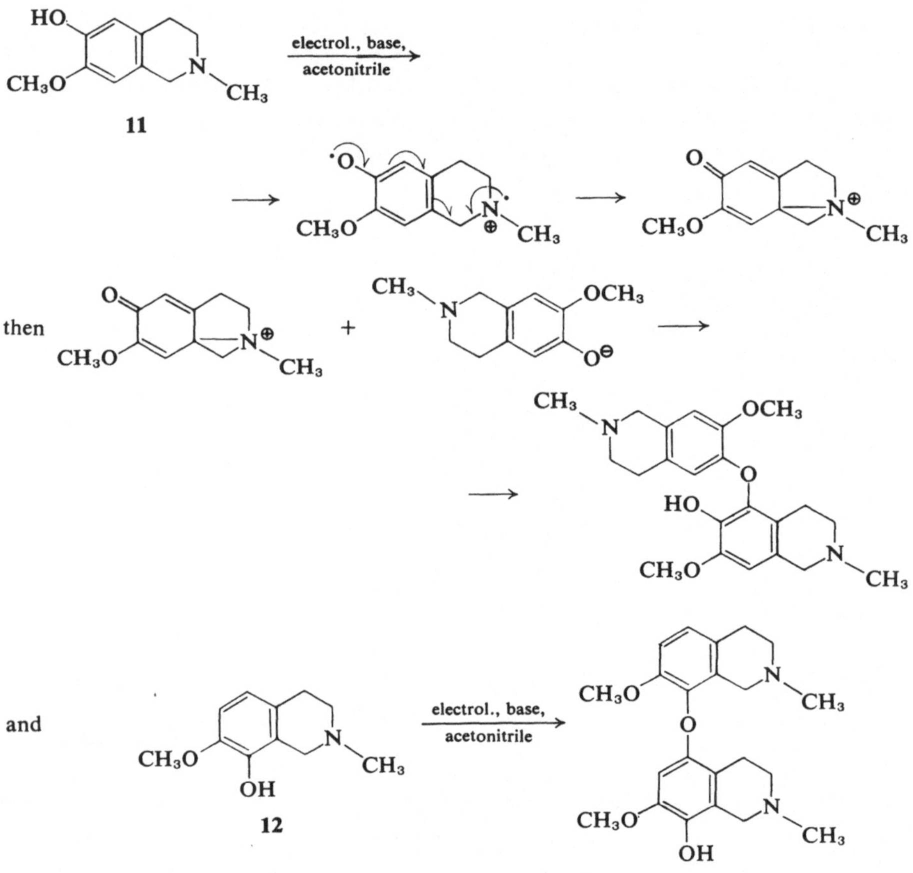

SCHEME 1.4

The stereochemistry and absolute configuration of a variety of dimers, prepared by electrolytic or chemical oxidation of (S)-$(-)$-1-methylcorypalline have been elucidated through careful analysis of their ORD (optical rotation. dispersion) and CD (circular dichroism) curves.[76]

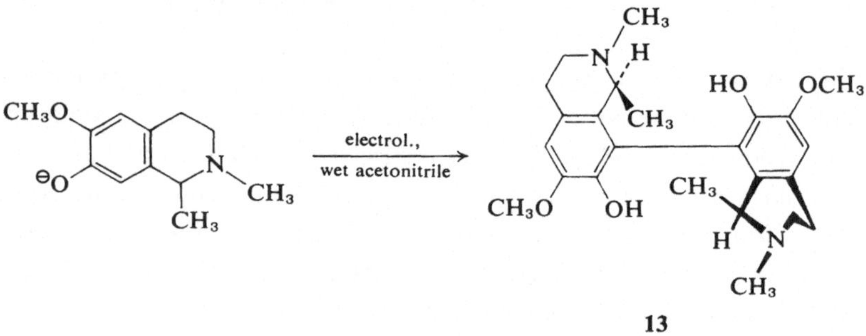

1.3.5. Rearrangements of 1,2-Dihydroisoquinolines

When 1-allyl-2-methyl-1,2-dihydroisoquinoline is heated with dilute mineral acid, a rearrangement occurs rapidly and in high yield to the 2-methyl-3-allyl-3,4-dihydroisoquinoline salt (14). This is an intramolecular suprafacial sigmatropic [3,3] shift analogous to the Claisen and Cope rearrangements, with a transition state as denoted below.[77-79]

A different mechanistic pathway is followed when the above immonium salt-14 is subjected to reflux with 2N HCl for 14 days, the product being tricyclic base 15. In addition, prolonged heating in acid of the enamine 16 provides the related base 17.

In these two instances, therefore, a carbonium ion mechanism probably applies in a two-step reaction[80]:

1.4. Biogenesis

An important series of *in vivo* experiments by Kapadia and his colleagues using [¹⁴C]-labeled precursors has led to a substantial advance in our understanding of the biogenesis of tetrahydroisoquinoline alkaloids.[81] *N*-Acetyl-3-demethylmescaline can act as a precursor for a variety of peyote alkaloids since administration of this amide, labeled as indicated, to peyote, *Lophophora williamsii* (Lemaire) Coult., generated radioactive anhalamine and anhalonidine.

N-Acetyl-3-demethylmescaline   Anhalamine   Anhalonidine

These results indicated that the administered *N*-acetyl compound was deacetylated to 3-demethylmescaline prior to its incorporation into the simple isoquinolines. 3-Demethylmescaline can itself act as an excellent precursor for anhalamine and anhalonidine.[81]

Attention was next turned to the two amino acids which have recently been found in peyote, namely peyoxylic and peyoruvic acids. Administration of labeled peyoxylic and peyoruvic acids to the plant led to the specific incorporation of the labels in anhalamine and anhalonidine, respectively.[81]

Peyoxylic acid   Anhalamine

Peyoruvic acid   Anhalonidine

Finally, the facile decarboxylation of peyoruvic and peyoxylic acid was observed when incubated with fresh peyote slices. In the case of labeled peyoruvic acid, the 3,4-dihydroisoquinoline **18** was isolated from the decarboxylation mixture.[81] It is interesting to note in this respect that Bobbitt and coworkers have recently found that simple tetrahydroisoquinoline-1-carboxylic acids containing at least one free phenolic function in the aromatic ring undergo *in vitro* oxidative decarboxylation when stirred in air, under basic conditions, to supply the corresponding 3,4-dihydroisoquinolines.[81a]

The above results confirm that simple tetrahydroisoquinolines such as anhalamine and anhalonidine are formed in nature from 3-demethylmescaline. 3-Demethylmescaline condenses with glyoxylic or pyruvic acid to give peyoxylic or peyoruvic acid. These amino acids undergo decarboxylation to supply 3,4-dihydroisoquinolines which are then enzymatically reduced to the tetrahydroisoquinolines.[81]

In a separate series of experiments, working with intact *Lophocereus schottii* (Engelm.) Britt. & Rose plants, it was demonstrated again by means of [^{14}C]-labeling that the five-carbon unit in lophocerine probably arises independently from both leucine and mevalonic acid by the pathways indicated below. In this instance, no α-keto acid is involved in the Mannich cyclization.[82]

Lophocerine

The recent isolation of a series of simple tetrahydroisoquinoline alkaloids monooxygenated in ring A, such as longimammatine and longimammidine, raises interesting and still unanswered questions concerning their biogenesis.[16]

1.5. Pharmacology

Debrisoquine (Hoffmann–LaRoche) is used in the treatment of hypertension. The simple carboxamidoximes 19 and 20, which bear structural similarities to debrisoquine, have been assayed for antihypertensive activity in rats and dogs at the Sterling–Winthrop Research Institute in Rensselaer, New York. Initial results were sufficiently promising to warrant further evaluation.[41]

Debrisoquine

19, R = H
20, R = CH$_3$

The analog 21 of bisobrin, an antithrombic agent, has shown appreciable fibrinolytic activity in rats.[83]

Bisobrin, $n = 4$
21, $n = 7$

Since the hypotensive properties of amiquinsin hydrochloride are established, the isomeric salt 22 was prepared and tested. It is hypotensive in anesthetized normotensive dogs, and it produced an antihypertensive effect in unanesthetized renal hypertensive dogs which is, however, of shorter duration than the effect of amiquinsin hydrochloride.[84]

Amiquinsin·HCl

22

4-Methyl-5-amino-1-formylisoquinoline thiosemicarbazone is a potential antitumor agent.[51] 7,8-Dichlorotetrahydroisoquinoline is an inhibitor of phenylethanolamine-N-methyltransferase, and is thus a potential antihypertensive agent.[84a]

The amino sulfoxide esproquin may be useful for the treatment of chronic heart failure when given orally to human subjects.[85]

Esproquin

Salsolinol, as well as the benzylisoquinoline tetrahydropapaveroline, are inhibitors of dopamine *O*-methylation *in vitro*, since they themselves are metabolized by *O*-methylation catalyzed by catechol *O*-methyltransferase (COMT). They are also *in vitro* inhibitors of rat brain monoamine oxidase. Both compounds were isolated in significant concentration from the urine of patients with Parkinson's disease during oral L-dopa treatment. The formation of these alkaloids probably originates from a Pictet–Spengler condensation of dopamine with the required aldehyde.[86]

Salsolinol

and

Dopamine

Tetrahydropapaveroline

They could, therefore, affect the metabolic disposition of endogenous neuroamines resulting in modification of the adrenergic function.[87] It should be stated, however, that no firm evidence has been presented thus far that salsolinol is a real alkaloid produced in the brain. This substance may be an artifact formed in the bladder or in the urine.[88]

Catecholamine-derived tetrahydroisoquinolines can be stored within adrenergic neurons,[89-92] and when released can either activate[91] or block[93] α-adrenergic receptors. A number of 6,7-dihydroxytetrahydroisoquinolines related to dopamine were, therefore, studied for their inhibitory effects on the accumulation of dopamine by rat brain slices. 6,7-Dihydroxytetrahydroisoquinoline and S-(−)-salsolinol were found to be effective inhibitors of dopamine accumulation, while R-(+)-salsolinol was less potent.[94] 6,7-Dihydroxytetrahydroisoquinoline is both a directly and indirectly acting sympathomimetic agent.[95]

S-(−)-Salsolinol

In an effort to identify those tetrahydroisoquinolines which will inhibit the action of COMT, but will not act as false neurotransmitters, a series of tetrahydroisoquinolines were evaluated as substrates and inhibitors of this enzyme, and were also tested for their ability to stimulate norepinephrine release from mouse hearts *in vivo*. Methyl substituents at C-2 and C-4 of 6,7-dihydroxytetrahydroisoquinolines had little effect in regard to COMT, but did eliminate the norepinephrine depleting activity. The interesting exception was 6,7-dihydroxy-2,2-dimethyltetrahydroisoquinolinium iodide which was an active depleter of norepinephrine from mouse hearts.[96]

The twin topics of the identification of isoquinoline alkaloids during alcohol intoxication,[97] and the pharmacology of isoquinoline alkaloids and ethanol interactions,[98] have recently been reviewed.

1-Benzyl-1-carboxy-6-hydroxytetrahydroisoquinoline and 1-methyl-1-carboxy-6-hydroxytetrahydroisoquinoline have shown anti-inflammatory activity in initial tests.[28a]

1.6. CMR Spectroscopy

The natural abundance CMR (carbon magnetic resonance) spectra of a number of isoquinoline alkaloids and model compounds have been recorded, and some of the data are summarized in Scheme 1.5.[99]

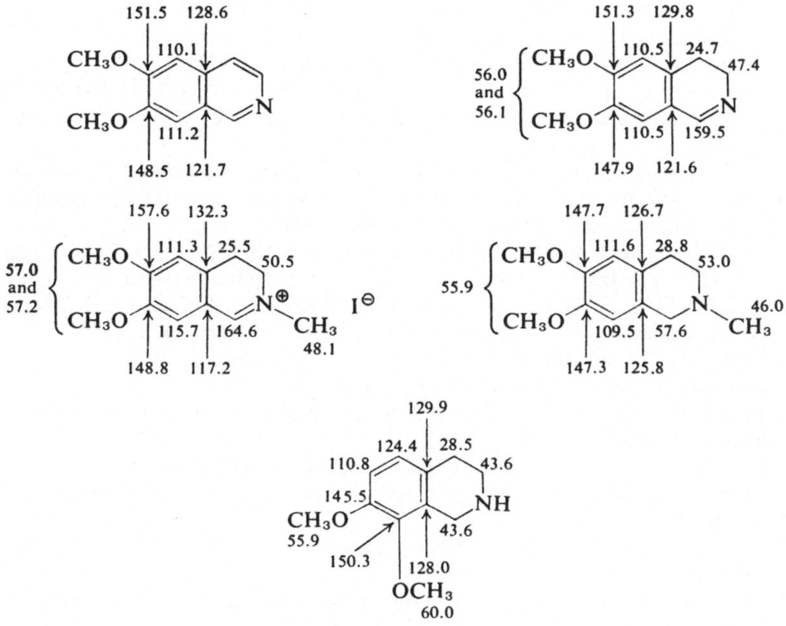

151.5 128.6

CH₃O 110.1

CH₃O 111.2   N

148.5 121.7

151.3 129.8

56.0 and 56.1 CH₃O 110.5 24.7 47.4

CH₃O 110.5 159.5 N

147.9 121.6

157.6 132.3

57.0 and 57.2 CH₃O 111.3 25.5 50.5

CH₃O 115.7 164.6 N⊕ CH₃ I⊖

148.8 117.2 48.1

147.7 126.7

55.9 CH₃O 111.6 28.8 53.0

CH₃O 109.5 57.6 N 46.0 CH₃

147.3 125.8

129.9

124.4 28.5

110.8 43.6

145.5

CH₃O NH

55.9 43.6

150.3 128.0

OCH₃

60.0

SCHEME 1.5

References and Notes

1. J. L. McLaughlin, *Lloydia*, 36, 1 (1973).
2. W. J. Keller, L. A. Spitznagle, L. R. Brady, and J. L. McLaughlin, *Lloydia*, 36, 397 (1973).
3. G. J. Kapadia and M. B. E. Fayez, *Lloydia*, 36, 9 (1973).
4. A. G. Paul, *Lloydia*, 36, 36 (1973).
5. A. T. Shulgin, *Lloydia*, 36, 46 (1973).
6. M. Fujita, H. Itokawa, J. Inoue, Y. Nozu, N. Goto, and K. Hasegawa, *J. Pharm. Soc. Japan*, 92, 482 (1972).
7. S. Naruto and H. Kaneko, *Phytochemistry*, 12, 3008 (1973).
8. T. H. Yang and C. M. Chen, *J. Chin. Chem. Soc., Ser. II*, 17, 54 and 235 (1970); see also H.-Y. Hsu and Y.-P. Chen, *Heterocycles*, 3, 265 (1975); and J.-E. Lindgren and J. G. Bruhn, *Lloydia*, 39, 464 (1976).
9. G. J. Kapadia, G. S. Rao, M. H. Hussain, and B. K. Chowdhury, *J. Heterocycl. Chem.*, 10, 135 (1973).
10. J. Lundström, *Acta Chem. Scand.*, 26, 1295 (1972).
11. J. G. Bruhn and J. Lundström, *Lloydia*, 39, 197 (1976).
12. E. A. Bell, J. R. Nulu, and C. Cone, *Phytochemistry*, 10, 2191 (1971).
13. M. E. Daxenbichler, R. Kleiman, D. Weisleder, C. H. VanEtten, and K. D. Carlson, *Tetrahedron Lett.*, 1801 (1972).
14. T. Ohta, H. Dato, T. Kurata, and M. Fujimaki, *Agric. Biol. Chem. (Japan)*, 39, 139 (1975).

15. P. Müller and H.-R. Schütte, *Z. Naturforsch.*, **23b**, 491 (1968).
16. R. L. Ranieri and J. L. McLaughlin, *J. Org. Chem.*, **41**, 319 (1976).
16a. R. L. Ranieri and J. L. McLaughlin, *Lloydia*, **40**, 173 (1977).
17. R. Tschesche, C. Spilles, and G. Eckhardt, *Chem. Ber.*, **107**, 1329 (1974). For the synthesis of amphibine-I, see R. Tschesche, J. Moch, and C. Spilles, *Chem. Ber.*, **108**, 2247 (1975).
18. J. E. Lindgren, S. Agurell, J. Lundström, and U. Svensson, *FEBS Lett.*, **13**, 21 (1971).
19. For a thorough discussion of the synthesis of 4-oxy- and 4-ketotetrahydroisoquinolines, see J. M. Bobbitt, *Adv. Heterocycl. Chem.*, **15**, 99 (1973).
20. S. Teitel, J. O'Brien, and A. Brossi, *J. Med. Chem.*, **15**, 845 (1972). For a compilation of the NMR spectra of monophenolic tetrahydroisoquinoline salts, see F. Schenker, R. A. Schmidt, T. Williams, and A. Brossi, *J. Heterocycl. Chem.*, **8**, 665 (1971).
21. S. Teitel, J. O'Brien, W. Pool, and A. Brossi, *J. Med. Chem.*, **17**, 134 (1974).
22. A. Brossi, A. Focella, and S. Teitel, *Helv. Chim. Acta*, **55**, 15 (1972).
23. S. Teitel and A. Brossi, *Lloydia*, **37**, 196 (1974).
24. H. Bruderer, A. Brossi, A. Focella, and S. Teitel, *Helv. Chim. Acta*, **58**, 795 (1975).
25. A. Brossi, *Heterocycles*, **3**, 343 (1975).
26. S. Yamada, K. Tomioka, and K. Koga, *Tetrahedron Lett.*, 61 (1976).
27. H. Akimoto, K. Okamura, M. Yui, T. Shioiri, and M. Kuramoto, *Chem. Pharm. Bull., Tokyo*, **22**, 2614 (1974).
28. M. R. Falco, J. X. de Vries, E. Marchelli, and H. Coda de Lorenzo, *Tetrahedron*, **28**, 5999 (1972).
28a. T. Kametani, K. Kigasawa, M. Hiiragi, H. Ishimaru, and K. Shiroyama, *J. Pharm. Soc. Japan*, **96**, 1031 (1976).
29. J. Sandrin, D. Soerens, L. Hutchins, E. Richfield, F. Ungemach, and J. M. Cook, *Heterocycles*, **4**, 1101 (1976).
30. T. Okawara and T. Kametani, *Heterocycles*, **2**, 571 (1974); and T. Kametani and T. Okawara, *J. Chem. Soc. Perkin I*, 579 (1977).
31. H. B. Kagan, N. Langlois, and T. P. Dang, *J. Organometal. Chem.*, **90**, 353 (1975).
32. L. Simon and G. Talpas, *Pharmazie*, **29**, 314 (1974). For the use of thioureas and thioamides in the Bischler–Napieralski cyclization, see A. A. B. Hazzaa, A. M. M. E. Omar, and M. S. Ragab, *Pharmazie*, **28**, 364 (1973); and M. S. Ragab, A. M. M. E. Omar, and A. A. B. Hazzaa, *Pharmazie*, **29**, 178 (1974). Alkyl or aryl thiocyanates can also be used in conjunction with phenacetyl chlorides, see M. A. Ainscough and A. F. Temple, *Chem. Commun.*, 695 (1976).
33. G. Fodor, J. Gal, and B. A. Phillips, *Angew. Chem. Int. Ed. Engl.*, **11**, 919 (1972).
34. P. E. Eaton, G. R. Carlson, and J. T. Lee, *J. Org. Chem.*, **38**, 4071 (1973).
35. E. F. Kiefer, *J. Med. Chem.*, **15**, 214 (1972). For the reduction of a β-nitrostyrene with aluminum amalgam, see E. McDonald and R. T. Martin, *Tetrahedron Lett.*, 1317 (1977).
36. J. H. Short, D. A. Dunnigan, and C. W. Ours, *Tetrahedron*, **29**, 1931 (1973).
37. M. A. Collins and F. J. Kernozek, *J. Heterocycl. Chem.*, **9**, 1437 (1972); and M. A. Collins, *Ann. N.Y. Acad. Sci.*, **215**, 92 (1973).
38. R. Sarges, *J. Heterocycl. Chem.*, **11**, 599 (1974).
39. S. F. Dyke, P. A. Bather, A. B. Garry, and D. W. Wiggins, *Tetrahedron*, **29**, 3881 (1973); P. Lejay, C. Viel, and G. Uchidaernouf, *Farm. Ed. Sci.*, **32**, 1 (1977); and P. Lejay and C. Viel, *Ann. Chim., Paris*, **2**, 87 and 127 (1977).
40. A. J. Birch, A. H. Jackson, and P. V. R. Shannon, *J. Chem. Soc. Perkin I*, 2185 (1974). See also G. A. Charnock, A. H. Jackson, J. A. Martin, and G. W. Stewart, *J. Chem. Soc. Perkin I*, 1911 (1974).
41. D. M. Bailey, C. G. DeGrazia, H. E. Lape, R. Frering, D. Fort, and T. Skulan, *J. Med. Chem.*, **16**, 151 (1973).
42. M. J. Bevis, E. J. Forbes, N. N. Naik, and B. C. Uff, *Tetrahedron*, **27**, 1253 (1971).

43. K. Kido and Y. Watanabe, *J. Pharm. Soc. Japan*, **95**, 1038 (1975).
44. M. Ikeda, K. Hirao, Y. Okuno, and O. Yonemitsu, *Tetrahedron Lett.*, 1181 (1974).
45. G. N. Dorofeenko, S. V. Krivun, and V. G. Korobkova, *Khim. Geterotsikl. Soedin.*, 1458 (1973); through *Curr. Abstr. Chem.*, **52**, issue 530, item 212844 (1974); and through *Chem. Abstr.*, **80**, 70649u (1974).
46. C. K. Bradsher and T. G. Wallis, *Tetrahedron Lett.*, 3149 (1972).
46a. S. D. Venkataramu, G. D. Mcdonell, W. R. Purdum, G. A. Dilbeck, and K. D. Berlin, *J. Org. Chem.*, **42**, 2195 (1977).
47. P. A. Wehrli and B. Schaer, *Synthesis*, 288 (1974).
48. J. Knabe, *Arch. Pharm.*, **292**, 416 (1959).
49. D. E. Schwartz and J. Rieder, *Clin. Chim. Acta*, **6**, 453 (1971).
50. J. Arthur, F. de Silva, N. Strojny, and N. Munno, *J. Pharm. Sci.*, **62**, 1066 (1973).
51. K. C. Agrawal, R. J. Cushley, S. R. Lipsky, J. R. Wheaton, and A. C. Sartorelli, *J. Med. Chem.*, **15**, 192 (1972); and K. C. Agrawal, P. D. Mooney, and A. C. Sartorelli, *J. Med. Chem.*, **19**, 970 (1976).
52. Y. Ogata and Y. Sawaki, *J. Am. Chem. Soc.*, **95**, 4692 (1973).
53. H. Stamm and J. Hoenicke, *Arch. Pharm.*, **307**, 340 (1974).
54. O. Hoshino, K. Ohyama, T. Taga, and B. Umezawa, *Chem. Pharm. Bull.*, *Tokyo*, **22**, 2587 (1974).
55. H. Hara, O. Hoshino, and B. Umezawa, *Heterocycles*, **3**, 123 (1975).
56. J. O'Brien, S. Teitel, and A. Brossi, *Synth. Commun.*, **2**, 171 (1972).
57. J. Kobor and K. Koczka, *Szegedi Tanakepzo Foiskola Tud. Kozl.* 179 (1969); through *Chem. Abstr.*, **78**, 4389s (1973).
58. A. Buzas, F. Cossais, J.-P. Jacquet, and A. Merour, *Bull. Soc. Chim. France*, 3476 (1973).
59. S. Queroix and J. Gardent, *C. R. Acad. Sci. Paris*, *C*, **276**, 703 (1973).
60. W. E. McEwen and R. L. Cobb, *Chem. Rev.*, **55**, 511 (1955).
61. F. D. Popp, *Adv. Heterocycl. Chem.*, **9**, 1 (1968).
62. F. D. Popp, *Heterocycles*, **1**, 165 (1973).
63. J. Knabe and A. Frie, *Arch. Pharm.*, **306**, 648 (1973). For a versatile new synthesis of Reissert compounds using an acid chloride and trimethylsilyl cyanide in methylene chloride, see S. Ruchirawat, N. Phadungkul, M. Chuankamnerdkarn, and C. Thebtaranonth, *Heterocycles*, **6**, 43 (1977).
64. H. W. Gibson, F. D. Popp, and A. Catala, *J. Heterocycl. Chem.*, **1**, 251 (1964).
65. T. Kametani, H. Nemoto, M. Takeuchi, M. Takeshita, and K. Fukumoto, *Heterocycles*, **4**, 921 (1976). This paper also gives references to earlier methods of alkylation at C-4. See also T. Kametani *et al.*, *J. Chem. Soc. Perkin I*, 386 (1977).
66. T.-K. Chen and C. K. Bradsher, *Tetrahedron*, **29**, 2951 (1973).
67. M. Natsume, S. Kumadaki, Y. Kanda, and K. Kiuchi, *Tetrahedron Lett.*, 2335 (1973). *N*-Methyl-1,2-dihydroisoquinolines can also be formylated at C-4 with $POCl_3$-DMF; see Ref. 39.
68. W. J. Gensler and K. T. Shamasundar, *J. Org. Chem.*, **40**, 123 (1975).
69. For other syntheses of C-4 alkylated isoquinolines, see W. J. Gensler, S. F. Lawless, A. L. Bluhm, and H. Dertouzos, *J. Org. Chem.*, **40**, 733 (1975).
69a. T. Shiraishi and H. Yamanaka, *Heterocycles*, **6**, 535 (1977).
70. L.-Å. Svensson, A. Björklund, and O. Lindvall, *Acta Chem. Scand. B*, **29**, 341 (1975).
71. G. F. Kirkbright, J. T. Stock, R. D. Pugliese, and J. M. Bobbitt, *J. Electrochem. Soc.*, **116**, 219 (1969).
72. J. M. Bobbitt, K. H. Weisgraber, A. S. Steinfeld, and S. G. Weiss, *J. Org. Chem.*, **35**, 2884 (1970).
73. J. M. Bobbitt, H. Yagi, S. Shibuya, and J. T. Stock, *J. Org. Chem.*, **36**, 3006 (1971).

74. J. M. Bobbitt, I. Noguchi, H. Yagi, and K. H. Weisgraber, *J. Am. Chem. Soc.*, **93**, 3551 (1971); and *J. Org. Chem.*, **41**, 845 (1976).
75. J. M. Bobbitt, *Heterocycles*, **1**, 181 (1973). For a study of the photo induced cyclization of bis-1,2-dihydroisoquinolines, see Y. Nakamura, J. Zsindely, and H. Schmid, *Heterocycles*, **5**, 427 (1976).
76. G. G. Lyle, *J. Org. Chem.*, **41**, 850 (1976).
77. R. G. Kinsman, S. F. Dyke, and J. Mead, *Tetrahedron*, **29**, 4303 (1973).
78. M. Sainsbury, S. F. Dyke, D. W. Brown, and R. G. Kinsman, *Tetrahedron*, **26**, 5265 (1970).
79. M. Sainsbury, D. W. Brown, S. F. Dyke, R. G. Kinsman, and B. J. Moon, *Tetrahedron*, **24**, 6695 (1968).
80. R. G. Kinsman and S. F. Dyke, *Tetrahedron Lett.*, 2231 (1975).
81. G. J. Kapadia, G. S. Rao, E. Leete, M. B. E. Fayez, Y. N. Vaishnav, and H. M. Fales, *J. Am. Chem. Soc.*, **92**, 6943 (1970).
81a. J. M. Bobbitt, C. L. Kulkarni, and P. Wiriyachitra, *Heterocycles*, **4**, 1645 (1976).
82. D. G. O'Donovan and E. Barry, *J. Chem. Soc. Perkin I*, 2528 (1974).
83. L. J. Fliedner, Jr., J. M. Schor, M. J. Myers, and I. J. Pachter, *J. Med. Chem.*, **14**, 580 (1971). See also R. L. Buchanan, V. Sprancmanis, T. A. Jenks, R. R. Crenshaw, G. M. Luke, H. M. Holava, and R. A. Partyka, *J. Med. Chem.*, **17**, 1241 (1974); and R. L. Buchanan, V. Sprancmanis, T. A. Jenks, R. R. Crenshaw, and G. M. Luke, *J. Med. Chem.*, **17**, 1248 (1974).
84. G. C. Wright and R. P. Halliday, *J. Pharm. Sci.*, **63**, 149 (1974).
84a. R. G. Pendleton, C. Kaiser and G. Gessner, *J. Pharmacol. Exp. Ther.*, **197**, 623 (1976).
85. B. Kotelanski, R. J. Grozman, and J. N. Cohn, *Clin. Pharmacol. Ther.*, **14**, 427 (1973); through *Chem. Abstr.*, **79**, 49340f (1973).
86. M. Sandler, S. Bonham-Carter, R. Hunter, and G. M. Stern, *Nature*, **241**, 439 (1973).
87. A. C. Collins, J. L. Cashaw, and V. E. Davis, *Biochem. Pharmacol.*, **22**, 2337 (1973).
88. A. Brossi, private communication.
89. S. Locke, G. Cohen, and D. Dembiec, *J. Pharmacol. Exp. Ther.*, **187**, 56 (1973).
90. V. M. Tennyson, G. Cohen, C. Mytilineou, and R. E. Heikkila, *Brain Res.*, **51**, 161 (1973).
91. C. Mytilineou, G. Cohen, and R. Barrett, *Eur. J. Clin. Pharmacol.*, **25**, 390 (1974).
92. R. S. Greenburg and G. Cohen, *J. Pharmacol. Exp. Ther.*, **184**, 119 (1973).
93. O. S. Lee, J. E. Mears, J. J. Bardin, D. D. Miller, and D. R. Feller, *Fed. Proc.*, 32,723 Abstr. (1973).
94. G. Cohen, R. E. Heikkila, D. Dembiec, D. Sang, S. Teitel, and A. Brossi, *Eur. J. Clin. Pharmacol.*, **29**, 292 (1974).
95. L. L. Simpson, *J. Pharmacol. Exp. Ther.*, **192**, 365 (1975).
96. E. E. Smissman, J. R. Reid, D. A. Walsh, and R. T. Borchardt, *J. Med. Chem.*, **19**, 127 (1976).
97. M. A. Collins, in *Alcohol and Opiates*, K. Blum, ed., Academic Press, New York (1977), p. 155.
98. M. Hirst, M. G. Hamilton, and A. M. Marshall, in *Alcohol and Opiates*, K. Blum, ed., Academic Press, New York (1977), p 167.
99. D. W. Hughes, H. L. Holland, and D. B. MacLean, *Can. J. Chem.*, **54**, 2252 (1976).

THE BENZYLISOQUINOLINES

2

2.1. Introduction

New benzylisoquinolines of interest are shown in Scheme 2.1.

(−)-Macrostomine is the first benzylisoquinoline alkaloid to possess an α-(N-methylpyrrolidine) substituent at C-4. The absolute configuration of this base was deduced by comparison of its CD curve with those of (−)-nicotine and (−)-brevicolline.[1] New sources of benzylisoquinolines are the Celastraceae,[7] the Fumariaceae,[8,9] and the Leguminosae.[10]

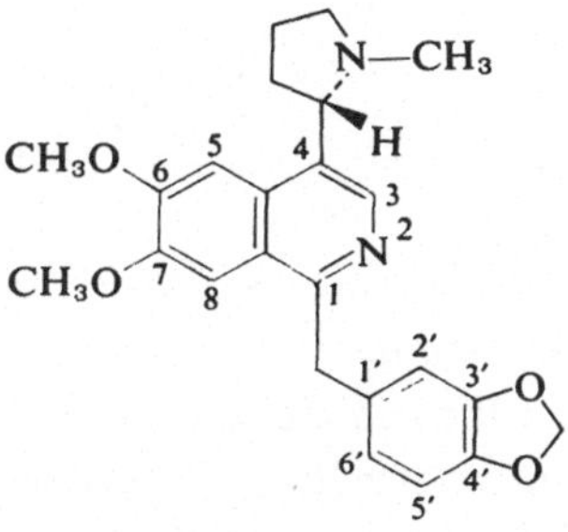

(−)-Macrostomine[1]
Papaver macrostomum
Boiss. et Huet

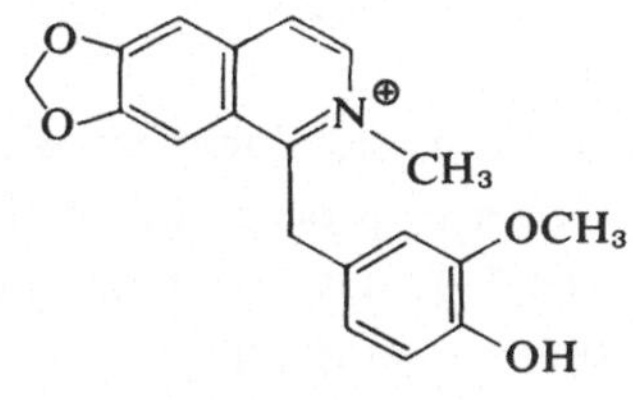

Escholamidine[2]
Eschscholtzia oregana
Greene?

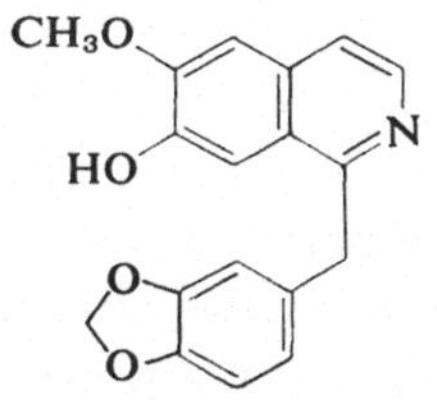

Sevanine[1,3]
P. macrostomum

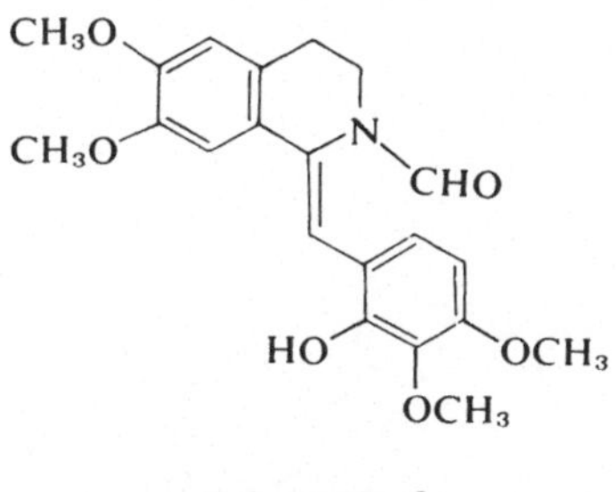

Polycarpine[3]
Enantia polycarpa
Engl. and Diels.

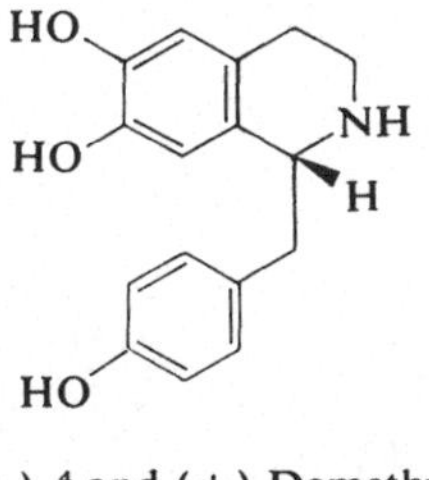

(+)-[4] and (±)-Demethyl-
coclaurine[5]
Nelumbo nucifera Gaertn.[4] and
Aconitum japonicum Decne.[5]

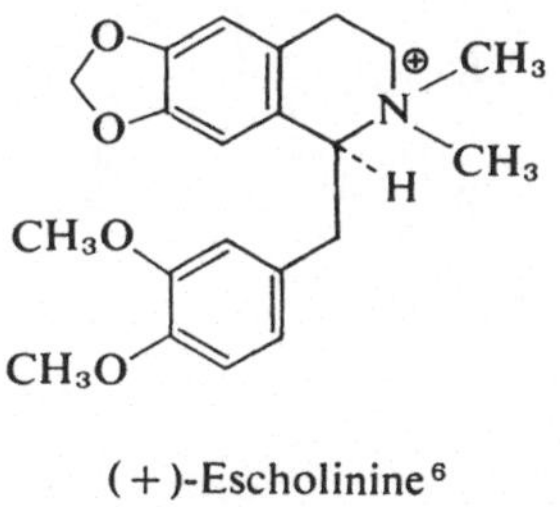

(+)-Escholinine[6]
Eschscholtzia californica Cham.

SCHEME 2.1

27

2.2. Synthesis[11]

2.2.1. The Use of Reissert Compounds[12]

A new synthesis of petaline proceeds through a Reissert derivative of 7-methoxy-8-hydroxyisoquinoline,[13] as shown below.

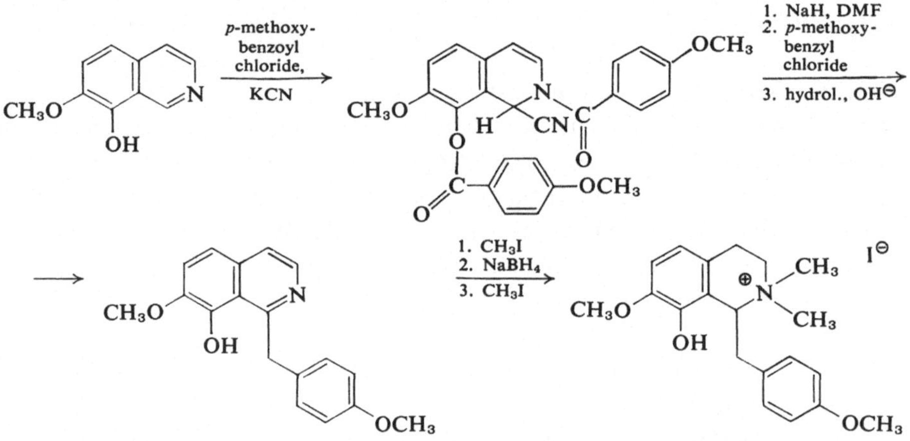

Petaline

The condensation of aromatic aldehydes with the anion of an *N*-benzoyl Reissert compound is known to proceed through an intramolecular rearrangement to give esters of type **1**. It has now been shown that condensation of the anion of the *N*-acetyl Reissert compound **2** with 2-nitrobenzaldehyde proceeds in a different manner to afford the derivative **3**,[14] the by-product being isoquinoline.[14]

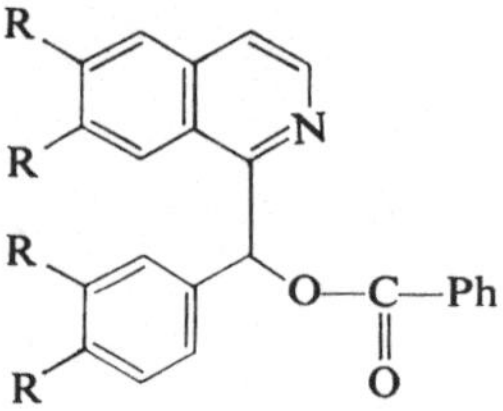

1

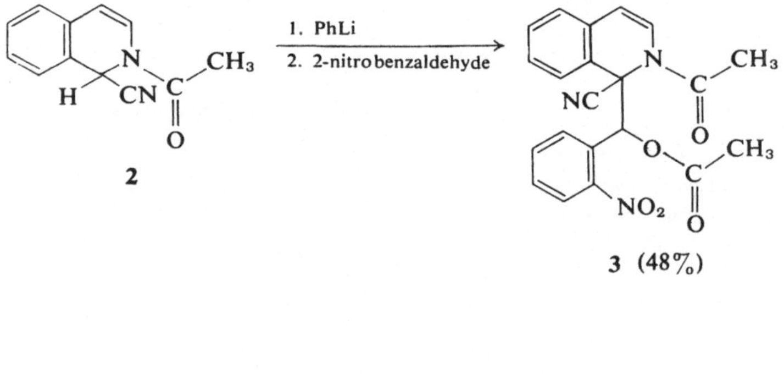

2

3 (48%)

2.2.2. Friedel–Crafts Alkylation and Acylation

A new preparation of 1-benzyltetrahydroisoquinoline starts with benzyl phenyl ketone and uses a Friedel–Crafts alkylation in the final step to form the tetrahydroisoquinoline.[15]

Friedel–Crafts acylation has been utilized in a new preparation of 1-benzyl-4-oxoisoquinolines. The ketone **4**, itself obtained by Friedel–Crafts acylation, gave under Leuckart conditions the formamide **5**. Hydrolysis under acidic or basic conditions generated the corresponding amine which was converted to the glycine ester. Hydrolysis, followed by cyclization of the carboxylic acid, provided a white crystalline material assumed to be the ketonic benzylisoquinoline **6**.[16]

Reaction of the keto acid **7**, obtained through Friedel–Crafts acylation, with ammonium acetate in acetic acid gives the yellow pyridone **8** as the major product. Compound **8** is amphoteric, forming both a sodium salt and a stable hydrochloride. Alkylation with methyl iodide in alkaline methanol affords a mixture of the N- and O-methylated products **9** and **10**.[17] The N-methylpyridone **9** is unstable and oxidized by air to the bicyclic keto imide **11**.[18,19]

It would not be surprising if **11** were to be found in nature as an oxidative degradation product of a benzylisoquinoline alkaloid.

The colorless minor product obtained from the cyclization of the keto acid **7** has been characterized as the aminonaphthol **12**.[18]

Another reaction sequence of the yellow pyridone **8** involves its catalytic reduction to the lactam **13** which was further reduced to norlaudanosine. A related sequence was then adopted to prepare laudanosine itself.[19]

Norlaudanosine
(Tetrahydropapaverine)

2.2.3. Pictet–Spengler Condensation

The synthesis of a variety of tetrahydroisoquinoline alkaloids by phenolic cyclization has been reviewed.[20] A recent example of such an approach resulted in a synthesis of isococlaurine.[21]

Isococlaurine

A synthesis of (+)-laudanosine was accomplished by condensation of the L-amino ester hydrochloride **14** with the glycidate salt **15** at pH 4 to provide a 44% yield of a separable diastereomeric mixture of **16** and **17**. Elimination of the chiral center at C-3 was accomplished by conversion of the methyl ester to a nitrile group which was then cleaved reductively[22] as shown in Scheme 2.2. A similar approach was used to prepare (+)-reticuline, the diphenolic analog of (+)-laudanosine.[22]

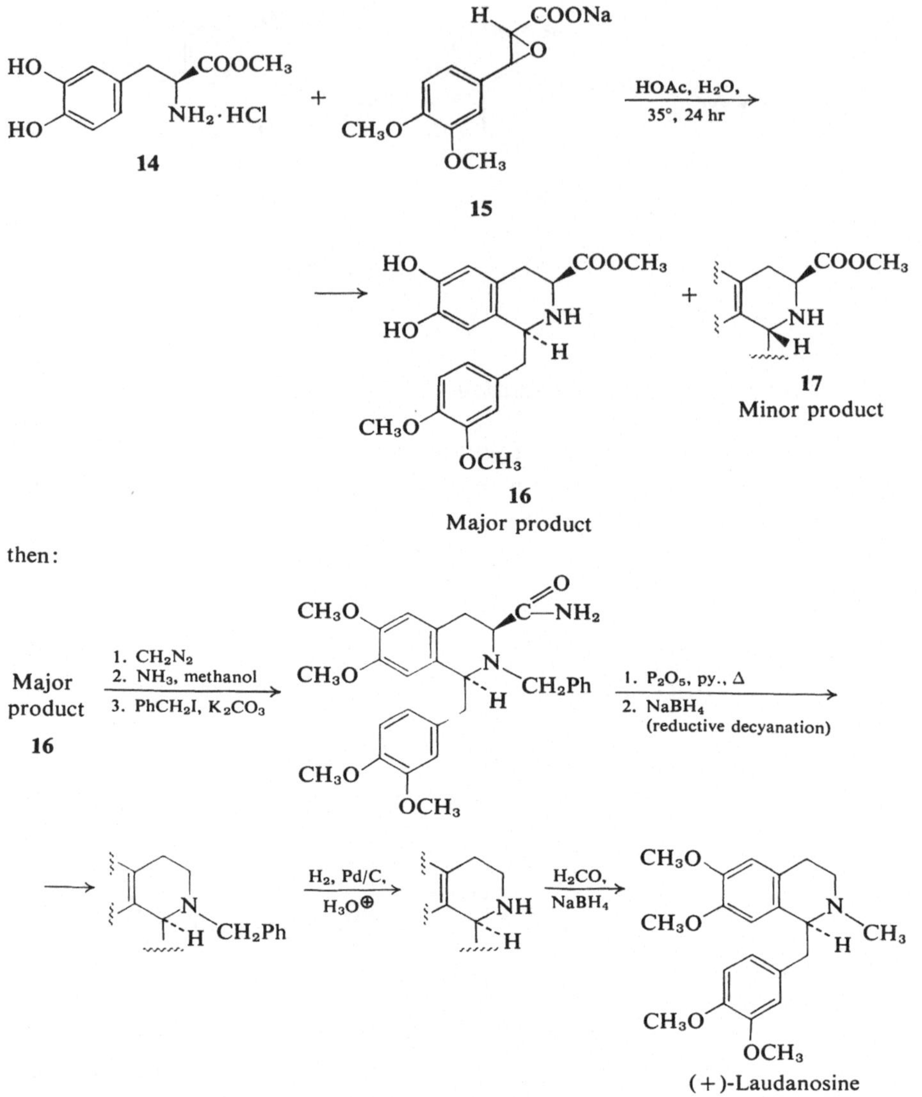

SCHEME 2.2

2.2.4. The Pomeranz–Fritsch/Reissert Approach

A practical route to the benzylisoquinolines utilizes the Jackson modification of the Pomeranz–Fritsch cyclization (see Sec. 1.2.3), followed by C-1 alkylation of the isoquinoline intermediate using a Reissert compound. For example, takatonine was prepared in 59% overall yield from 3,4,5-trimethoxybenzaldehyde by the sequence shown in Scheme 2.3.[23]

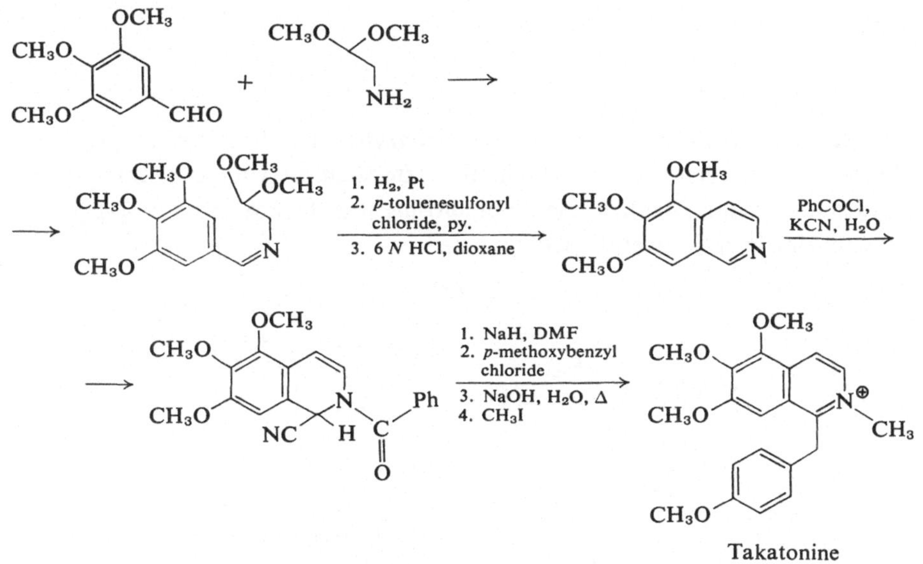

SCHEME 2.3

2.2.5. Cyclization of N-Sulfonylphenethylamines

N-Sulfonylphenethylamines have been readily condensed with aldehydes under acidic conditions to provide N-sulfonyltetrahydroisoquinolines or N-sulfonyltetrahydrobenzylisoquinolines. Nororientaline, found as a natural product in *Erythrina* X *bidwillii* (Leguminosae), was prepared by this approach. In the last step, N-detosylation, as well as O-debenzylation, was accomplished with sodium in liquid ammonia.[23a]

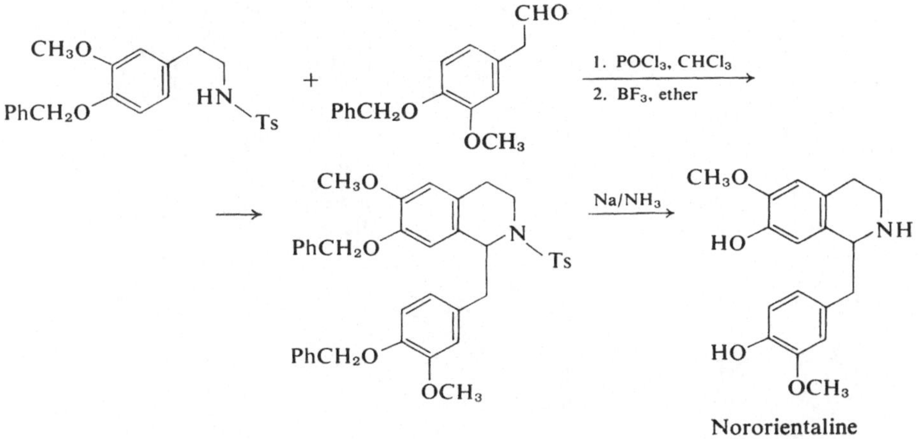

Nororientaline

2.3. Reactions of Benzylisoquinolines

2.3.1. Oxidation

Selective oxidation of a tetrahydrobenzylisoquinoline can be performed through the sequence monobromination, metallation with butyllithium, and oxygenation using nitrobenzene. Laudanosine was thus oxidized to 6'-hydroxylaudanosine in 65% overall yield.[24]

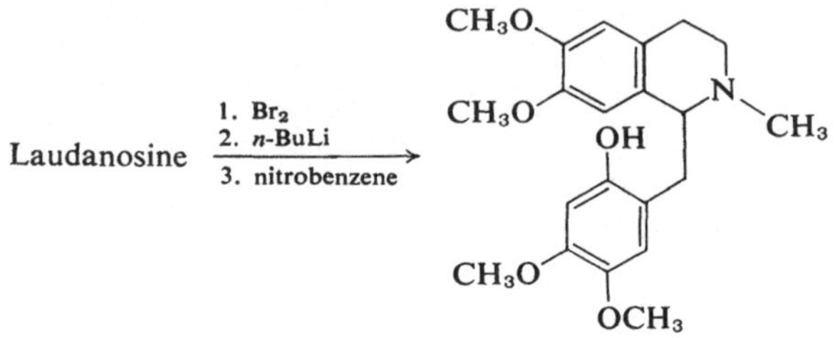

6'-Hydroxylaudanosine

The oxidation of benzylisoquinolines and 3,4-dihydrobenzylisoquinolines at C-α to 1-aroylisoquinolines can be conveniently achieved using oxygen and triton-B in pyridine.[25]

2.3.2. Reduction and the Chemistry of the Reduction Products

Some tetrahydrobenzylisoquinolines hydroxylated at C-α, e.g., **18**, could not be efficiently obtained by direct reduction of the corresponding α-hydroxylated benzylisoquinolines. Instead, it was found necessary to proceed through the intermediacy of the 4-oxazolin-2-one system as a protecting group. The sequence, shown in Scheme 2.4, was also adapted to the preparation of ketone **19**.[26]

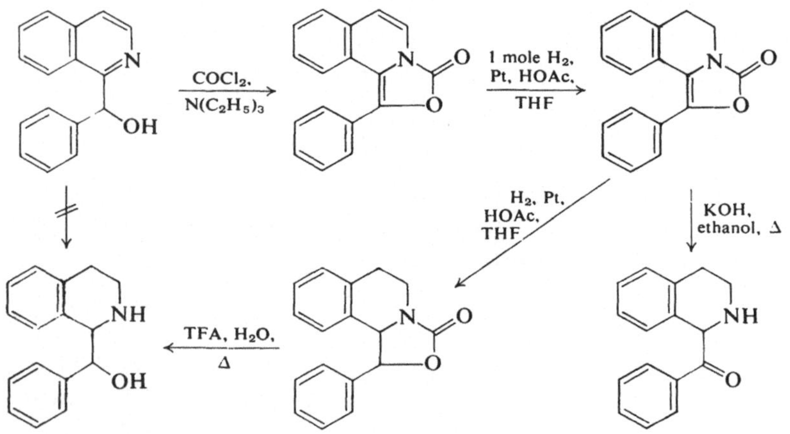

SCHEME 2.4

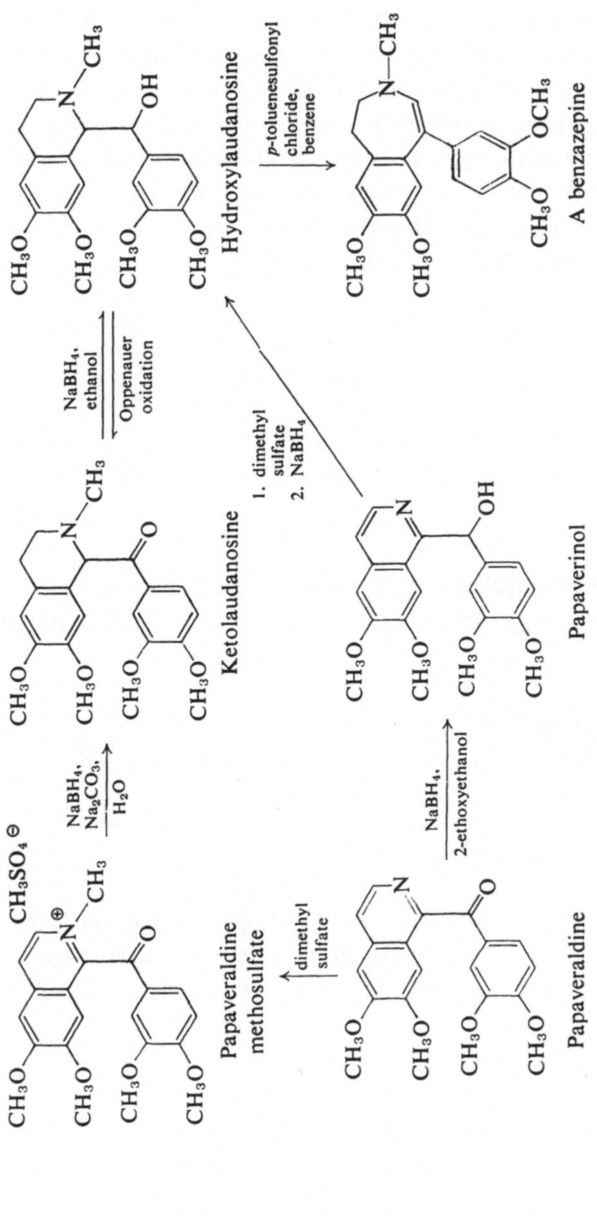
Papaveraldine methosulfate
dimethyl sulfate
Papaveraldine
NaBH4, Na2CO3, H2O
NaBH4, 2-ethoxyethanol
Ketolaudanosine
Papaverinol
1. dimethyl sulfate 2. NaBH4
NaBH4, ethanol
Oppenauer oxidation
Hydroxylaudanosine
p-toluenesulfonyl chloride, benzene
A benzazepine

SCHEME 2.5

On the other hand, papaveraldine methosulfate can be reduced to keto-laudanosine which, in turn, can be converted to hydroxylaudanosine. Alternatively, papaveraldine itself can be reduced first to papaverinol; N-methylation followed by reduction provides hydroxylaudanosine.[27] Subsequent rearrangement of this β-amino alcohol induced by p-toluenesulfonyl chloride in pyridine was found to give a small yield of a benzazepine (see Scheme 2.5).[28]

Reaction of papaverinol with glacial acetic acid produces papaverine and papaveraldine through hydride transfer. However, treatment of papaverinol with 90% sulfuric acid results in some O-demethylation of one of the methoxyl groups in ring C, together with some cleavage of the C-α to C-1′ bond to form a substituted isoquinoline-1-carboxaldehyde.[29]

When a tetrahydrobenzylisoquinoline is subjected to Clemmensen conditions, benzylic cleavage occurs with formation of a simple tetrahydroisoquinoline.[30]

A chiral rhodium complex with (+)-diop as ligand has been used for an asymmetric reduction of papaverine to its tetrahydro derivative.[31]

2.3.3. Photolysis

Oxidation of papaverine with m-chloroperbenzoic acid yields papaverine N-oxide. Its irradiation in acetone results in the formation of a small amount of papaverine, together with the 1,3-benzoxazepine **20** and its unstable valence tautomer **21** which upon acid work-up of the reaction mixture leads to furans **22** and **23** (see Scheme 2.6).[32]

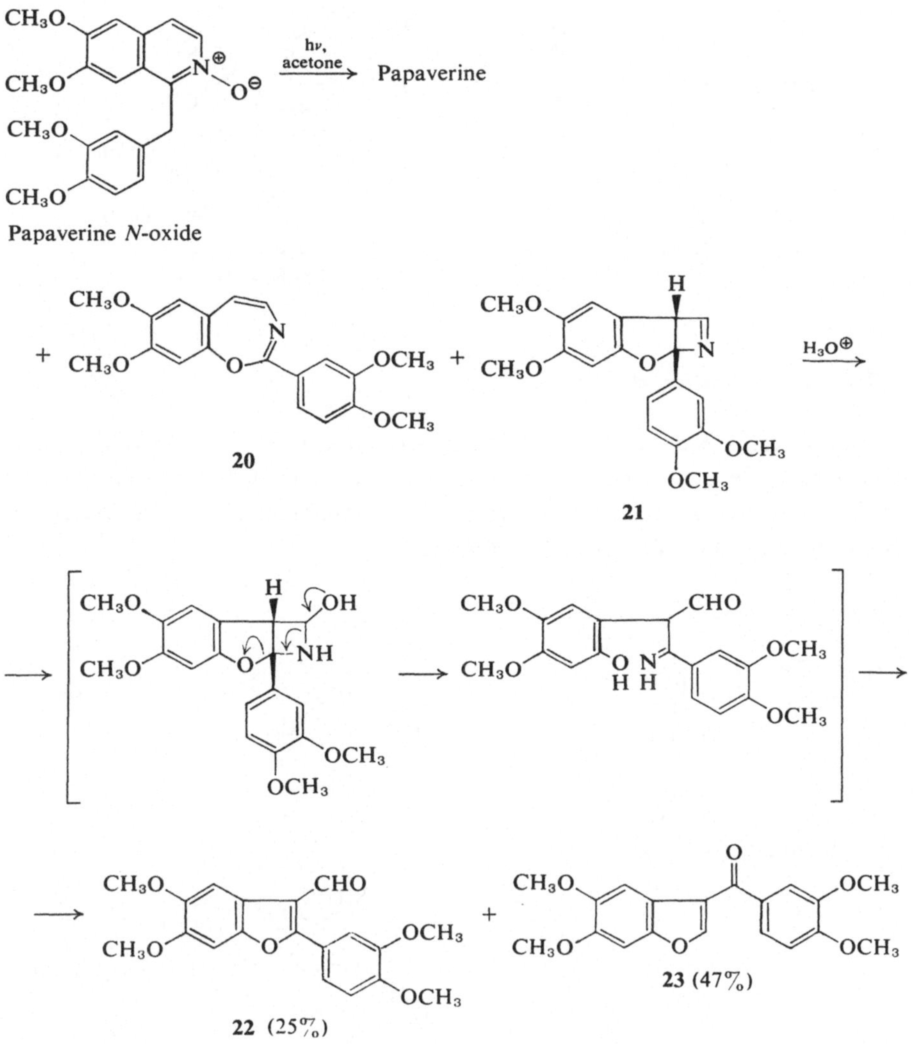

Papaverine N-oxide

20   21

22 (25%)   23 (47%)

SCHEME 2.6

Irradiation of laudanosine methiodide in methanol gives the ether 24 in 80% yield. In aqueous acid, the photolysis reaction provides instead a 54% yield of the alcohol 25.[33]

In chloroform solution, papaverine decomposes in sunlight to a mixture of papaveraldine, papaverinol, papaverine N-oxide, and 6,7-dimethoxyisoquinoline N-oxide.[34]

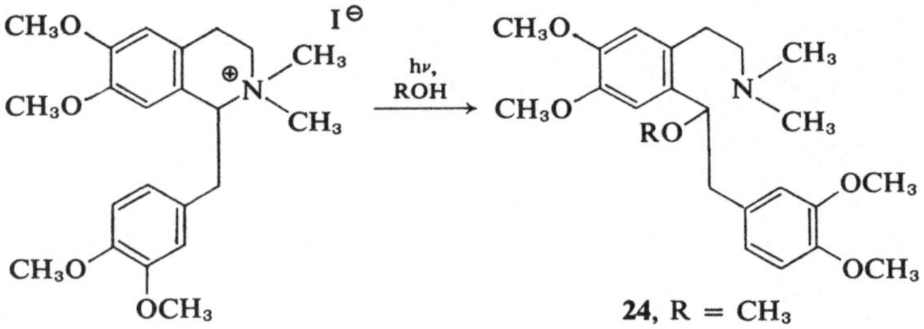

24, R = CH$_3$
25, R = H

2.3.4. Acetonylation at C-6'

Papaverine can be alkylated or acylated at C-6'.[35,36] π-(2-Methoxyallyl)-nickel bromide has recently been introduced as a reagent for acetonylation, and 6'-acetonylpapaverine was prepared by the reaction of the known 6'-bromopapaverine with this nickel complex.[37]

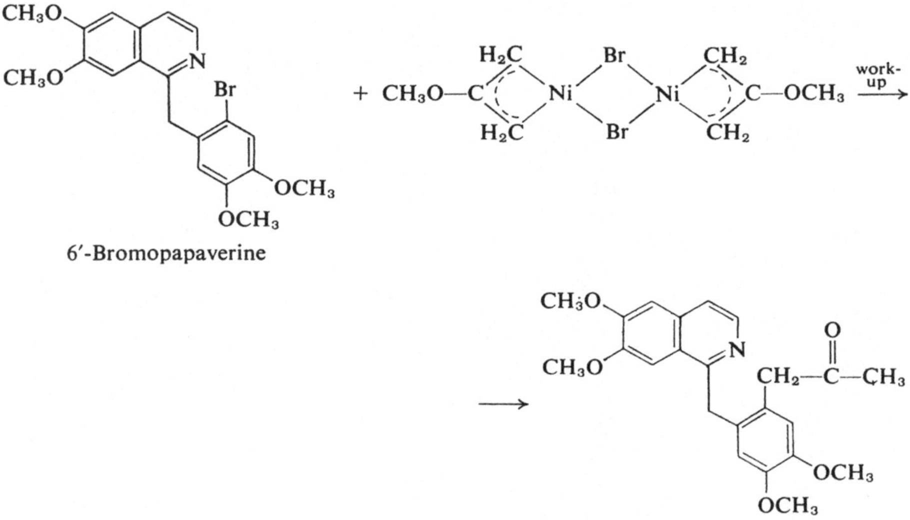

6'-Bromopapaverine

6'-Acetonylpapaverine

2.3.5. Fission of Ring B[38]

Reaction of the known 6'-hydroxymethyllaudanosine with ethyl chloroformate results in opening of ring B with simultaneous formation of a six-membered cyclic ether (see Sec. 20.1.4)[39]:

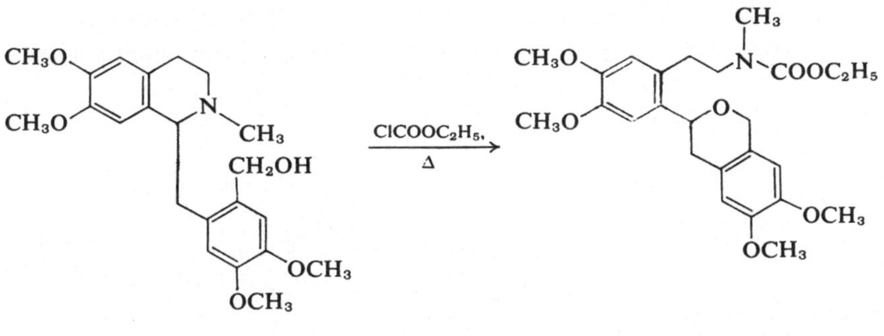

CH₃O figure

6′-Hydroxymethyllaudanosine

Another method for cleavage of ring B involves treatment of 3,4-dihydro-papaverine methiodide with ethyl chloroformate in base, giving the keto urethan **26**. A parallel reaction occurs when the 6′-acetyl derivative of dihydro-papaverine methiodide is used.[40] Such cleavage of ring B is a well-known transformation of 3,4-dihydroisoquinolines (see Sec. 26.2.3).

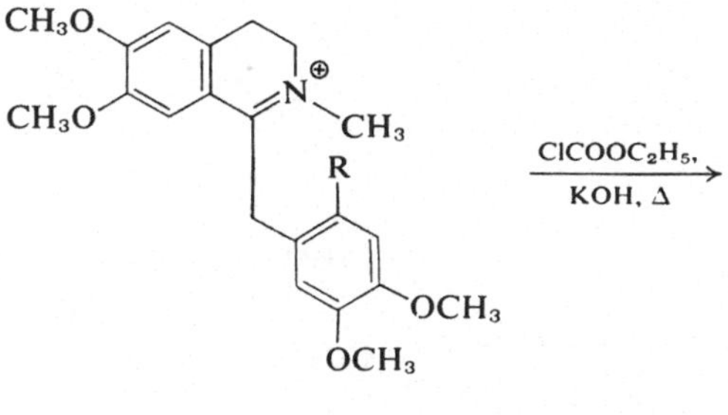

R = H or acetyl

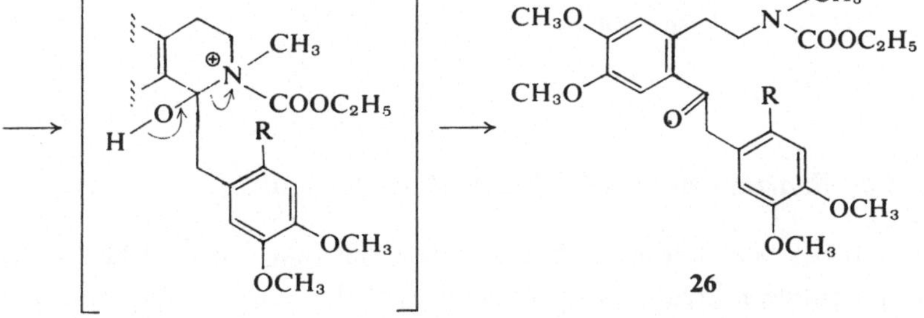

26

R = H or acetyl
(> 75%)

If, however, 6′-acetyl-3,4-dihydropapaverine methiodide is treated with an excess of ethyl chloroformate and base, the product is the naphthalene ester **27**.[40]

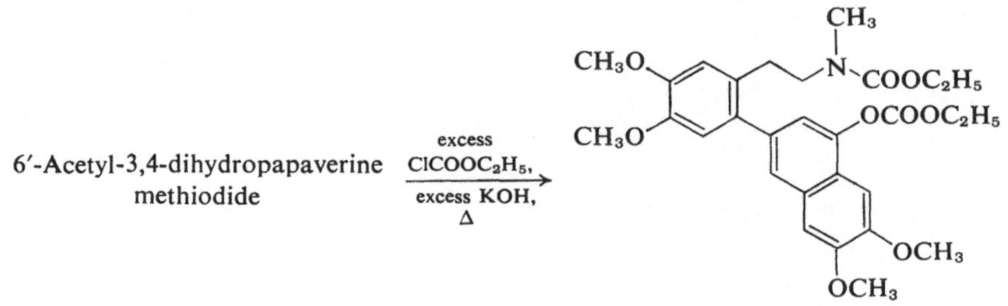

6′-Acetyl-3,4-dihydropapaverine
methiodide

excess
ClCOOC$_2$H$_5$,
⟶
excess KOH,
Δ

27
(58%)

Reaction of laudanosine with benzyl chloroformate in the presence of sodium hydroxide gave a *trans*-stilbene which was epoxidized with *m*-chloroperbenzoic acid. A benzazepine was then isolated upon acid treatment.[41]

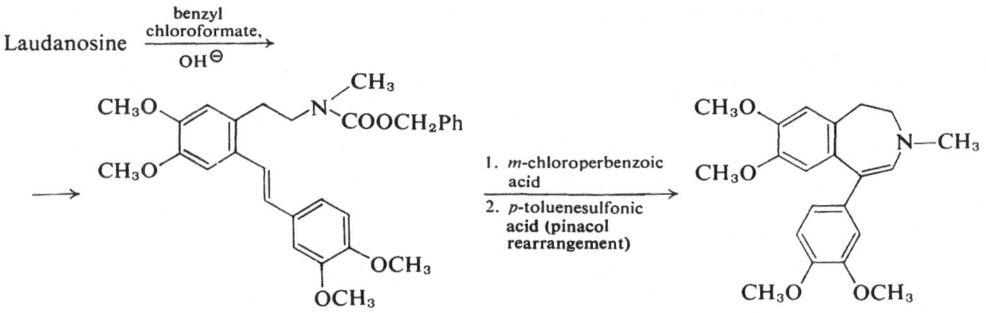

Laudanosine

benzyl
chloroformate,
⟶
OH$^\ominus$

1. *m*-chloroperbenzoic acid
2. *p*-toluenesulfonic acid (pinacol rearrangement)

A *trans*-stilbene

A benzazepine

2.3.6. Rearrangement of 1,2-Dihydroisoquinolines

It has been known since the early 1960s that when a 1-benzyl-1,2-dihydroisoquinoline is treated with mineral acid the benzyl group can migrate from C-1 to C-3.[41a] The reaction is intermolecular, and an increase in the size of the nitrogen substituent results in a decrease in the yield of rearrangement product. The transformation involves initial protonation of the enamine at C-4 to form a 1,4-dihydroisoquinoline intermediate, e.g., **29**.

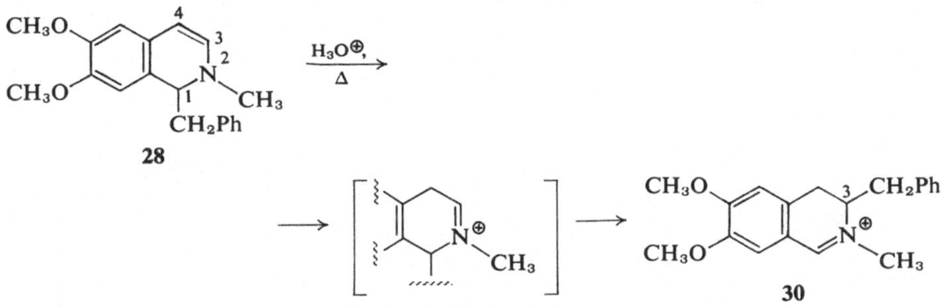

It has now been shown by Knabe, Dyke, and their co-workers, that this interesting rearrangement occurs by a bimolecular exchange mechanism. Cyclization of an optically active (*) sample of the amino acetal **31** provides optically active **30**, most probably through transition state **32**.[42]

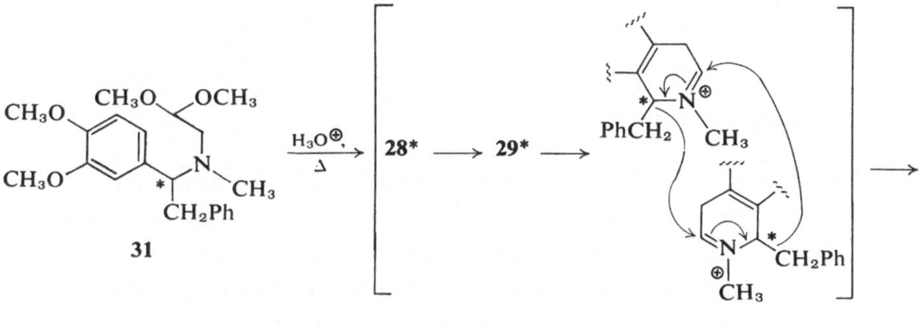

Salt **30**,
optically active
at C-3

2.3.7. Transformation of α-Ketobenzylisoquinolines

Gardent and co-workers have found that α-keto-3,4-dihydroisoquinolines, such as **33**, slowly undergo self-condensation in acid solution, the product being a highly substituted imidazolinium salt.[43] Compound **34** will also condense with methyl vinyl ketone (MVK) in protic solvents to yield a substituted pyrrole, **34**.[44]

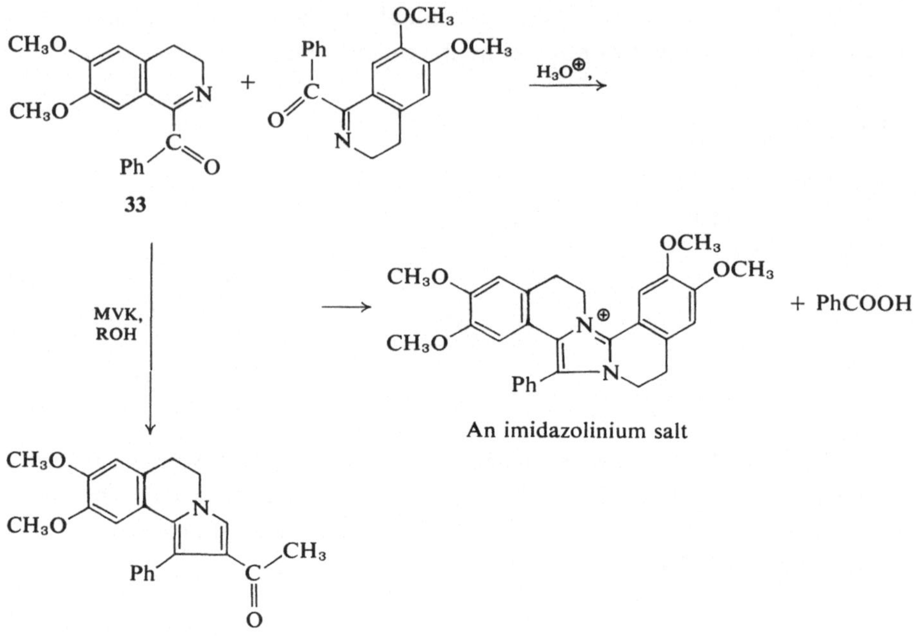
33
MVK, ROH
+ PhCOOH
An imidazolinium salt

34

2.3.8. *Electrolytic Oxidation*

The electrochemical intramolecular coupling of both nonphenolic[45] and diphenolic[46] tetrahydrobenzylisoquinolines to supply morphinandienones is now well established (see the two equations below).

A series of oxygenated tetrahydroisoquinoline-1-carboxylic acids has been decarboxylated by electrochemical oxidation to 3,4-dihydroisoquinolines, and these were reduced with sodium borohydride to their tetrahydro analogs. The ease of decarboxylation could be correlated with the electron density of the

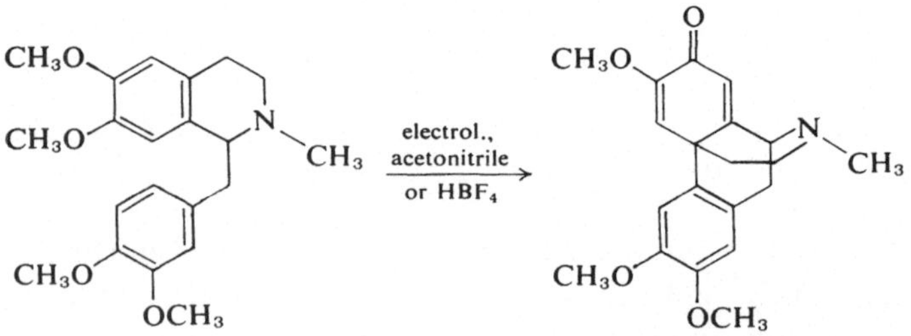
electrol., acetonitrile or HBF_4

O-Methylflavinantine[45]
(52%, CH_3CN)
(94%, HBF_4)

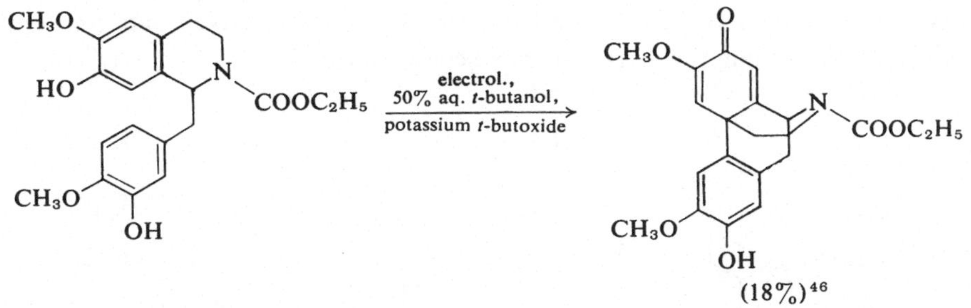

(18%)[46]

aromatic ring. This decarboxylation parallels the biogenetic process since it has been demonstrated that simple isoquinolines and benzylisoquinolines are formed *in vivo* by a similar pathway. An intriguing possibility is that more complex isoquinoline alkaloids, for instance the bisbenzylisoquinolines, could be formed *in vivo* through phenolate addition to the dihydroisoquinoline intermediate in its quinoid form. If such is the case, no oxidation would be required in the dimerization step.[47]

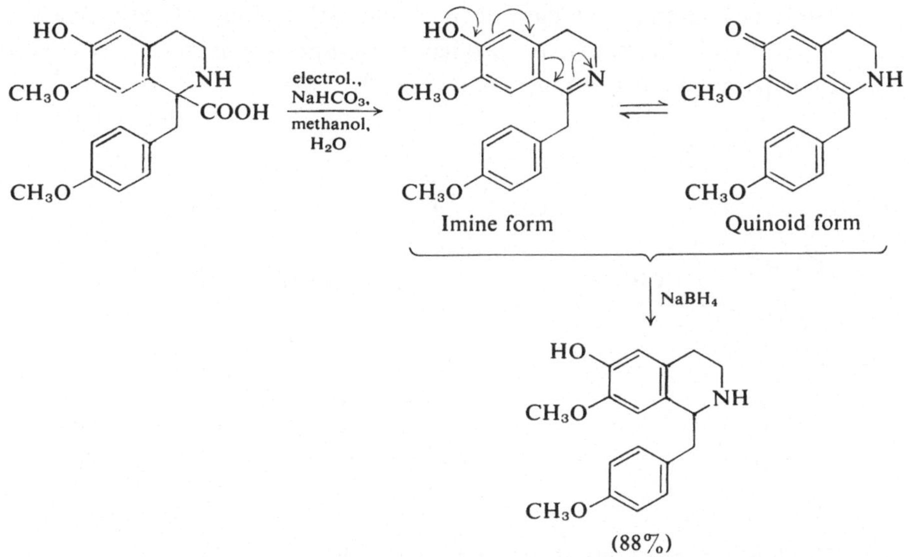

Imine form   Quinoid form

(88%)

No oxidative decarboxylation occurred when the nitrogen was *N*-acetylated or when no free phenolic group was present in ring A of the starting carboxylic acid.[47]

2.3.9. Oxidative Dimerization

Papaverine can be dimerized by oxidation with vanadium oxytrifluoride in trifluoroacetic acid. Such condensation does not usually occur with tetrahydrobenzylisoquinolines which instead tend to undergo intramolecular coupling (see Sec. 10.2.3).[48] The topic of phenolic oxidative coupling in the synthesis of isoquinoline alkaloids has been reviewed.[49]

Papaverine

2.4. Conversion of Benzylisoquinolines to Morphine

The many attempts at para-ortho oxidative coupling of reticuline have usually resulted in the para–para product isosalutaridine in small yield, and/or the ortho–para coupling product isoboldine.[50]

Reticuline   Isosalutaridine   Isoboldine

The long sought laboratory analogy for the *in vivo* para–ortho coupling of reticuline to salutaridine has now been achieved by Schwartz and Mami using the thallium tristrifluoroacetate (TTFA) coupling of *N*-ethoxycarbonyl-norreticuline (see Scheme 2.7). The product was the salutaridine derivative **35** isolated in a remarkable 23% yield. Reduction of **35** with lithium aluminum hydride in THF provided a 1:1 mixture of diastereomeric salutaridinols which upon treatment with 1 *N* HCl afforded thebaine. Since thebaine had previously been converted to codeinone, and the latter to codeine and thence to morphine, this work constitutes a formal synthesis of all these alkaloids in their racemic

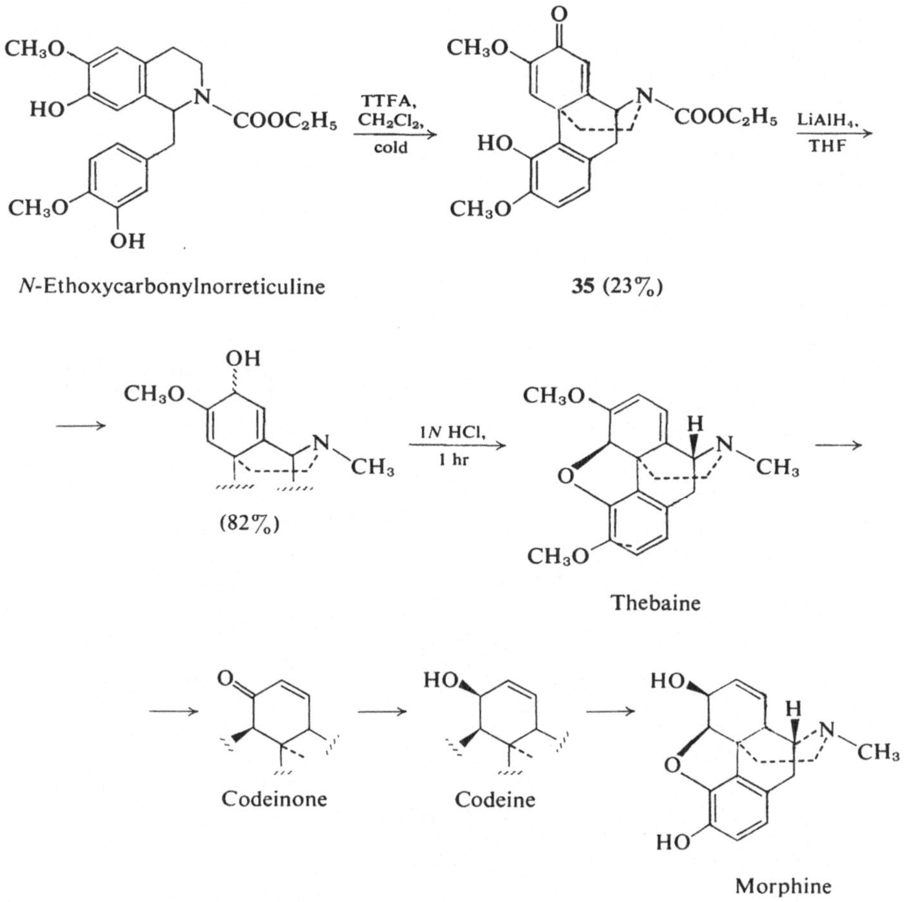

N-Ethoxycarbonylnorreticuline   35 (23%)

(82%)

Thebaine

Codeinone   Codeine

Morphine

SCHEME 2.7

forms. The influence of TTFA as compared with other reagents, e.g., $K_3Fe(CN)_6$, MnO_2–SiO_2, Ag_2CO_3–Celite, $VOCl_3$, $FeCl_3$, in directing para–ortho coupling in this system is quite likely a coordination phenomenon.[51]

In another efficient approach to morphine (see Scheme 2.8), the 3,4-dihydro-benzylisoquinoline **36** is reduced with lithium in liquid ammonia, followed by N-formylation to give the amide **37**. Acid-catalyzed cyclization of **37** gives the tetracyclic ketoamide **38**. The 2-hydroxyl substituent is removed by the Musliner–Gates procedure, and the resulting material reduced to dihydrothebainone.[52] The conversion of (−)-dihydrothebainone into (−)-morphine has already been recorded.

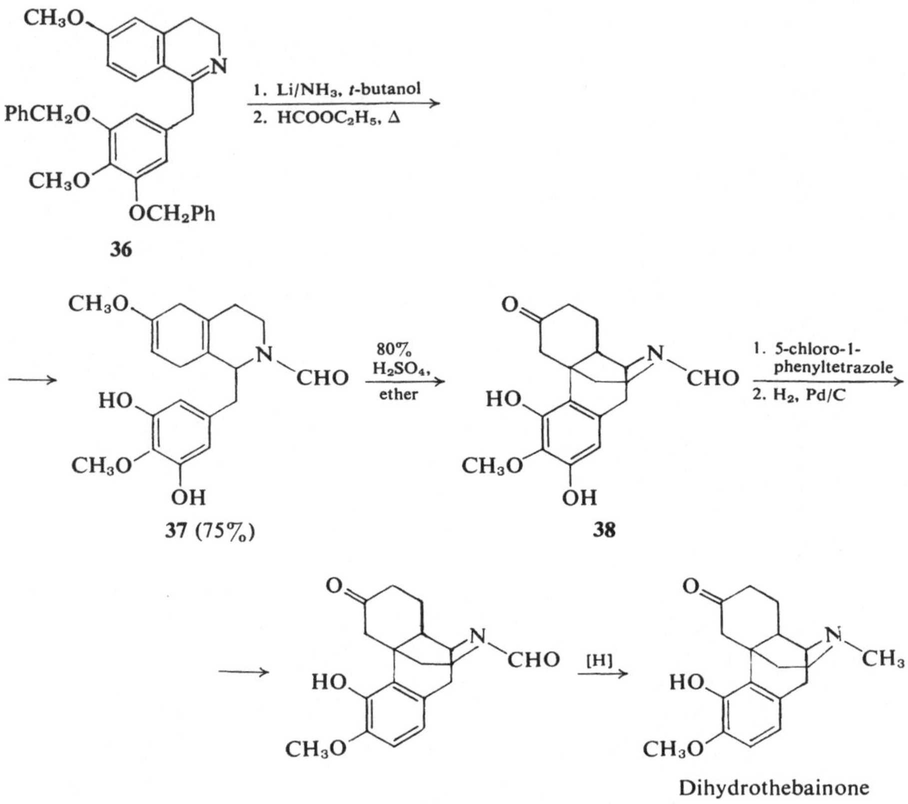

SCHEME 2.8

2.5. Cryptopleurospermine, an Unusual Natural α-Dione

Johns, Lamberton, and their co-workers have found the unusual alkaloid cryptopleurospermine in an Australian member of the Lauraceae, *Cryptocaria pleurosperma* White & Francis. Cryptopleurospermine, $C_{20}H_{21}NO_6$, is colorless and optically inactive, and shows ν_{max} 1660 cm^{-1} (6.03 μ), λ_{max}^{EtOH} 233, 284, and 326 nm (4.45, 4.13, and 4.16, respectively). Sodium borohydride reduction to the diol, followed by oxidation with periodic acid furnished two aldehydes, isovanillin and 2-dimethylaminoethyl-4,5-methylenedioxybenzaldehyde (see Scheme 2.9). The IR, UV, PMR, and mass spectra of the alkaloid are completely consonant with the structure assigned.[53] The biogenesis of cryptopleurospermine probably involves oxidation of a tetrahydrobenzylisoquinoline precursor to an α-ketoimmonium salt which, upon conversion to its pseudobase and further N-alkylation, undergoes cleavage of ring B.

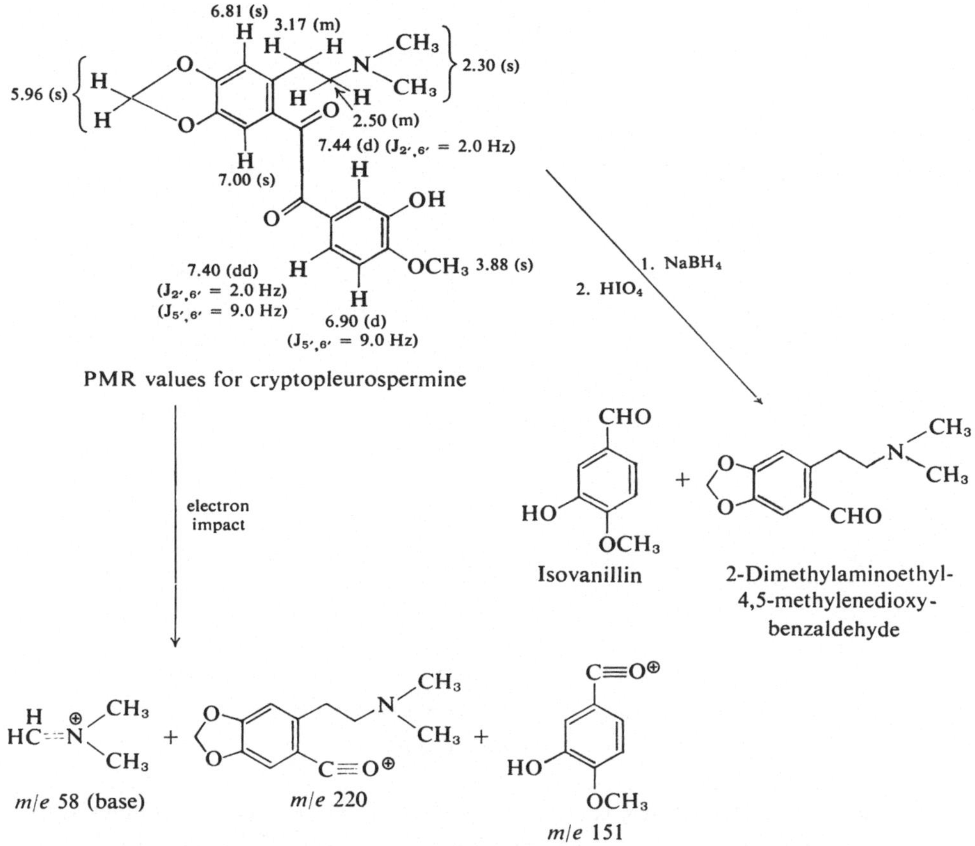
6.81 (s)
3.17 (m)
2.30 (s)
5.96 (s)
2.50 (m)
7.44 (d) (J2',6' = 2.0 Hz)
7.00 (s)
OH
7.40 (dd)
(J2',6' = 2.0 Hz)
(J5',6' = 9.0 Hz)
OCH3 3.88 (s)
6.90 (d)
(J5',6' = 9.0 Hz)
PMR values for cryptopleurospermine
1. NaBH4
2. HIO4
electron impact
Isovanillin
2-Dimethylaminoethyl-4,5-methylenedioxy-benzaldehyde
m/e 58 (base)
m/e 220
m/e 151

SCHEME 2.9

2.6. Biogenesis[53a]

In an important series of experiments involving ^{14}C and ^{3}H doubly labeled precursors, Wilson and Coscia have shown that in *Papaver orientale* L., dopa may be metabolized in at least two ways to yield dopamine and 3,4-dihydroxyphenylpyruvic acid. Decarboxylation of dopa to yield dopamine is actually favored over transamination to the substituted pyruvic acid. Norlaudanosoline carboxylic acid is the first tetrahydrobenzylisoquinoline alkaloid to be formed, and it is readily converted to norlaudanosoline through the presumed intermediacy of an imine (see Scheme 2.10).[54]

For these experiments to be successful, it is imperative that the labeled precursors be fed directly to the latex in the poppy capsules, otherwise the added labeled dopa fails to penetrate to the site of the appropriate aminotransferase

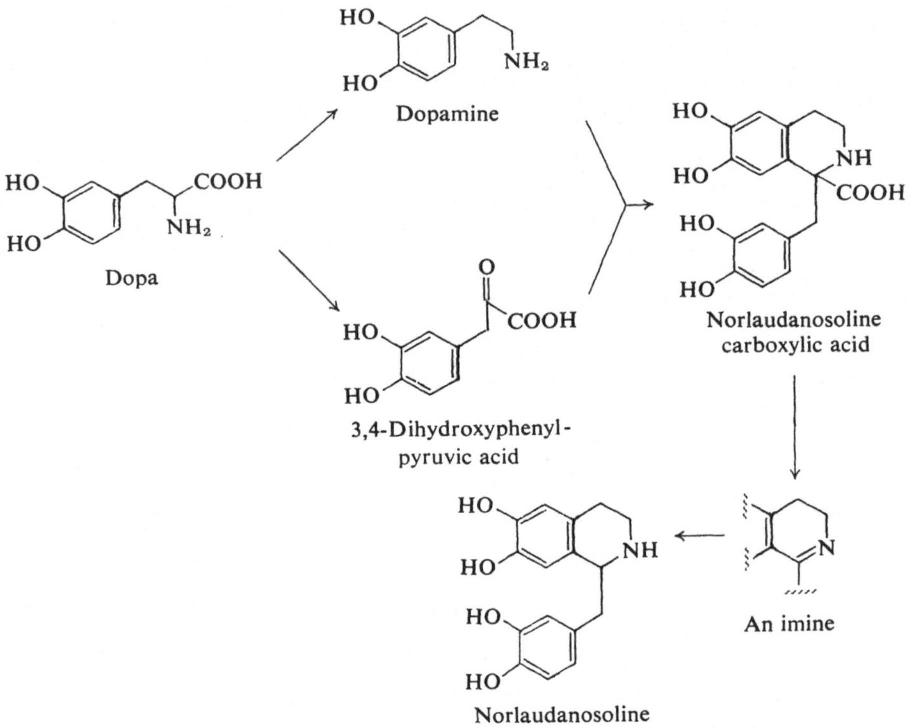

SCHEME 2.10

enzyme.[54–56] The pathway for the biogenesis of the benzylisoquinolines thus bears a strong similarity to that previously established[57] for the simple isoquinolines (see Sec. 1.4). Significantly, analogs of norlaudanosoline carboxylic acid still bearing a phenolic function at C-6 undergo facile decarboxylation to the corresponding imine when stirred in air under basic conditions.[57a]

Working with *Litsea glutinosa* (Lauraceae), and concentrating on the biosynthesis of reticuline, Bhakuni and co-workers have shown that dopa and dopamine contribute only to the formation of the phenethylamine portion of reticuline. The benzylic portion is biosynthesized from 3,4-dihydroxyphenylpyruvic acid not derived from dopa. Tyrosine, 4-hydroxy-, and 3,4-dihydroxyphenylpyruvic acid all participate in the formation of both halves of the molecule. Norlaudanosoline carboxylic acid, norlaudanosoline, and didehydronorlaudanosoline are intermediates in the biosynthetic sequence; *O*-methylation precedes *N*-methylation.[57b]

The biosynthesis of papaverine in *P. somniferum* L. has been investigated using [14]C- and ^{3}H-labeled precursors and follows the sequence shown in Scheme 2.11.[58,59]

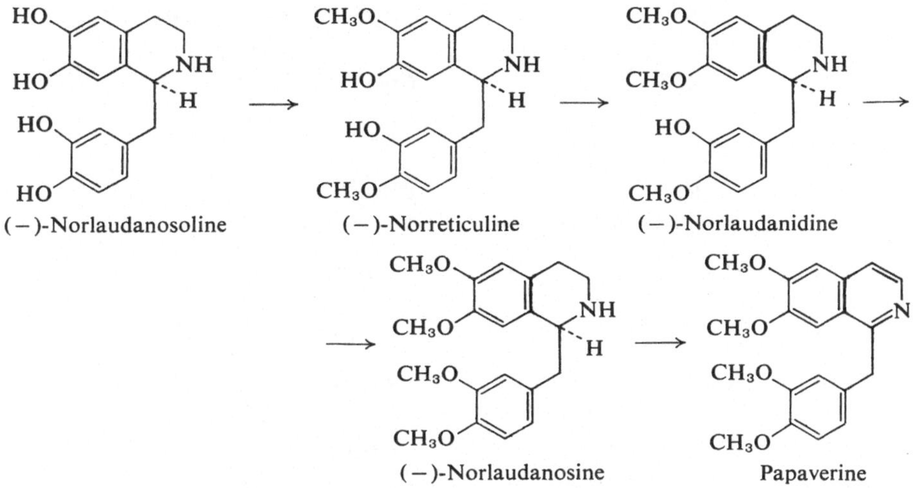

(−)-Norlaudanosoline (−)-Norreticuline (−)-Norlaudanidine

(−)-Norlaudanosine Papaverine

SCHEME 2.11

(+)-Norreticuline is not a precursor for papaverine.[58,59] The problem of the ease of N-demethylation in *Papaver* species is still not completely solved since labeled (±)-reticuline was not incorporated in papaverine while (+)-laudanosoline, which possesses the same absolute configuration as (−)-norlaudanosoline, was incorporated.[58] It is known that in some cases N-demethylation definitely does occur since morphine is metabolized to normorphine.[60]

It has been claimed that norlaudanosoline is present in nanogram quantities in rat brain following chronic administration of (−)-dopa.[61] The laboratory synthesis of papaverine derivatives specifically labeled with ^{14}C has been discussed.[62]

2.7. Pharmacology

(+)-Demethylcoclaurine has been isolated from the seed embryos of *Nelumbo nucifera* Gaertn. and exhibits smooth muscle and uterine relaxation activity.[4] Its racemate, called "higenamine," is found in *Aconitum japonicum* Decne. The levorotatory form has shown much stronger cardiac activity in the frog's heart test than its enantiomer.[5]

The four benzylisoquinolines of clinical interest are papaverine, used as an antispasmodic and peripheral vasodilator; dioxyline, also a peripheral and coronary vasodilator; the closely related ethaverine, which is another smooth muscle relaxant; and trimetoquinol, which is a bronchodilator. Several new analogs of these compounds have been screened for pharmacological activity.[63]

(+)-Demethylcoclaurine

Dioxyline

Ethaverine

Trimetoquinol

Pyridones, such as **39** and **40,** which bear a structural relationship to papaverine, have shown enhanced activity and/or duration of peripheral vaso-dilator activity in dogs.[16] (−)-Trimetoquinol has more potent β-adrenergic activity than its enantiomer in a variety of systems.[64] The intact tetrahydro-isoquinoline nucleus appears to be necessary for the β-adrenoceptor action of this synthetic diphenolic tetrahydrobenzylisoquinoline.[65]

A series of quaternary tetrahydropapaverine salts have been evaluated for their neuromuscular blocking potency relative to the bisbenzylisoquinoline alkaloid (+)-tubocurarine. Salt **41** was found to be more potent than the isomer with the reverse stereochemistry at the nitrogen.[66]

39, R = CH₃
40, R = C₂H₅

41

Another example of the importance of stereochemistry in determining physiological activity is in the finding that (−)-norlaudanosoline is a very effective lipolytic agent with isolated fat cells, whereas the (+)-form is essentially inactive.[67]

The enzyme phosphodiesterase catalyzes the apparently irreversible hydrolysis of the 3'-bond in the cyclic nucleotides to produce their acyclic

5'-monophosphate derivatives. It is known that papaverine is an effective inhibitor of this enzyme. The 6,7-dimethoxy substituents of papaverine contribute more to its effectiveness as an inhibitor than do the methoxyl functions in the benzyl side chain.[68]

Metabolism of papaverine in rats gives mostly 7-demethylpapaverine, whereas in guinea pigs 6-demethyl- and 4'-demethylpapaverine were found in approximately equal amounts. The last two compounds also appear as metabolites of papaverine in microorganisms.[69]

Despite claims to the contrary,[61] no hard evidence has been presented thus far that norlaudanosoline is a real alkaloid produced in the brain. Rather, it is felt that this substance could be an artifact formed in the bladder or the urine.[70] Following direct administration of norlaudanosoline of unspecified stereochemistry into the cerebral ventricle of rats, the animals developed a need for ethanol in increasingly excessive amounts. This result has been taken as evidence that an abnormal metabolite in the brain may produce the addictive state caused by alcoholic beverages.[71]

The benzylisoquinoline **42** has shown greater antiarrhythmic potency and is considerably less toxic than lidocaine. It is thus a candidate for further evaluation as an antiarrhythmic drug.[71a]

42

2.8. NMR Spectroscopy

It has been firmly established that in most N-methyltetrahydrobenzylisoquinolines, rings A and C lie on the same side of the molecule. Further studies of PMR spectra have now shown that in aromatic benzylisoquinolines such as papaverine and its methiodide salt, ring C is preferentially located on the same side as the tertiary or quaternary nitrogen. For papaverine methiodide in TFA, H-5 is at $\delta7.68$, and H-8 is at $\delta7.49$. The C-6 and C-7 methoxyls are found at $\delta4.12$ and 4.00, respectively. Only minor differences in chemical shifts appear if an additional methoxyl is present at C-8.[72]

The ^{13}C chemical shifts for laudanosine have been assigned. It will be noted that the C-6' signal comes at 121.5 ppm since this carbon atom is de-shielded with respect to its ortho and meta methine neighbors by its lacking the strongly shielding influence of ortho methoxy groups. (The asterisked values are interchangeable.[73])

2.9. UV Spectroscopy

Escholamidine iodide[2]	λ_{max}^{EtOH}	254, 285 and 314 nm (4.80, 3.90 and 4.05)
	λ_{min}^{EtOH}	275, 297 and 345 nm (3.87, 3.87 and 3.92)
Sevanine[1]	λ_{max}^{EtOH}	241, 277 sh, 285, 291 sh, 321 and 332 nm (4.74, 3.82, 3.87, 3.85, 3.69 and 3.73)

References and Notes

1. V. A. Mnatsakanyan, V. Preininger, V. Šimánek, A. Klašek, L. Dolejš, and F. Šantavý, *Tetrahedron Lett.*, 851 (1974); and V. A. Mnatsakanyan, V. Preininger, V. Šimanek, J. Juřina, A. Klašek, L. Dolejš, and F. Šantavý, *Collect. Czech. Chem. Commun.*, **42**, 1421 (1977).
2. J. Slavík, L. Slavíková, and L. Dolejš, *Collect. Czech. Chem. Commun.*, **40**, 1095 (1975).
3. V. Šimánek, V. Preininger, A. Klásek, and J. Juřina, *Heterocycles*, **4**, 1263 (1976).
3a. A. Jossand, M. Leboeuf, A. Cavé, M. Damak, and C. Riche, *C. R. Acad. Sci., Ser. C*, **284**, 467 (1977).

4. H. Koshiyama, H. Ohkuma, H. Kawaguchi, H. Y. Hsu, and J. P. Chen, *Chem. Pharm. Bull.*, *Tokyo*, **18**, 2564 (1970).
5. T. Kosuge and M. Yokota, *Chem. Pharm. Bull.*, *Tokyo*, **24**, 176 (1976). For the crystal structure of higenamine hydrobromide, see N. Masaki, H. Iisuka, M. Yokota, and A. Ochiai, *J. Chem. Soc. Perkin I*, 717 (1977).
6. J. Slavík and L. Dolejš, *Collect. Czech. Chem. Commun.*, **38**, 3514 (1973).
7. D. W. Bishay, Z. Kowalewski, and J. D. Philippson, *J. Pharm. Pharmacol.*, **23** (Suppl.), 233(S) (1971).
8. G. Nonaka and I. Nishioka, *Phytochemistry*, **13**, 2620 (1974).
9. C. Tani, N. Nagakura, S. Hattori, and M.-T. Kao, *J. Pharm. Soc. Japan*, **94**, 844 (1974); through *Chem. Abstr.*, **81**, 166363 (1974).
10. K. Ito, H. Haruna, and H. Furukawa, *J. Pharm. Soc. Japan*, **95**, 358 (1975); and references cited therein.
11. For a review on the synthesis of benzylisoquinolines and phenethylisoquinolines, see T. Kametani and K. Fukumoto, *MTP Int. Rev. Sci. Alkaloids, Org. Chem.*, Series One, Vol. 9, K. Wiesner, ed., Butterworths, London (1973), p. 181.
12. For a review on the use of Reissert compounds in the synthesis of isoquinoline alkaloids, see F. D. Popp, *Heterocycles*, **1**, 165 (1973).
13. B. C. Uff, J. R. Kershaw, and S. R. Chhabra, *J. Chem. Soc. Perkin I*, 479 (1972). For related syntheses, see B. C. Uff, J. R. Kershaw, and J. L. Neumeyer, *Org. Syn.*, **54**, 1892 (1974); and F. R. Stermitz and D. K. Williams, *J. Org. Chem.*, **38**, 1761 (1973).
14. H. W. Gibson and F. D. Popp, *Heterocycles*, **2**, 5 (1974).
15. L. W. Deady, N. H. Pirzada, R. D. Topsom, and J. W. Bobbitt, *Aust. J. Chem.*, **26**, 2063 (1973).
16. M. Srinivasan and J. B. Rampal, *Chem. Ind. (London)*, 89 (1975).
17. W. E. Kreighbaum, W. F. Kavanaugh, W. T. Comer, and D. Deitchman, *J. Med. Chem.*, **15**, 1131 (1972).
18. W. E. Kreighbaum, W. F. Kavanaugh, and W. T. Comer, *J. Heterocycl. Chem.*, **10**, 317 (1973).
19. I. W. Elliott, Jr., *J. Heterocycl. Chem.*, **9**, 853 (1972).
20. T. Kametani, K. Fukumoto, K. Kigasawa, M. Hiiragi, and H. Ishimaru, *Heterocycles*, **3**, 311 (1975).
21. M. R. Falco, J. X. DeVries, and G. Mann, *Z. Chem.*, **13**, 56 (1973); *Chem. Abstr.*, **78**, 148096s (1973).
22. M. Konda, T. Shioiri, and S. Yamada, *Chem. Pharm. Bull.*, *Tokyo*, **23**, 1025 and 1063 (1975). For an asymmetric synthesis of (−)-laudanosine from L-3(3,4-dihydroxyphenyl)alanine, see M. Konda, T. Ohishi, and S. Yamada, *Chem. Pharm. Bull.*, *Tokyo*, **25**, 69 (1977).
23. A. J. Birch, A. H. Jackson, and P. V. R. Shannon, *J. Chem. Soc. Perkin I*, 2190 (1974). For other examples of the Reissert approach, see Ref. 11 above; J. Knabe and A. Frie, *Arch. Pharm. (Weinheim)*, **306**, 648 (1973), which discusses the preparation of 1-acylisoquinolines; and S. F. Dyke, A. W. C. White, and D. Hartley, *Tetrahedron*, **29**, 857 (1973).
23a. K. Ito and H. Tanaka, *Chem. Pharm. Bull.*, *Tokyo*, **25**, 1732 (1977).
24. P. Wiriyachitra and M. P. Cava, *J. Org. Chem.*, **42**, 2274 (1977).
25. S. Ruchirawat, V. Somchitman, N. Tongpenyai, W. Lertwanawatana, S. Issarayangyuen, N. Prasitpan, and P. Prempree, *Heterocycles*, **4**, 1917 (1976).
26. J. L. Neumeyer and C. B. Boyce, *J. Org. Chem.*, **38**, 2291 (1973).
27. K. W. Bentley and A. W. Murray, *J. Chem. Soc.*, 2487 (1963).
28. T. Kametani, S. Hirata, S. Shibuya, and K. Fukumoto, *J. Chem. Soc.*, C, 1927 (1971).
29. S. Ruchirawat, N. Tongpenyai, N. Prasitpan, and P. Prempree, *Heterocycles*, **4**, 1893 (1976).
30. M. Shamma and G. S. Jayatilake, unpublished results.

31. H. B. Kagan, N. Langlois, and T. P. Dang, *J. Organometal. Chem.*, **90**, 353 (1975).
32. J. B. Bremner and P. Wiriyachitra, *Aust. J. Chem.*, **26**, 437 (1973).
33. J. B. Bremner and Le van Thuc, *Chem. Ind. (London)*, 453 (1976). I. R. C. Bick, J. B. Bremner, and P. Wiriyachitra, *Tetrahedron Lett.*, 4795 (1971).
34. S. Pfeifer, G. Behnson, L. Küehn, and R. Kraft, *Pharmazie*, **27**, 734 (1972); through *Chem. Abstr.*, **78**, 84607z (1973).
35. P. Mathieu and J. Gardent, *C. R. Acad. Sci. Ser. C.*, **267**, 1416 (1968).
36. W. Wiegrebe, H. Reinhart, H. Budzikiewicz, and U. Krüger, *Tetrahedron*, **25**, 2899 (1969).
37. L. S. Hegedus and R. K. Stiverson, *J. Am. Chem. Soc.*, **96**, 3250 (1974).
38. It is well established that sodium in liquid ammonia reduction of tetrahydrobenzyl-isoquinoline methiodides leads to cleavage of the C-1 to N-2 bond; see K. W. Bentley and A. W. Murray, *J. Chem. Soc.*, 2501 (1963).
39. W. Wiegrebe, H. Reinhart, and J. Fricke, *Pharm. Acta Helv.*, **48**, 420 (1973).
40. W. Wiegrebe, J. Fricke, H. Budzikiewicz, and L. Pohl, *Tetrahedron*, **28**, 2849 (1972).
41. T. Kametani, H. Nemoto, K. Suzuki, and K. Fukumoto, *J. Org. Chem.*, **41**, 2988 (1976).
41a. J. Knabe and J. Kubitz, *Arch. Pharm. (Weinheim)*, **297**, 129 (1964).
42. J. Knabe, R. Dörr, S. F. Dyke, and R. G. Kinsman, *Tetrahedron Lett.*, 5373 (1972); and J. Knabe and R. Dörr, *Arch. Pharm. (Weinheim)*, 784 (1973); and R. G. Kinsman, A. W. C. White, and S. F. Dyke, *Tetrahedron*, **31**, 449 (1975).
43. P. Cauwel, J. Chazerain, and J. Gardent, *Tetrahedron Lett.*, 1023 (1971).
44. P. Cauwel and J. Gardent, *Tetrahedron Lett.*, 2781 (1972).
45. L. L. Miller, F. R. Stermitz, and J. R. Falck, *J. Am. Chem. Soc.*, **95**, 2651 (1973); J. Y. Becker, L. L. Miller, and F. R. Stermitz, *J. Electroanal. Chem. Interfacial Electrochem.*, **68**, 181 (1976); S. M. Kupchan, and C.-K. Kim, *J. Am. Chem. Soc.*, **97**, 5623 (1975); and references cited therein.
46. J. M. Bobbitt, I. Noguchi, R. S. Ware, K. N. Chiong, and S. J. Huang, *J. Org. Chem.*, **40**, 2924 (1975).
47. J. M. Bobbitt and T. Y. Cheng, *J. Org. Chem.*, **41**, 443 (1976).
48. S. M. Kupchan, A. J. Liepa, V. Kameswaran, and R. F. Bryan, *J. Am. Chem. Soc.*, **95**, 6861 (1973).
49. T. Kametani and K. Fukumoto, *Synthesis*, 657 (1972).
50. An important and early exception to this statement is the isolation of salutaridine in small yield from the ferricyanide oxidation of reticuline; see D. H. R. Barton, D. S. Bhakuni, R. James, and G. W. Kirby, *J. Chem. Soc.*, C, 128 (1967).
51. M. A. Schwartz and I. S. Mami, *J. Am. Chem. Soc.*, **97**, 1239 (1975).
52. H. C. Beyerman, T. S. Lie, L. Maat, H. H. Bosman, E. Buurman, E. J. M. Bijsterveld, and H. J. M. Sinnige, *Rec. Trav. Chim. Pays-Bas*, **95**, 24 (1976). See also H. C. Beyerman, E. Buurman, T. S. Lie, and L. Maat, *Rec. Trav. Chim. Pays-Bas*, **95**, 43 (1976); and H. C. Beyerman, E. Buurman, and L. Maat, *Chem. Commun.*, 918 (1972). For a practical method of carrying out a Birch reduction using lithium in liquid ammonia in the presence of *t*-butanol, see H. C. Beyerman, F. F. van Leeuwen, T.S. Lie, L. Maat, and C. Olieman, *Rec. Trav. Chim. Pays-Bas*, **95**, 238 (1976).
53. S. R. Johns, J. A. Lamberton, A. A. Sioumis, and R. I. Willing, *Aust. J. Chem.*, **23**, 353 (1970).
53a. For some speculations on the biogenesis and metabolism of benzylisoquinoline alkaloids, see S. F. Dyke, *Heterocycles*, **6**, 1441 (1977).
54. M. L. Wilson and C. J. Coscia, *J. Am. Chem. Soc.*, **97**, 431 (1975).
55. A. R. Battersby, R. C. F. Jones, and R. Kazlaukas, *Tetrahedron Lett.*, 1873 (1975).
56. S. Tewari, D. S. Bhakuni, and R. S. Kapil, *Chem. Commun.*, 554 (1975).
57. G. J. Kapadia, G. S. Rao, E. Leete, M. B. E. Fayez, Y. N. Vaishnav, and H. M. Fales, *J. Am. Chem. Soc.*, **92**, 6943 (1970).
57a. J. M. Bobbitt, C. L. Kulkarni, and P. Wiriyachitra, *Heterocycles*, **4**, 164 (1976).

57b. D. S. Bhakuni, A. N. Singh, S. Tewari, and R. S. Kapil, *J. Chem. Soc. Perkin I*, 1662 (1977).
58. E. Brochmann-Hanssen, C. Chen, C. R. Chen, H. Chiang, A. Y. Leung, and K. McMurtrey, *J. Chem. Soc. Perkin I*, 1531 (1975).
59. H. Uprety, D. S. Bhakuni, and R. S. Kapil, *Phytochemistry*, 14, 1535 (1975).
60. R. J. Miller, C. Jolles, and H. Rapoport, *Phytochemistry*, 12, 597 (1973).
61. A. J. Turner, K. M. Baker, S. Algeri, A. Frigerio, and S. Garattini, *Life Sci.*, 14, 2247 (1974).
62. S. D. Ithakissios, G. Tsatsas, J. Nikokavouras, and A. Tsolis, *J. Labelled Compd.*, 10, 369 (1974).
63. For a GLC procedure for the determination of papaverine in blood plasma or urine, see D. E. Guttman, H. B. Kostenbauder, G. R. Wilkinson, and P. H. Dubé, *J. Pharm. Sci.*, 63, 1625 (1974).
64. D. D. Miller, P. Osei-Gyimah, J. Bardin, and D. R. Feller, *J. Med. Chem.*, 18, 454 (1975). See also D. D. Miller, P. Osei-Gyimah, R. V. Raman, and D. R. Feller, *J. Med. Chem.*, 20, 1502 (1977).
65. D. D. Miller, P. F. Kador, R. Venkatraman and D. R. Feller, *J. Med. Chem.*, 19, 763 (1976); and references cited therein.
66. J. B. Stenlake, W. D. Williams, N. C. Dhar, and I. G. Marshall, *Eur. J. Med. Chim. Ther.*, 9, 233 (1974).
67. G. Cohen, R. E. Heikkila, D. Dembiec, D. Sang, S. Teitel, and A. Brossi, *Eur. J. Clin. Pharmacol.*, 29, 292 (1974).
68. M. S. Amer and W. E. Kreighbaum, *J. Pharm. Sci.*, 64, 1 (1975); and references cited therein. See also S. F. Berndt and H. U. Schultz, *J. Neural Transm., Suppl.* 1974, p. 187.
69. J. P. Rosazza, M. Kammer, L. Youel, R. V. Smith, P. W. Erhardt, D. H. Troung and S. W. Leslie, Proc. of the ACS Div. Med. Chem., 172nd National Meeting, San Francisco, Cal., Aug.–Sept. 1976, item 43. Laudanosine has also been microbially O-demethylated to pseudocodamine; see P. J. Davis and J. P. Rosazza, *J. Org. Chem.*, 41, 2548 (1976).
70. A. Brossi, private communication.
71. R. D. Myers and C. L. Melchior, *Science*, 196, 554 (1977). For a review on the pharmacology of isoquinoline alkaloids and ethanol interactions, see M. Hirst, M. G. Hamilton, and A. M. Marshall, in *Alcohol and Opiates*, K. Blum, ed., Academic Press, New York (1977), p. 167.
71a. J. L. Neumeyer, C. Perianayagam, S. Ruchirawat, H. S. Feldman, B. H. Takman, and P. A. Tenthorey, *J. Med. Chem.*, 20, 894 (1977).
72. T. Tomimatsu, S. Yamada, and R. Yuasa, *J. Pharm. Soc. Japan*, 97, 217 (1977).
73. E. Wenkert, B. L. Buckwalter, I. R. Burfitt, M. J. Gašič, H. E. Gottlieb, E. W. Hagaman, F. M. Schell, and P. M. Wovkulich, in *Topics in C-13 NMR Spectroscopy*, Vol. 2, G. C. Levy, ed., Wiley-Interscience, New York (1976), p. 105.

3

THE ISOQUINOLONES

3.1. Introduction

Siamine[1] and doryphornine[2] are two isoquinolones discovered recently.

Siamine is unusual, incorporating two phenolic groups in a meta relationship and a methyl group at C-3; thus it may not be derived biogenetically from tyrosine.

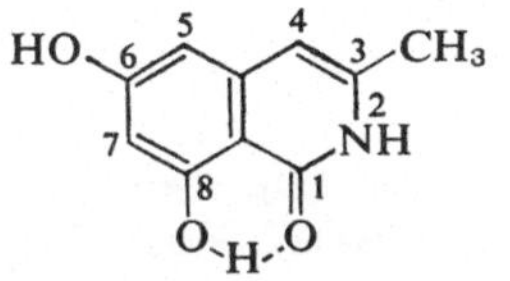

Siamine
Cassia siamea (Leguminosae)

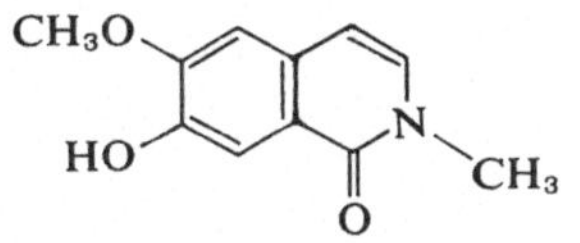

Doryphornine
Doriphora sassafras Endlicher
(Monimiaceae)

3.2. Siamine

Siamine, $C_{10}H_9NO_3$, ν_{max} 1635 cm^{-1} (6.12 μ), λ_{max}^{MeOH} 245, 262 sh, 272 sh, 282, 294, 321, and 330 nm, exhibits the PMR values shown below. Very tentative CMR chemical shift assignments were also made.[1]

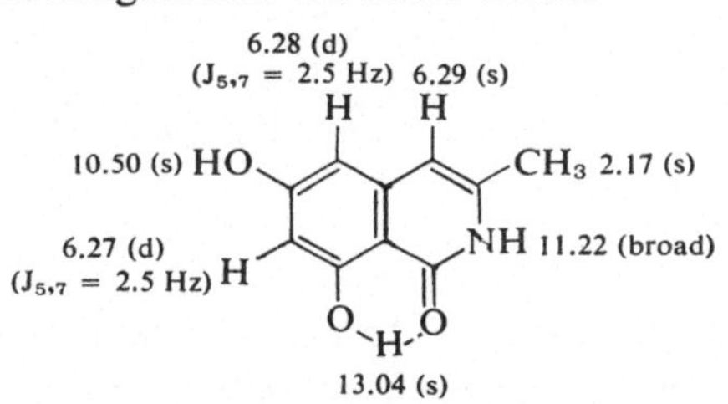

PMR chemical shifts for siamine

The structure was settled by synthesis from the diphenolic homophthalic acid 1.[1]

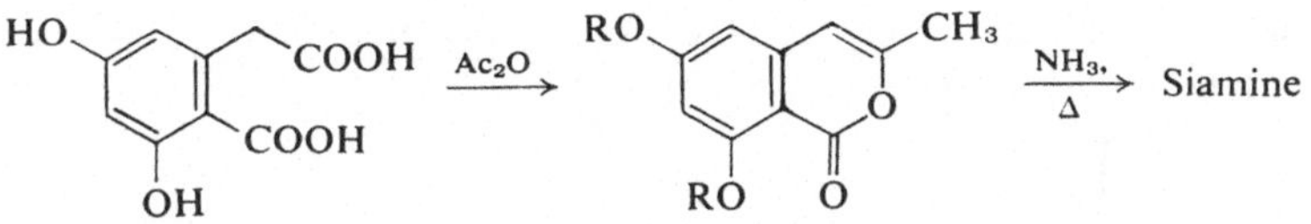

R = Acetyl or H

3.3. Synthesis and Reactions of Isoquinolones

Aromatic isoquinolones can be obtained by the thermal rearrangement of isoquinoline *N*-oxides[3]; thus it was possible to derive the isoquinoline **3** from the *N*-oxide **2**. Compound **3** could be further functionalized through bromination and subsequent treatment with a mixture of cuprous cyanide and sodium cyanide.[4]

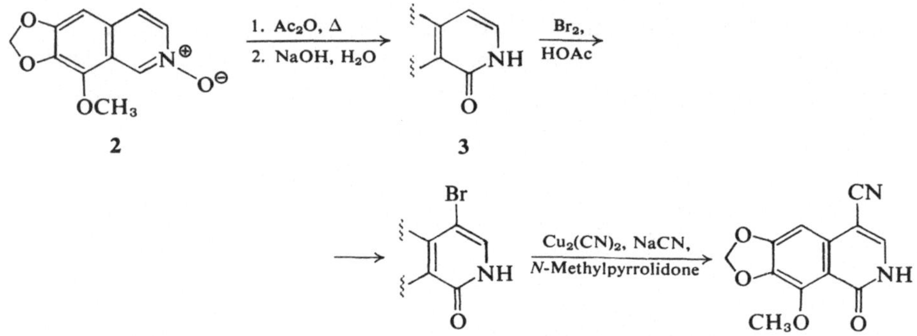

A new synthesis of isoquinolones originates with homophthalic acid, and is described in the following sequence[5]:

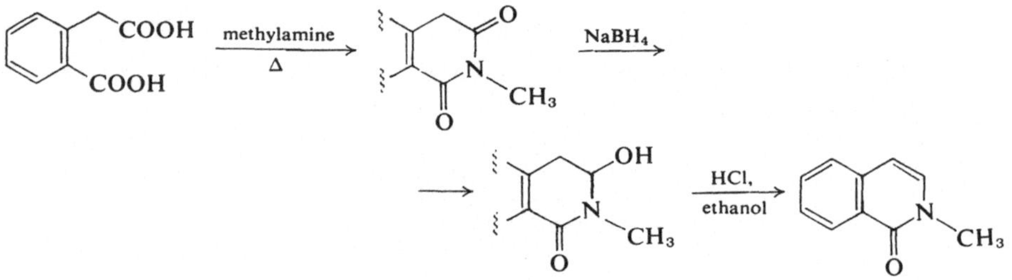

This sequence was adapted[5] to provide the protoberberine lactam **4**; the starting material was obtained from reduction of the imide derived from homoveratrylamine and homophthalic acid.

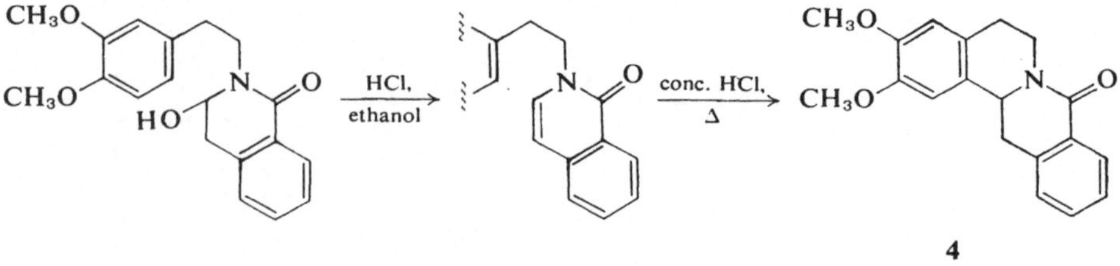

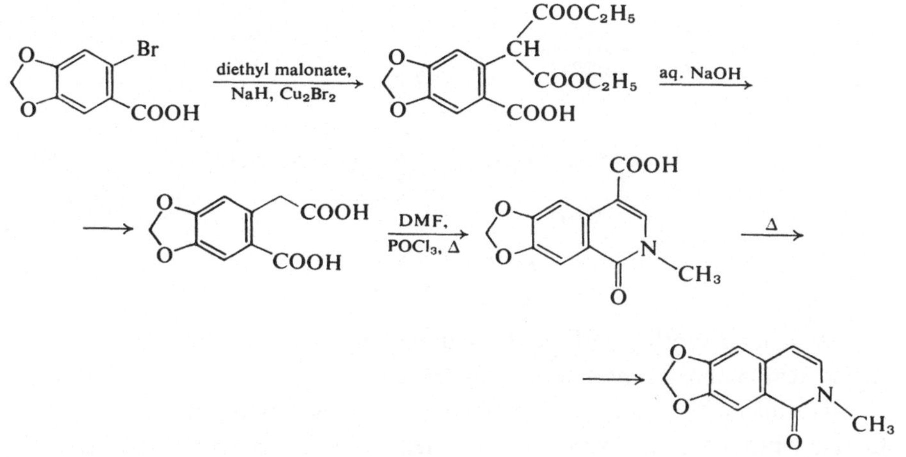

Doryanine

SCHEME 3.1

An attractive novel preparation of isoquinolones proceeds through reaction of a homophthalic acid with dimethyl formamide and phosphorus oxychloride, as described in Scheme 3.1. The alkaloid doryanine was prepared in good yield by this method.[5a]

It is known that tetrahydrobenzylisoquinolines can be cleaved to isoquinolones using potassium permanganate. A method has now been developed for the oxidation of isoquinolinium salts. Taking papaverine methiodide as an example, this salt is first treated with aqueous alkali to supply the enamine **5**. Treatment of a benzene solution of the enamine with oxygen in the presence of cuprous chloride provides the isoquinolone **6** in 80% yield.[6] (For the oxidation of bisbenzylisoquinolines to isoquinolones, see Secs. 5.4 and 6.1.)

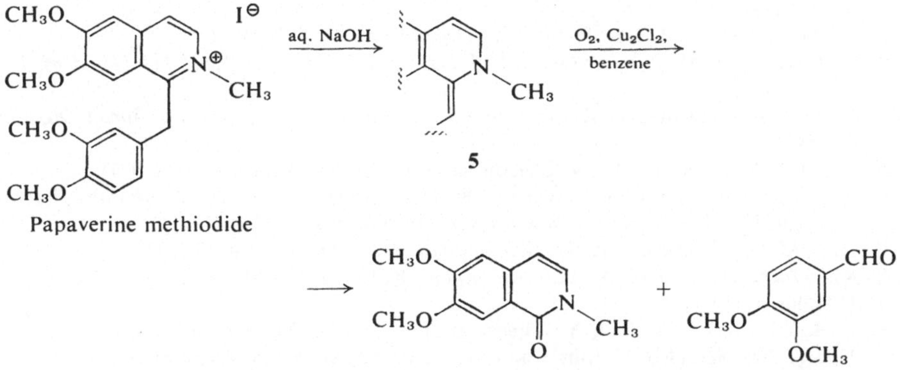

Papaverine methiodide

5

6

Veratraldehyde

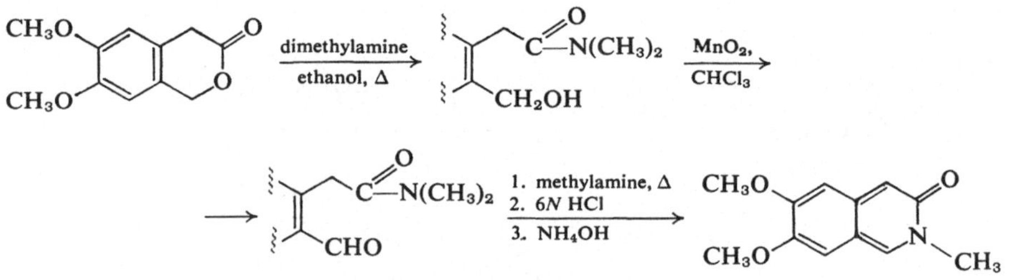

SCHEME 3.2

A general synthesis of 3(2H)-isoquinolinones from 2-hydroxymethylaryl-acetic acid lactones is presented in Scheme 3.2.[7]

The introduction of a 3,4-double bond in an isoquinolone is often difficult. In the synthesis of thalactamine, this dehydrogenation was performed using palladium on carbon.[8]

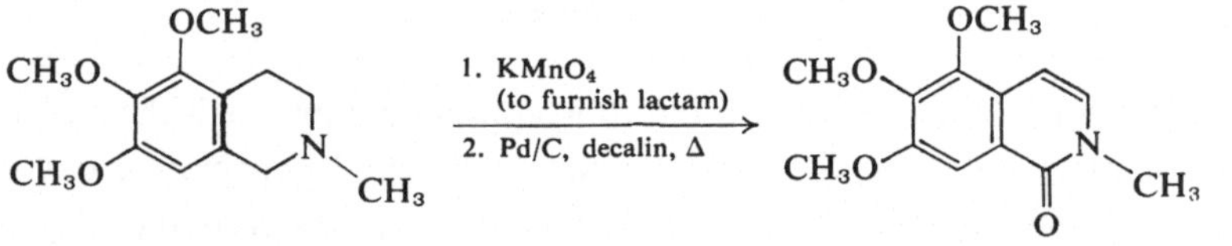

Thalactamine

Isoquinolinium salts can be readily converted to isoquinolones through air oxidation of the corresponding pseudobases.[9]

References

1. B. Z. Ahn and F. Zymalkowski, *Tetrahedron Lett.*, 821 (1976).
2. C. R. Chen, J. L. Beal, R. W. Doskotch, L. A. Mitscher, and G. H. Svoboda, *Lloydia*, **37**, 493 (1976).
3. M. M. Robison and B. L. Robison, *J. Org. Chem.*, **21**, 1337 (1956).
4. S. Passannanti, M. P. Paternostro, F. Piozzi, and G. Savona, *Chem. Ind. (London)*, 791 (1975).
5. H. Iida, K. Kawano, T. Kikuchi, and F. Yoshimizu, *J. Pharm. Soc. Japan*, **96**, 176 (1976).
5a. V. H. Belgaonkar and R. N. Usgaonkar, *J. Chem. Soc. Perkin I*, 702 (1977).
6. S. Ruchirawat, U. Borvornvinyanant, K. Hantawong, and Y. Thebtaranonth, *Heterocycles*, **6**, 1119 (1977); and S. Ruchirawat, *Heterocycles*, **6**, 1724 (1977).
7. N. J. McCorkindale and A. W. McCullough, *Tetrahedron*, **27**, 4653 (1971).
8. Kh. Duchevska and N. Mollov, *Izv. Khim.*, **8**, 134 (1975); through *Chem. Abstr.*, **84**, 150809u (1976).
9. S. Ruchirawat, S. Sunkul, Y. Thebtaranonth, and N. Thirasasna, *Tetrahedron Lett.*, 2335 (1977). See also A. Albert and W. L. F. Armarego, in *Advances in Heterocyclic Chemistry*, Vol. 4, A. R. Katritzky, ed., Academic Press, New York (1965), p. 13.

4

THE PAVINES AND ISOPAVINES

Some recently isolated pavines and isopavines are shown below:

Pavines

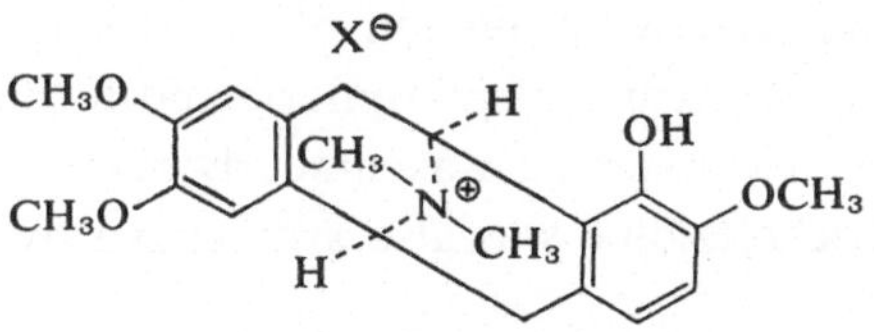

(−)-Platycerine N-metho salt[1]
Argemone platyceras Link & Otto

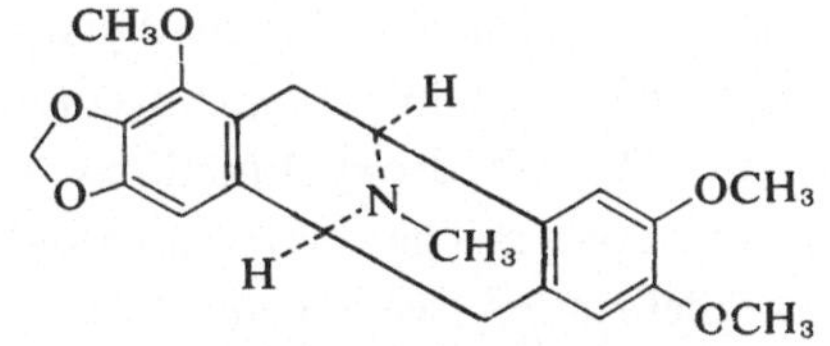

(−)-2,3-Methylenedioxy-
4,8,9-trimethoxypavinane[2]
Thalictrum strictum Lebed.

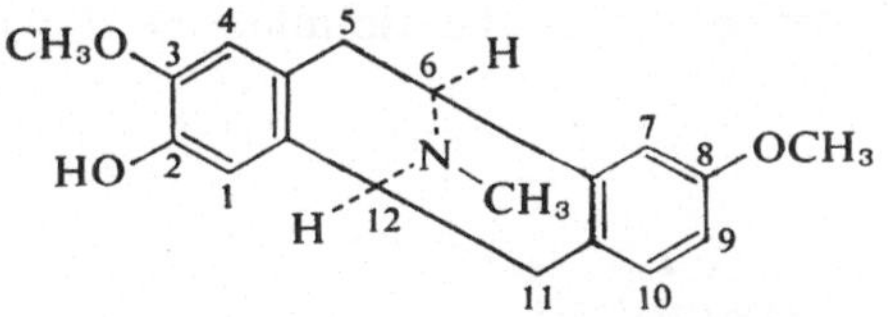

(−)-2-Hydroxy-3,8-dimethoxypavinane[3]
Argemone munita Dur. & Hilg.
subsp. *rotundata* (Rydb.) G. B. Owmb.

Isopavine

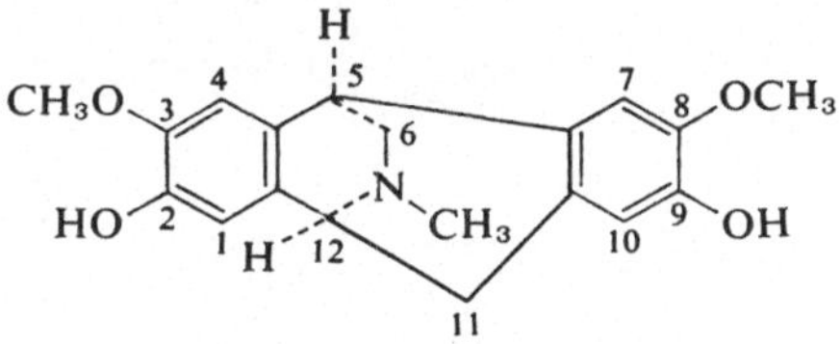

(−)-Thalidine[4]
Thalictrum dioicum L.

4.1. Synthesis

4.1.1. Pavines

Reduction of papaverine methiodide with tin and hydrochloric acid produces, in addition to laudanosine, N-methylpavine formed through cyclization of the transitory 1,4-dihydroisoquinolinium intermediate.[5] Recent syntheses in the pavine series have utilized 1,4-dihydroisoquinolinium intermediates under acidic but nonreductive conditions. This was accomplished by the construction of appropriately substituted benzylisoquinoline methiodides followed by selective reduction of the immonium bond to the corresponding 1,2-dihydrobenzylisoquinolines. Subsequent acid treatment generated and cyclized the 1,4-dihydro intermediates.

Pai and Natarajan have thus confirmed the structure of (−)-caryachine by a total synthesis of the racemic base.[6] The requisite 1,2-dihydrobenzylisoquinoline, formed by *N*-methylation and selective reduction of the Bischler–Napieralski product **1**, was cyclized using a mixture of phosphoric and formic acids. The 8-hydroxy-9-methoxy analog of caryachine was similarly prepared in this study.[6] The related alkaloid eschscholtzidine was also synthesized by a parallel route[7] (see Scheme 4.1).

Stermitz and his collaborators[9,10] constructed the benzylisoquinoline methiodides necessary for conversion to the unsymmetrically substituted pavines platycerine and munitagine by way of a Reissert compound.[8] Alkylation of the Reissert compound derived from the isoquinoline **2** with the appropriately

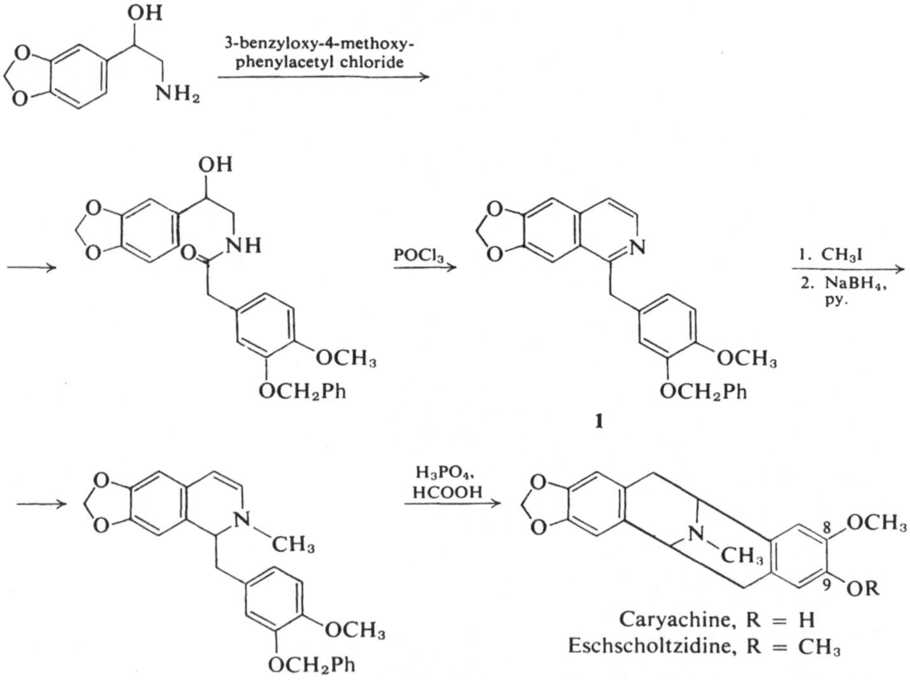

SCHEME 4.1

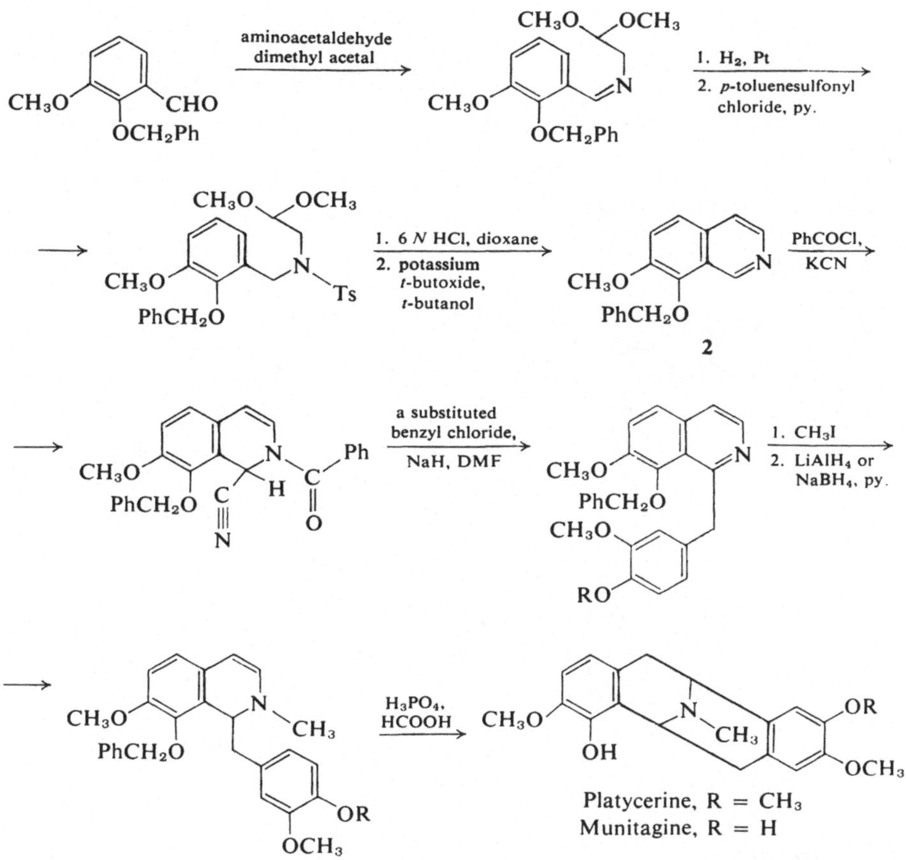

SCHEME 4.2

substituted benzyl halides, followed by *N*-methylation, selective reduction and acid-catalyzed cyclization supplied platycerine[9] and munitagine[10] (see Scheme 4.2[11]).

An interesting synthesis of pavines along totally different lines was developed by Ito's group. Canadine methiodide was the starting material. The B ring of this salt was contracted by an established route to the quaternary salt **3**, which upon treatment with phenyllithium underwent a Stevens-type rearrangement to the pavinane system. Selective cleavage of the methylenedioxy group with boron trichloride, followed by diazomethane methylation then provided the unsymmetrically substituted di-*O*-methylmunitagine.[12] A by-product formed in this rearrangement was the tetracyclic base **4**:

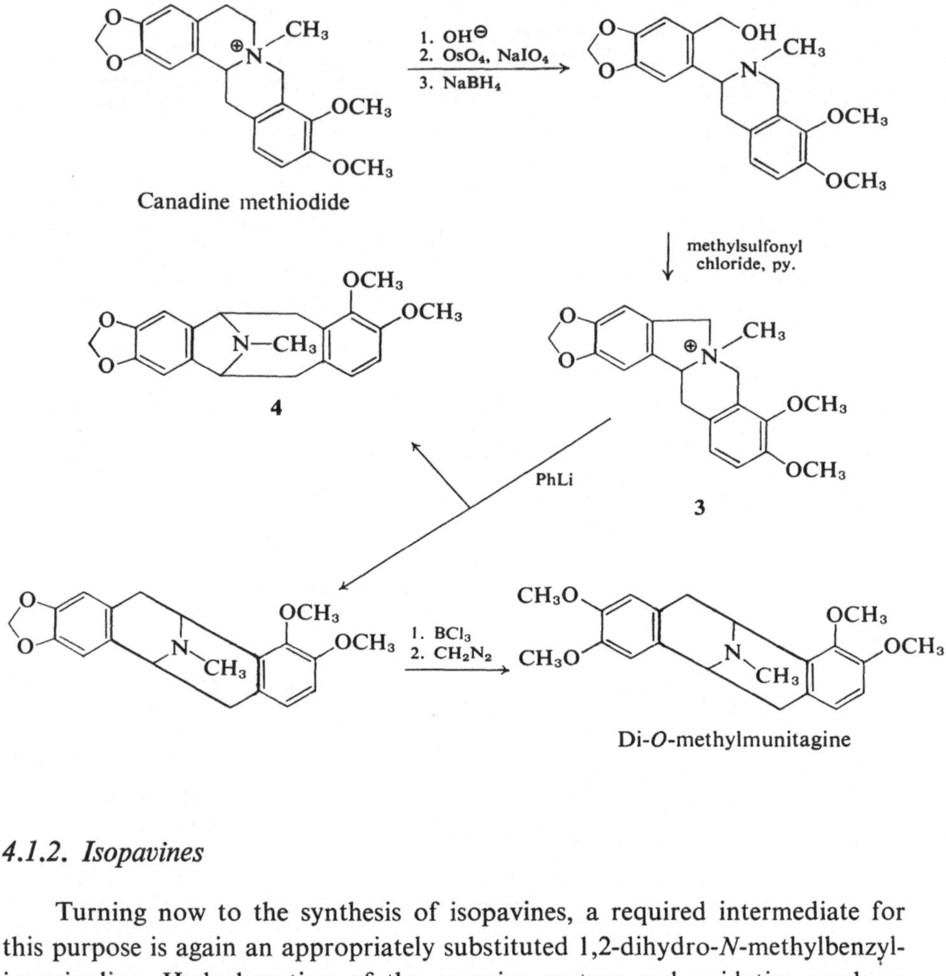

Canadine methiodide

1. OH⊖
2. OsO₄, NaIO₄
3. NaBH₄

methylsulfonyl
chloride, py.

4

3

PhLi

1. BCl₃
2. CH₂N₂

Di-*O*-methylmunitagine

4.1.2. Isopavines

Turning now to the synthesis of isopavines, a required intermediate for this purpose is again an appropriately substituted 1,2-dihydro-*N*-methylbenzyl-isoquinoline. Hydroboration of the enamine system and oxidative work-up induce formation of a 4-hydroxybenzylisoquinoline which can be cyclized in acid to the isopavine skeleton. A recent example of such a synthesis leads to the new isopavine thalidine[4]:

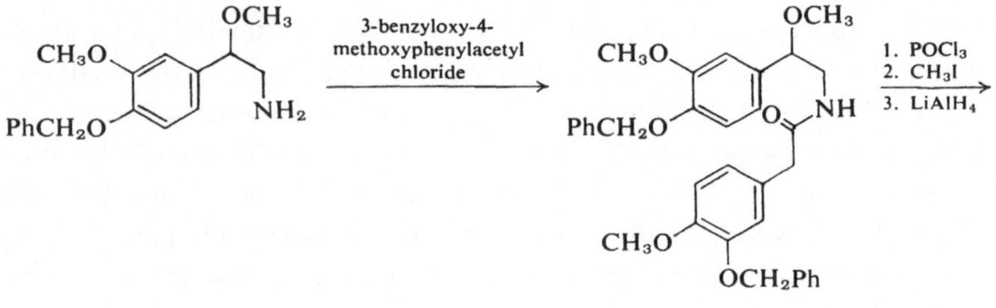

3-benzyloxy-4-
methoxyphenylacetyl
chloride

1. POCl₃
2. CH₃I
3. LiAlH₄

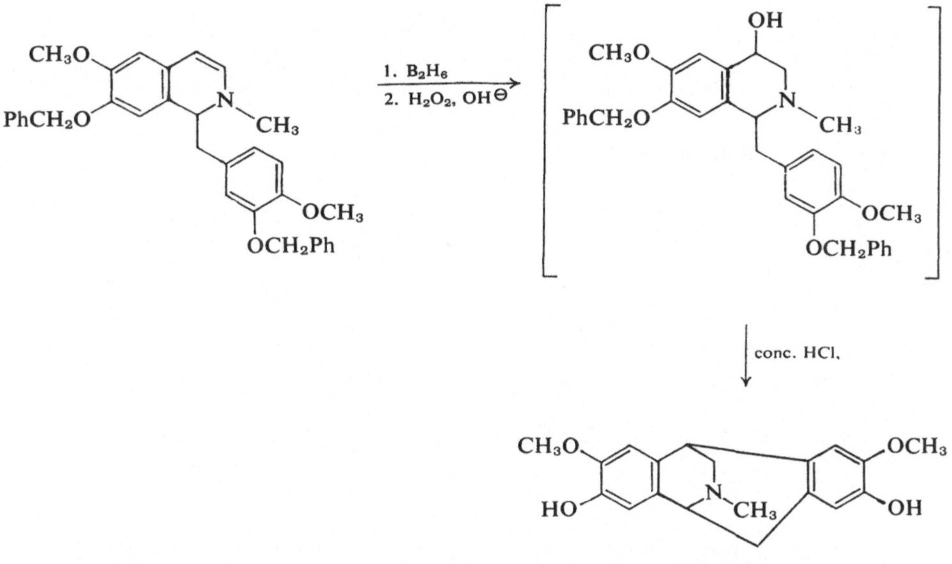

Thalidine

The methylenedioxy substituted isopavine amurensine was prepared by a parallel method,[13] as were the related bases reframine and reframoline.[14] In the latter study, undertaken to confirm the structure of reframoline, the two positional isomers with respect to the hydroxyl and methoxyl groups, namely, 5 and 6, were prepared synthetically, but could not be distinguished by spectral means. They were, however, easily characterized by way of the UV spectra of their respective methine bases, 7 and 8. The 3-hydroxy methine base exhibited a large bathochromic shift upon the addition of base, while little alteration of the spectrum of the 2-hydroxy isomer occurred in base:

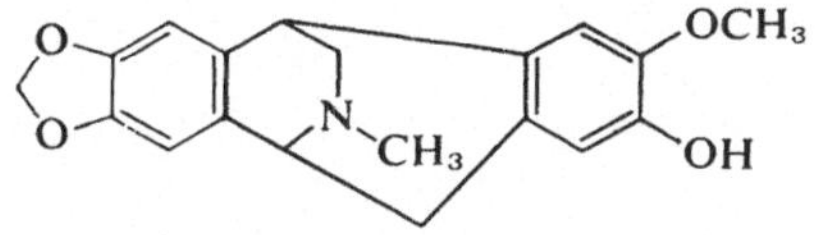

Amurensine

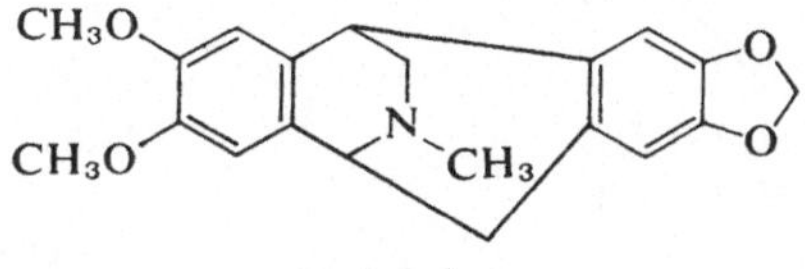

Reframine

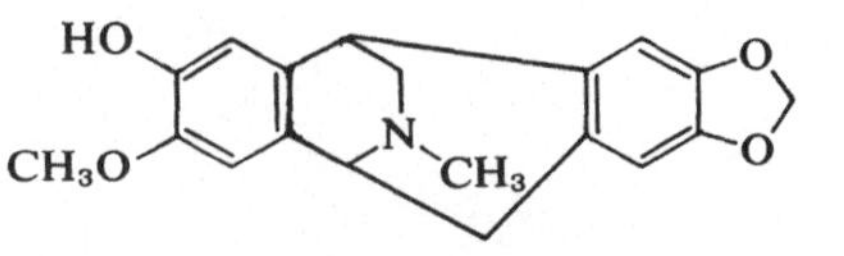

Reframoline, **5**

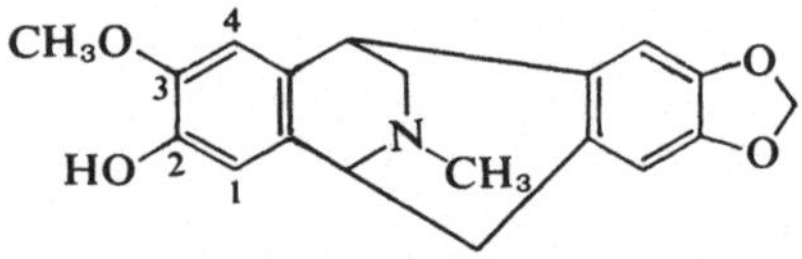

1. CH₃I
2. OH⊖

6

1. CH₃I
2. OH⊖

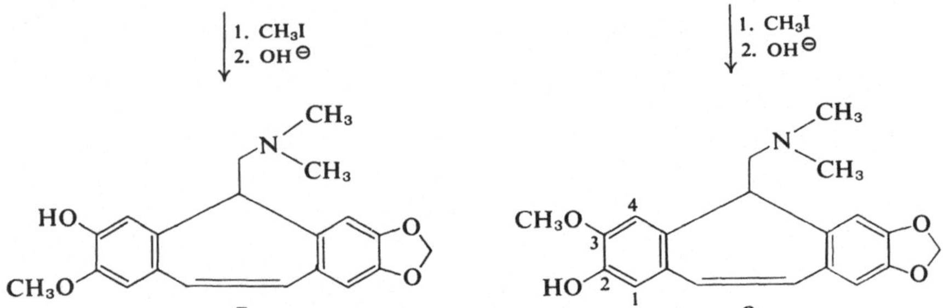

7

8

Umezawa and co-workers have succeeded in improving the yields in the transformation of benzylisoquinolines to isopavines.[15] Lead tetraacetate oxidation of 6-hydroxy substituted tetrahydrobenzylisoquinolines generates C-4 acetoxylated products which in acid cyclize to isopavines in over 90% yield. The syntheses of reframine and *O*-methylthalisopavine are examples of this method. By contrast, the yields in the acid-catalyzed cyclization of 4,7-dihydroxy-tetrahydrobenzylisoquinolines to isopavines rarely exceed 30–35%:

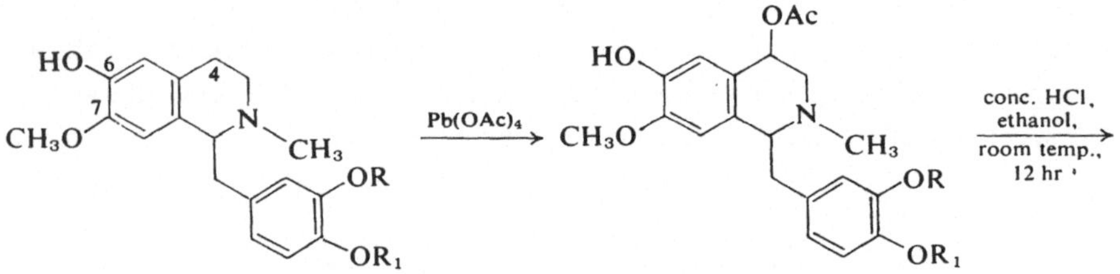

$\xrightarrow{\text{Pb(OAc)}_4}$

conc. HCl,
ethanol,
room temp.,
12 hr

R = R₁ = CH₃ (76%)
R + R₁ = CH₂ (60%)

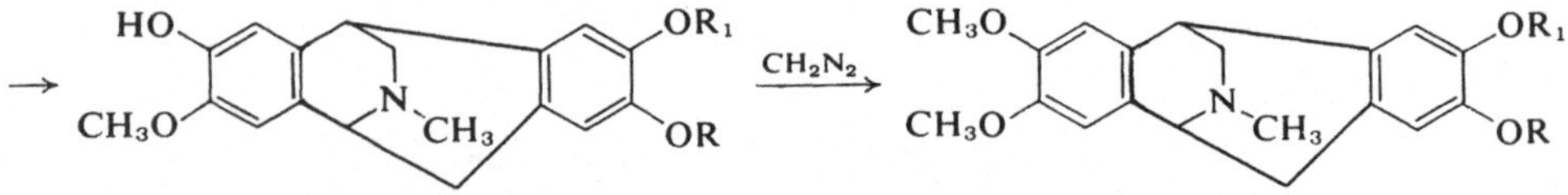

R = R₁ = CH₃ (92%)
R + R₁ = CH₂ (90%)

$\xrightarrow{\text{CH}_2\text{N}_2}$

O-Methylthalisopavine, R = R₁ = CH₃
(66%)
Reframine, R + R₁ = CH₂
(73%)

In an extension of these studies, trifluoroacetic acid (TFA) treatment of a series of variously substituted 8-chloro-*p*-quinol acetates such as **9** and **10** produced isopavine **11** and homoisopavine **12**, but in reduced yields.[16]

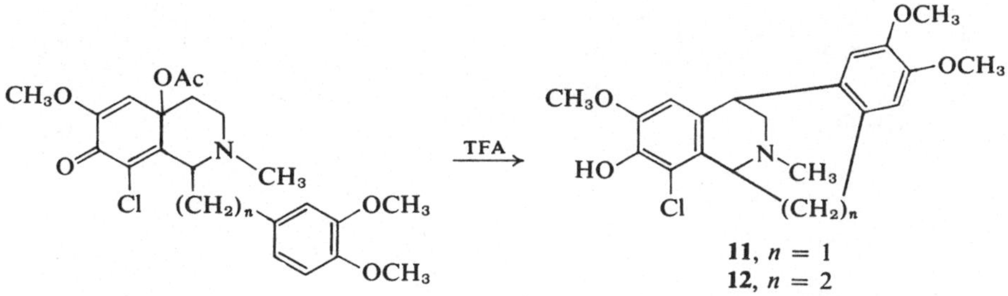

9, $n = 1$
10, $n = 2$

11, $n = 1$
12, $n = 2$

Kametani's group has applied the diazomethane ring expansion reaction of 3,4-dihydroisoquinoline methiodides to the preparation of isopavines.[17] The carbon insertion reaction with the 3-aryl-3,4-dihydroisoquinoline methiodide **13** supplied a crude aziridinium iodide which upon standing in 6*N* hydrochloric acid for a week underwent successive ring expansion and closure, probably through the quinone methide **14**, to furnish reframidine in 20% yield from the aziridine salt:

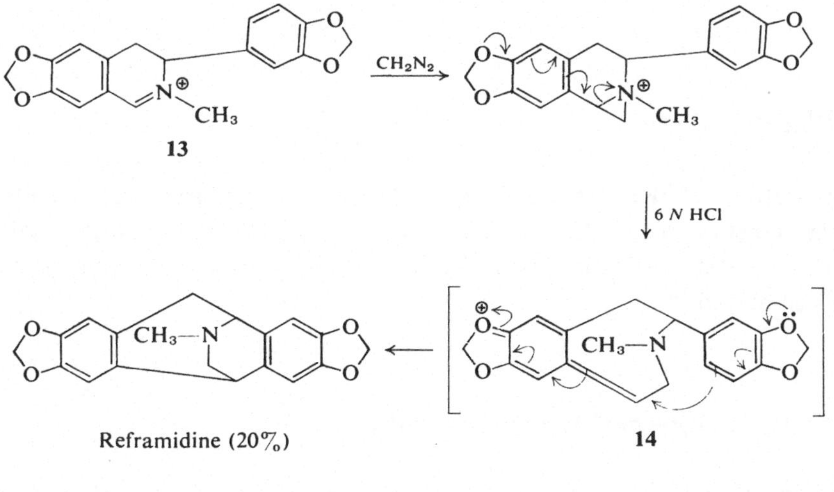

13

6 *N* HCl

Reframidine (20%)

14

The angularly methylated isopavinane base **16** was unexpectedly obtained

when the propynylbenzylamine **15** was treated with polyphosphoric acid
(PPA).[18]

4.2. Biogenesis

The isolation of the trioxygenated alkaloid (−)-2-hydroxy-3,8-dimethoxy-pavinane[3] poses the interesting question of whether this compound is derived from (+)-coclaurine or from (+)-reticuline. In either case, it is possible that a propavinane such as **17** could be involved in the biogenetic sequence.

17, R = H or OCH$_3$

4.3. Pharmacology

In a study of the stereochemical preferences for curarimimetic neuro-muscular junction blockade, the *R,R*-isomer of *N*-methylpavine methiodide showed a modest statistical superiority in potency in two of the four assay systems employed.[19]

4.4. Spectroscopy and Crystallography

The CMR chemical shifts of the symmetrically substituted pavine alkaloid argemonine are as indicated below. The *N*-methyl signal at δ40.6 is shielded with respect to its chemical shift in laudanosine, cularine, and a variety of aporphines.[20]

The absolute configuration of (−)-argemonine methiodide has been confirmed by an X-ray analysis.[21]

References and Notes

1. J. Slavík, L. Slavíková, and K. Haisová, *Collect. Czech. Chem. Commun.*, **38**, 2513 (1973).
2. S. K. Maekh, S. Y. Yunusov, and P. G. Gorovoi, *Khim. Prir. Soedin.*, 116 (1976).
3. R. M. Coomes, J. R. Falck, D. K. Williams, and F. R. Stermitz, *J. Org. Chem.*, **38**, 3701 (1973). The absolute configuration and numbering system of this alkaloid as given here are different from those in the original paper, but are consistent with the evidence reported, especially the strong negative specific rotation.
4. M. Shamma, A. S. Rothenberg, S. S. Salgar, and G. S. Jayatilake, *Lloydia*, **39**, 395 (1976).
5. For details of early synthetic work in this series, see M. Shamma, *The Isoquinoline Alkaloids*, Academic Press, New York (1972), p. 101.
6. S. Natarajan and B. R. Pai, *Indian J. Chem.*, **12**, 550 (1974).
7. M. S. Premila and B. R. Pai, *Indian J. Chem.*, **11**, 1084 (1973).
8. For a review of the application of Reissert compound chemistry to the synthesis of isoquinoline alkaloids, see F. D. Popp, *Heterocycles*, **1**, 165 (1973).
9. F. R. Stermitz and D. K. Williams, *J. Org. Chem.*, **38**, 1761 (1973).
10. F. R. Stermitz, D. K. Williams, S. Natarajan, M. S. Premila, and B. R. Pai, *Indian J. Chem.*, **12**, 1249 (1974).
11. For a synthesis of an *N*-methylhomopavinane base, see F. R. Stermitz and D. K. Williams, *J. Org. Chem.*, **38**, 2099 (1973).
12. K. Ito, H. Furukawa, T. Iida, K. H. Lee, and T. O. Soine, *Chem. Commun.*, 1037 (1974).
13. S. F. Dyke and A. C. Ellis, *Tetrahedron*, **28**, 3999 (1972).
14. S. F. Dyke, A. C. Ellis, R. G. Kinsman, and A. W. C. White, *Tetrahedron*, **30**, 1193 (1974).
15. O. Hoshino, M. Taga, and B. Umezawa, *Heterocycles*, **1**, 223 (1973).
16. H. Hara, O. Hoshino, and B. Umezawa, *Heterocycles*, **5**, 713 (1976).
17. T. Kametani, S. Hirata, and K. Ogasawara, *J. Chem. Soc. Perkin I*, 1466 (1973). See also T. Kametani and K. Ogasawara, *Chem. Pharm. Bull. Tokyo*, **21**, 893 (1973).
18. J. R. Brooks, D. N. Harcourt, and R. D. Waigh, *J. Chem. Soc. Perkin I*, 2588 (1973).

19. A. A. Genenah, T. O. Soine, and N. A. Shaath, *J. Pharm. Sci.*, **64**, 62 (1975); and P. W. Erhardt and T. O. Soine, *J. Pharm. Sci.*, **64**, 53 (1975).
20. E. Wenkert, B. L. Buckwalter, I. R. Burfitt, M. J. Gasic, H. E. Gottlieb, E. W. Hagaman, F. M. Schell, and P. M. Wovkulich in *Topics in Carbon-13 NMR Spectroscopy*, Vol. 2, G. C. Levy, ed., Wiley-Interscience, New York (1976), p. 110.
21. T. Kaneda, N. Sakabe, and J. Tanaka, *Bull. Chem. Soc., Tokyo*, **49**, 1263 (1976).

THE BISBENZYLISOQUINOLINES

5

5.1. Introduction

Bisbenzylisoquinolines have[1,2] been classified into 26 types.[3] The recent characterization of the new alkaloids tiliamosine[4] and thalistyline[4a] brings this total to 28. These types can best be described by a numerical system obeying the following rules.[3]

1. The numbering system for a benzylisoquinoline half of a dimer is shown in structure **1** (see page 81).

2. Each benzylisoquinoline half of the dimer is described in terms of its oxygenation pattern, since only oxygenated positions are indicated. The more highly oxygenated benzylisoquinoline half constitutes the left-hand side of the dimer, and is listed first. The two sets of numerical values are separated by a hyphen. In the case of head-to-tail coupling, the more highly oxygenated benzylisoquinoline is placed on top, and is listed first.

3. The symbols *, †, and ‡ indicate the shared oxygens of diaryl ethers, and are placed at the upper right of the appropriate numbers.

4. Numbers between parentheses, appearing after the listing of the oxygenated sites, denote a direct carbon–carbon bond for the presence of a biphenyl linkage. Numbers between squared brackets, appearing after the listing of the oxygenated sites, indicate the terminals of a methylenoxy bridge.

Using the above simple rules, the 28 types of bisbenzylisoquinolines can be denoted as shown in Table 5.1.[3] This classification system allows the precise description of the oxygenated skeleton of a bisbenzylisoquinoline without having resort to a structural diagram. In the case where both sides of a dimer bear an equal number of oxygens, the moiety appearing at the left or on top is that possessing the higher numbers, for example, for thalicrine 6*,7,11†,12-6,7,8*,12†, rather than the reverse notation 6,7,8*,12†-6*,7,11†,12. Whenever the numerical-type description shows any position between 5 and 8 bonded to one between 9 and 14 in the other half of the dimer, this *ipso facto* can be taken as an indication that head-to-tail coupling is involved. It should be borne in mind that in nature no head-to-head coupling is encountered without accompanying tail-to-tail coupling. In instances where a large ring has been formed through diaryl ether, biphenyl, or methylenoxy bridging, this ring is usually 18-membered, although rings as small as 16 or as large as 20 are also known.[3]

Table 5.1. The 28 Types of Bisbenzylisoquinolines

Dimer	Type	Alkaloids
	5,6,7,11*,12-5,6,7,12*	Thalistyline
	6,7,11*,12-6,7,12*	Dauricine, daurinoline, dauricinoline, dauricoline, magnoline, berbamunine, espinine, espinidine, lindoldhamine, thalibrine, cuspidaline, grisabine
	6,7,11*,12-6,7*,12	Liensinine, isoliensinine, neferine

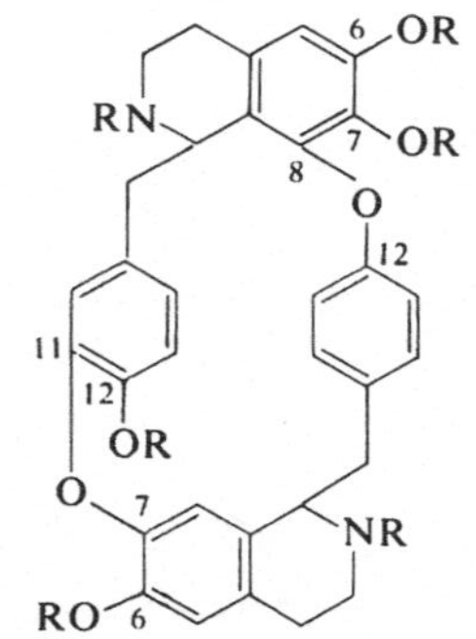

6,7,8*,11†,12-6,7†,12*

Curine (chondodendrine), tubocurine (chondocurine), chondocurarine, chondrofoline, tubocurarine, hayatinine, hayatidine, cycleacurine, toxicoferine, hayatine

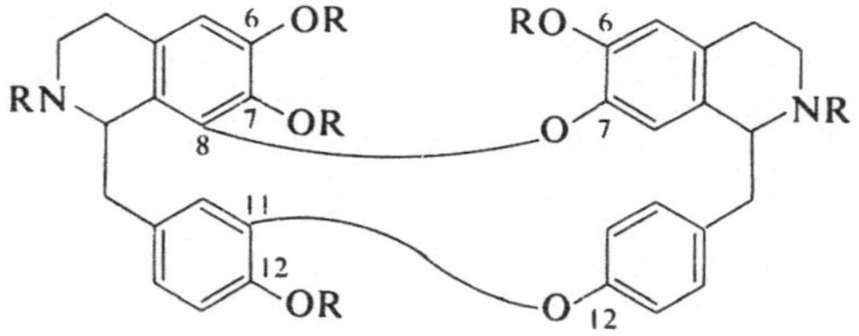

6,7,8*,11†,12-6,7*,12†

Berbamine, pycnamine, tetrandrine, isotetrandrine, phaeanthine, fangchinoline, limacine, obamegine, thalrugosine, penduline, atherospermoline, cycleadrine, cycleahomine, cycleanorine, krukovine, phaeantharine, menisidine, menisine, peinamine

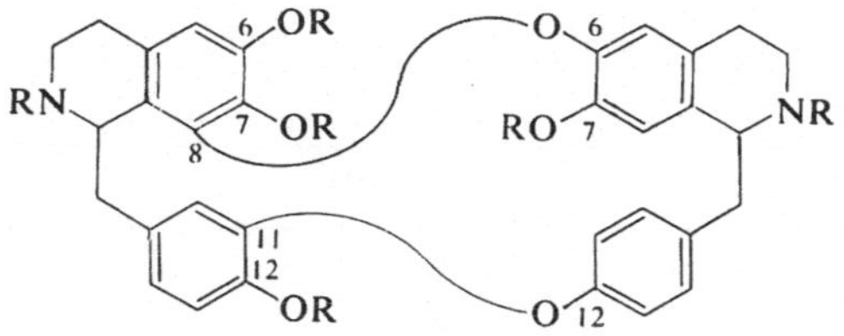

6,7,8*,11†,12-6*,7,12†

Thalicberine, O-methylisothalicberine, thalmetine, belarine

(continued)

Table 5.1 —continued

Dimer	Type	Alkaloids
	5*,6,7,11†,12-6*,7,12†	Nemuarine
	5*,6,7,11†,12-6,7*,12†	Panurensine
	5,6,7,8*,11†,12-6,7*,12†	Hernandezine, thalsimine, thalidezine, isothalidezine, thalisamine, thalsimidine

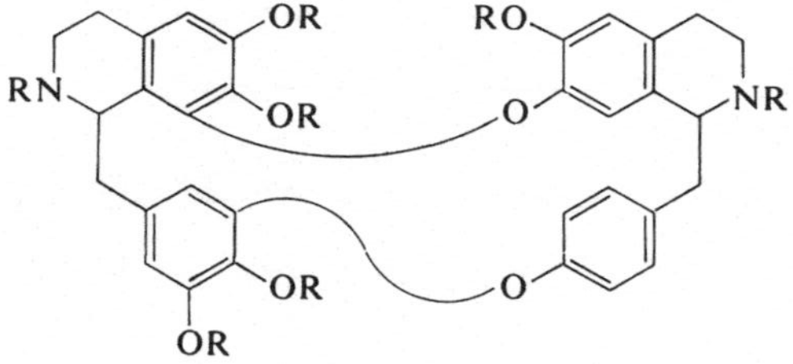

Tenuipine, isotenuipine, repandinine

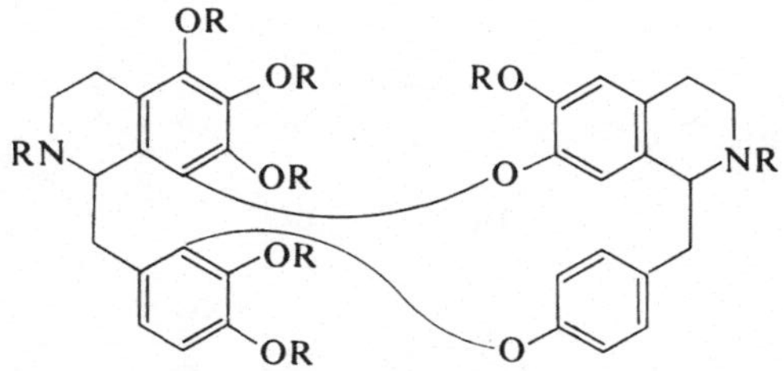

6,7,8*,11†,12,13-6,7*,12†

Thalibrunine, thalibrunimine

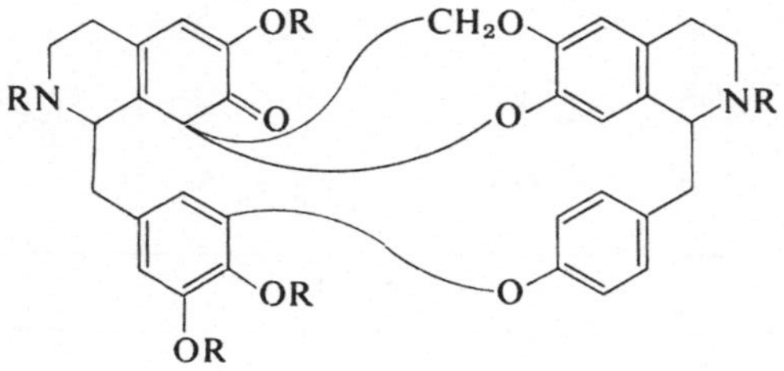

5,6,7,8*,10†,11,12-6,7*,12†

Repanduline, pseudorepanduline

6,7,8*,11†,12,13-6,7*,12†[8-6]

(continued)

Table 5.1 —continued

Dimer	Type	Alkaloids
	6,7,10*,12,13-6,7,12*	Magnolamine
	6,7,8*,12-6,7*,12(11-11)	Rodiasine, funiferine, ocotosine, tiliageine, phlebicine ocotine, dirosine
	6,7*,8†,12-6*,7†,12(11-11)	Tiliacorine, tiliacorinine, nortiliacorine-A, nortiliacorinine-A and -B

5,6,7*,8†,12-6*,7†,12(11-11)   Tiliamosine[4]

6,7,8*,12†-6,7,8†,12*   Isochondodendrine, cycleanine, protocuridine, sciadenine, sciadoline

6,7,8*,12†-6,7,8†,12*[11-7]   Insularine, insulanoline

(continued)

Table 5.1 —continued

Dimer	Type	Alkaloids
	6,7,*11†,12-6,7,8*,12†	Oxyacanthine, ocoteamine, epistephanine, repandine, obaberine, demerarine, stebisimine, limacusine, aromoline, homoaromoline, thalmine, coclobine, thalrugosamine, daphnoline, cepharanoline, cycleapeltine, daphnandrine, sepeerine, cepharanthine, hypoepistephanine, oblongamine, faralaotrine, macolidine, macoline
	6*, 7†, 11‡,12-6,7*,8†,12‡	Trilobine, isotrilobine, micranthine, cocsuline, telobine, tricordatine, cocsoline, trigilletimine, apateline
	6,7*,8†,11‡,12-6,7†,8*,12‡	Menisarine, cocsulinine
	6*,7,11†,12-6,7,8*,12†	Thalicrine

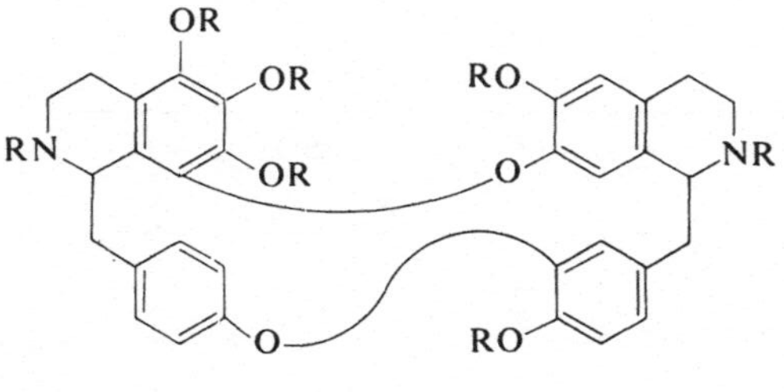

Thalisopine, thalisopidine, thalrugosaminine

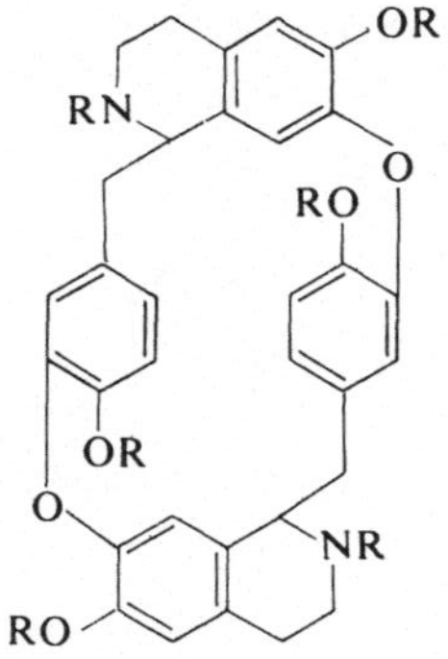

5,6,7,8*,12†-6,7*,11†,12

Skeleton formerly assigned to hayatine

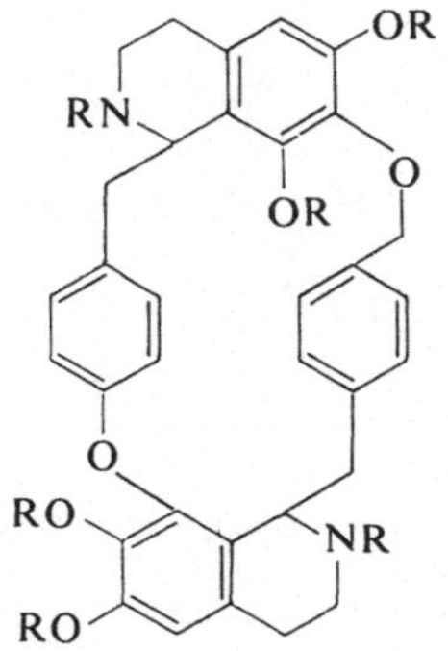

6,7*,11†,12-6,7†,11*,12

Cissampareine, warifteine

6,7,8,12*-6,7,8*[7-12]

(continued)

Table 5.1 —continued

Dimer	Type	Alkaloids
	6,7*,11†,12-5*,6,7,12†	Thalmine, dryadine, lauberine, dryadodaphnine, thalictine
	6,7,8*,11†,12-5*,6,7,12†	Thalidasine, thalfoetidine, thalrugosidine
	5*,6,7,11†,12-5,6,7,8*,12†	Thalfine, thalfinine

The building blocks for the bisbenzylisoquinolines are thus (+)- and (−)-O-demethylcoclaurine (**2**), followed by the less common (+)-laudanosoline (**3a**), (+)-thalinosoline (**4**), and (+)-5-hydroxylaudanosoline (**3b**). It appears, therefore, as if a benzylisoquinoline possessing a secondary amino function can occur either in the R- or the S-configuration. No cases are known where two

1

(+)-**2**, β H-1
(−)-**2**, α H-1

3a, R = H
3b, R = OH

4

laudanosoline or 5-hydroxylaudanosoline units have dimerized since the tendency is for these monomers to oxidize further to aporphine or protoberberine bases.[3]

An apparently new source of bisbenzylisoquinolines is the plant family Umbelliferae.[5]

5.2. Stepinonine, an Unusual Bisbenzylisoquinoline

The chemistry of the monophenolic yellow alkaloid (−)-stepinonine, $C_{36}H_{34}N_2O_7$, isolated from *Stephania japonica* Miers, has been investigated by Ibuka, Konoshima, and Inubushi.[6] The alkaloid shows a conjugated carbonyl band at 1663 cm^{-1} (6.02 μ), and can be reduced with sodium borohydride to tetrahydrostepinonine whose IR spectrum is devoid of carbonyl absorption. N-Methylation of this derivative furnished N-methyltetrahydrostepinonine whose N-methyl PMR signal appeared unusually upfield at δ2.16.

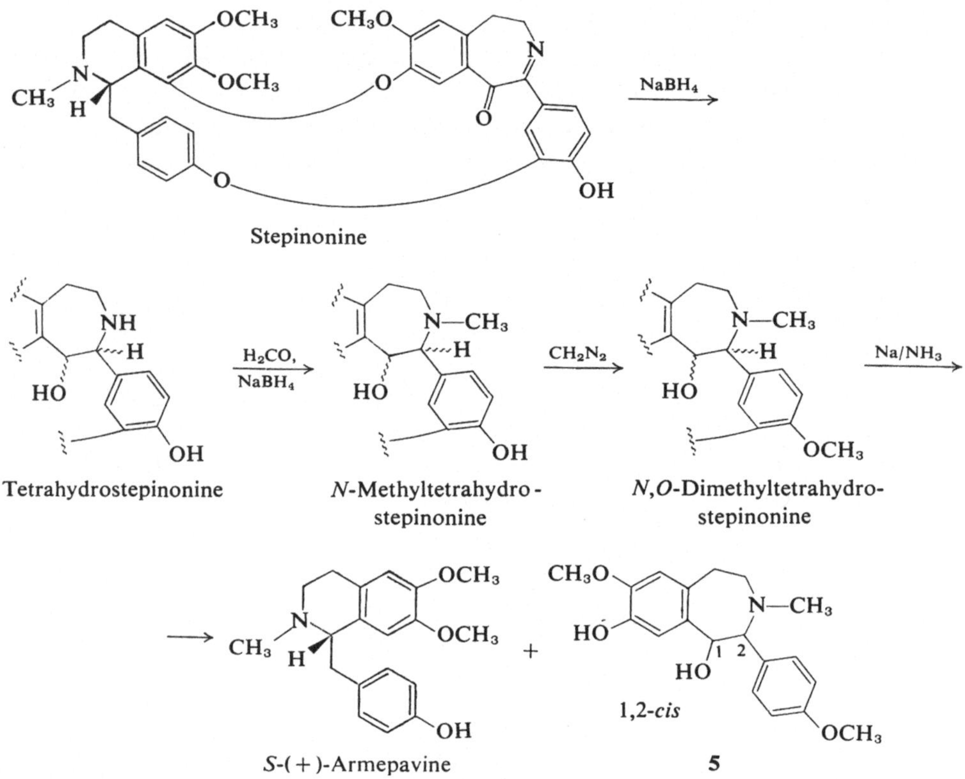

Stepinonine

Tetrahydrostepinonine   *N*-Methyltetrahydro-stepinonine   *N,O*-Dimethyltetrahydro-stepinonine

S-(+)-Armepavine   **5**

1,2-*cis*

SCHEME 5.1

This compound could in turn be *O*-methylated to *N,O*-dimethyltetrahydro-stepinonine. Reductive fission of the latter with sodium in liquid ammonia produced *S*-(+)-armepavine, together with the benzazepine **5** (see Scheme 5.1).

    O-Ethylation of *N*-methyltetrahydrostepinonine with diazoethane followed by permanganate in acetone oxidation generated the dicarboxylic acid **6**, thus establishing the position of the phenolic function as well as the nature of the lower diaryl linkage in stepinonine. The exact position of the upper diaryl ether was settled by treatment of the *O*-ethyl derivative of *N*-methyltetrahydro-stepinonine with $C_2H_5OD-D_2O-3\%$ DCl, conditions under which deuteration occurs only at unsubstituted aromatic sites ortho to methoxyl groups. The resulting dideuterio derivative **7** was then subjected to reductive fission to afford *S*-(+)-[5-D]-armepavine and a benzazepine deuterated at C-6 (see Scheme 5.2).[6]

    For characterization purposes, the *O*-ethyl derivative of the benzazepine degradation product **5** was synthesized as shown in Scheme 5.3.

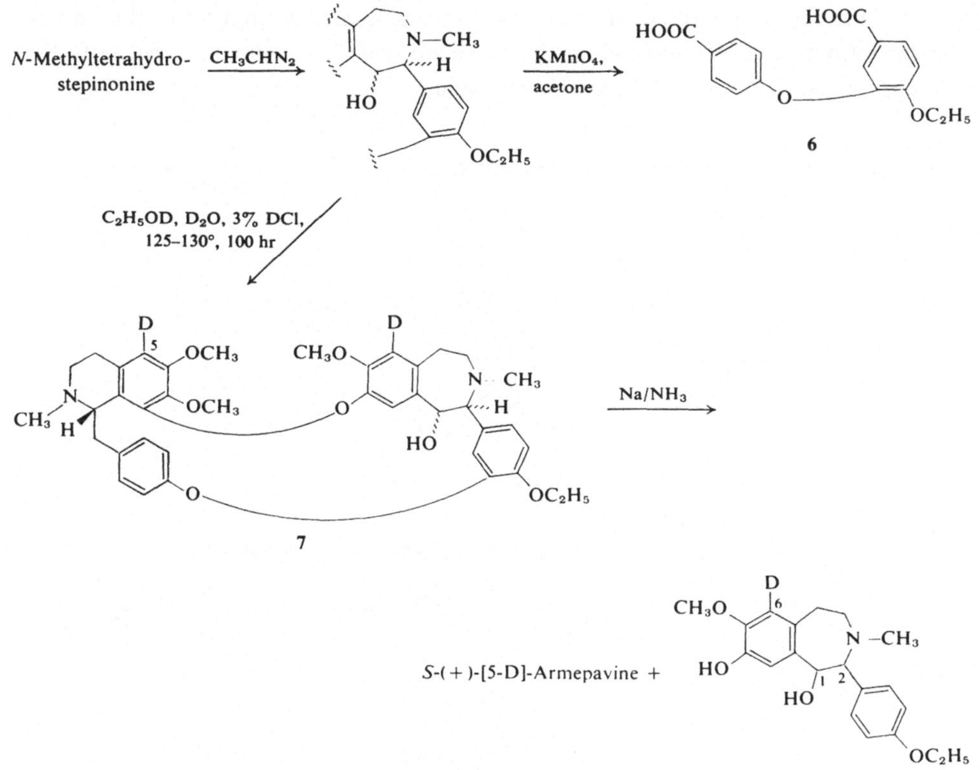
N-Methyltetrahydro-
stepinonine
CH3CHN2
N—CH3
H
HO
OC2H5
KMnO4,
acetone
HOOC
HOOC
O
OC2H5
6
C2H5OD, D2O, 3% DCl,
125–130°, 100 hr
D
5
OCH3
OCH3
N
CH3
H
CH3O
D
N CH3
H
O
HO
O
OC2H5
7
Na/NH3
S-(+)-[5-D]-Armepavine +
D
6
CH3O
N —CH3
HO
1 2
HO
OC2H5

SCHEME 5.2

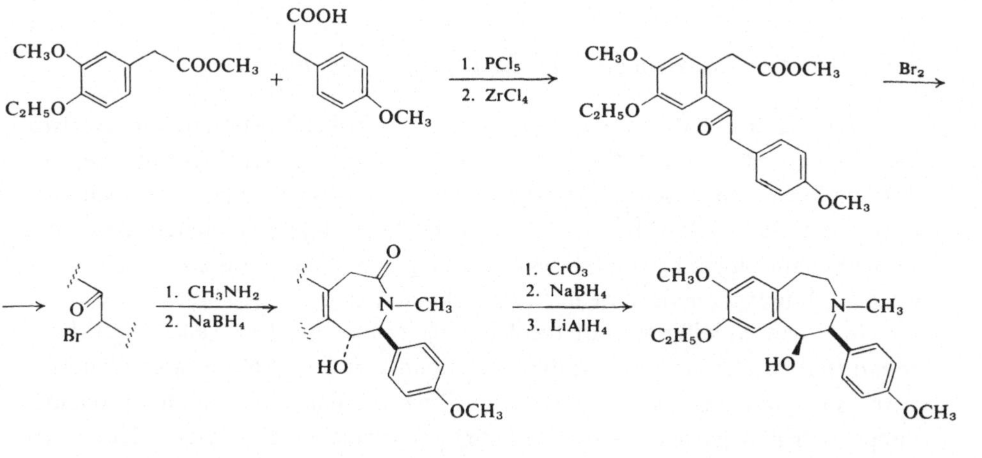
CH3O
C2H5O
COOCH3
+
COOH
OCH3
1. PCl5
2. ZrCl4
CH3O
C2H5O
COOCH3
O
OCH3
Br2
O
Br
1. CH3NH2
2. NaBH4
O
N—CH3
HO
OCH3
1. CrO3
2. NaBH4
3. LiAlH4
CH3O
C2H5O
N—CH3
HO
OCH3

SCHEME 5.3

Finally, stepinonine was converted into a separable mixture of the known *O*-methylrepandine and *O*-methyloxyacanthine by rearrangement of the appropriate ketone with zinc in acetic acid.[7]

Stepinonine $\xrightarrow[\substack{see \\ Scheme\ 5.1}]{several\ steps,}$ *N,O*-Dimethyltetrahydrostepinonine $\xrightarrow[HOAc]{CrO_3,}$

A ketone

O-Methylrepandine,   C-1′ α-H
O-Methyloxyacanthine, C-1′ β-H

5.3. Synthesis

5.3.1. The Cava Modification of the Ullmann Reaction

The classical Ullmann-type synthesis of a bisbenzylisoquinoline involves the direct coupling of a phenolic benzylisoquinoline with a halogenated benzylisoquinoline in the presence of copper or one of its salts or oxides. The advantage of this approach is that the two halves of the dimer may be prepared separately as pure enantiomers before the final coupling step. This feature is usually outweighed, however, by low yields in the Ullmann step.

In a new modification of the Ullmann procedure, Cava and Afzali have found that condensation of equimolar amounts of *S*-(+)-6′-bromolaudanosine with *S*-(+)-armepavine in pyridine in the presence of pentafluorophenylcopper (PFPC) gave (+)-*O*-trimethylmagnolamine in 42% yield. This reaction was then extended to the aporphine–benzylisoquinoline series.[8]

S-(+)-6′-Bromolaudanosine   S-(+)-Armepavine

(+)-O-Trimethylmagnolamine

It should be noted in this context that the old structure for (+)-magnol-amine which incorporated four phenolic groups, at C-7, 12, 13, and 7′, has lately been modified. Oxidative degradative studies have shown this alkaloid to be only a triphenol, with hydroxyls at C-7, 13, and 7′. The correct molecular formula is then $C_{37}H_{42}N_2O_7$ rather than $C_{36}H_{40}N_2O_7$.[8a]

5.3.2. The Inubushi Synthesis of Obaberine and Trilobine

The previously known Inubushi syntheses of (+)-isotetrandrine and (−)-phaeanthine[9] have now been adapted to preparations of (+)-obaberine and (+)-trilobine.[10] Advantage was taken of the facts that (a) the t-butoxycarbonyl group is resistant to catalytic hydrogenation but is readily hydrolyzed, and (b) amides are readily formed from the reaction of an amine with a carboxylic ester of p-nitrophenol. The formation of the third diaryl ether bridge present in trilobine by reaction of the appropriate polyphenol with hot aqueous HBr had previously been carried out on other bisbenzylisoquinolines (see Schemes 5.4 and 5.5).[11]

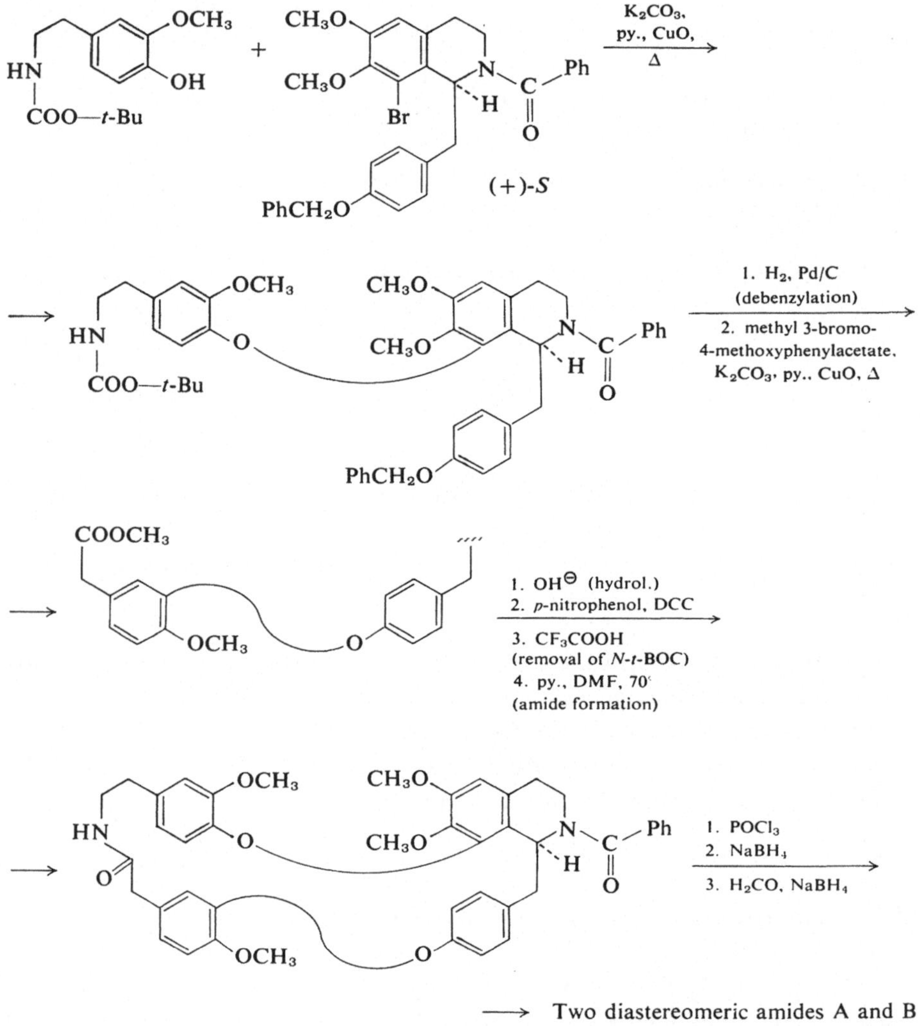

OCH3
HN
OH
COO—t-Bu
+
CH3O
CH3O
Br
N
H
C
Ph
O
(+)-S
PhCH2O
K2CO3,
py., CuO,
Δ
OCH3
HN
O
COO—t-Bu
CH3O
CH3O
N
H
C
Ph
O
PhCH2O
1. H2, Pd/C
(debenzylation)
2. methyl 3-bromo-
4-methoxyphenylacetate,
K2CO3, py., CuO, Δ
COOCH3
OCH3
O
1. OH⊖ (hydrol.)
2. p-nitrophenol, DCC
3. CF3COOH
(removal of N-t-BOC)
4. py., DMF, 70°
(amide formation)
OCH3
HN
O
O
CH3O
CH3O
N
H
C
Ph
O
OCH3
O
1. POCl3
2. NaBH4
3. H2CO, NaBH4
Two diastereomeric amides A and B

SCHEME 5.4

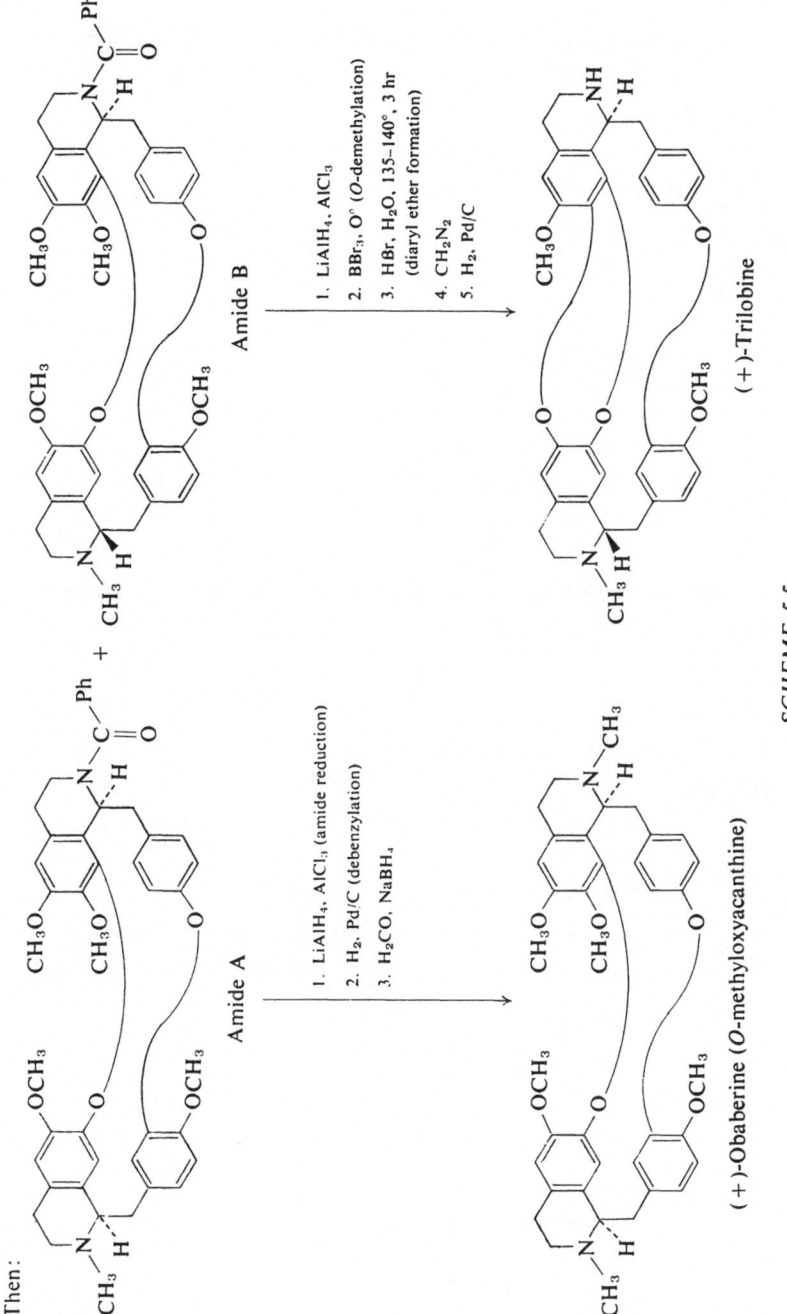

SCHEME 5.5

5.3.3. The Use of Reissert Compounds

A series of bisbenzylisoquinolines has been prepared through alkylation of 6,7-disubstituted isoquinoline Reissert compounds with diaryl ethers such as **8, 9,** and **10**.[12]

The sequence involving diaryl ether **9** and the Reissert derivative from 6,7-methylenedioxyisoquinoline is represented here.

5.4. Controlled Oxidation

Potassium permanganate in acetone oxidation of a bisbenzylisoquinoline results in selective cleavage at the C-1 to C-α benzylic bond of the isoquinoline moiety which is unsubstituted at C-8. Relative stereochemistry does not affect the site of the oxidation. The product is an aldehydo lactam showing $\nu_{max}^{CHCl_3}$ 1640–1645 cm^{-1} and 1690–1720 cm^{-1} (5.82–5.92 and 6.08–6.17 μ) which can be used for characterization purposes. The C-8 hydrogen peri to the new lactam carbonyl appears in the NMR spectrum as a singlet between δ7.20 and 7.42. To exemplify, in the case of the oxidation of (+)-hernandezine, the C-8 peri proton singlet is found at δ7.23. The yields range from 5 to 35%; and it is preferable to protect phenolic functions by acetylation prior to oxidation.[13]

(+)-Hernandezine

Secohernandezine aldehydo lactam

This oxidation has been utilized in the determination of the location of the phenolic function in (+)-tiliacorine, which had been formulated as either **11** or **12**. Permanganate in acetone oxidation of its O-acetyl derivative, followed by ester hydrolysis, yielded secotiliacorine aldehydo lactam, λ_{max}^{EtOH} 212, 282, and 310 sh (4.10, 3.44, and 2.94). This product showed a strong bathochromic shift in base, $\lambda_{max}^{EtOH-OH^-}$ 230, 292, and 340 nm (4.00, 3.27, and 3.21), indicative of a phenolic function para to an aromatic aldehyde, so that tiliacorine must be represented by structure **11**.[14]

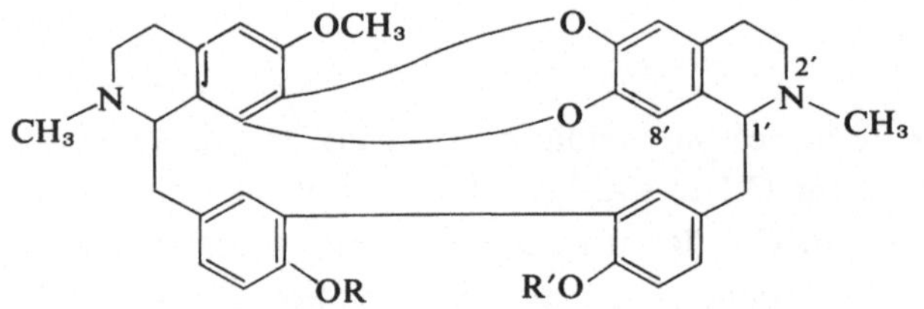

11, R = CH$_3$, R' = H
12, R = H, R' = CH$_3$

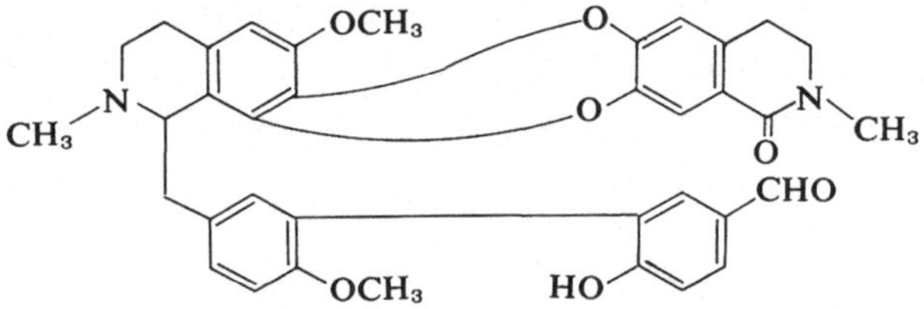

Secotiliacorine aldehydo lactam

Hernandezine has been prepared from tetrandrine through the sequence of selective monobromination, metal–bromine exchange, and oxygenation of the organometallic intermediate with nitrobenzene followed by O-methylation. Oxidations of this type have also been extended to the tetrahydrobenzylisoquinoline and the aporphine series.[15]

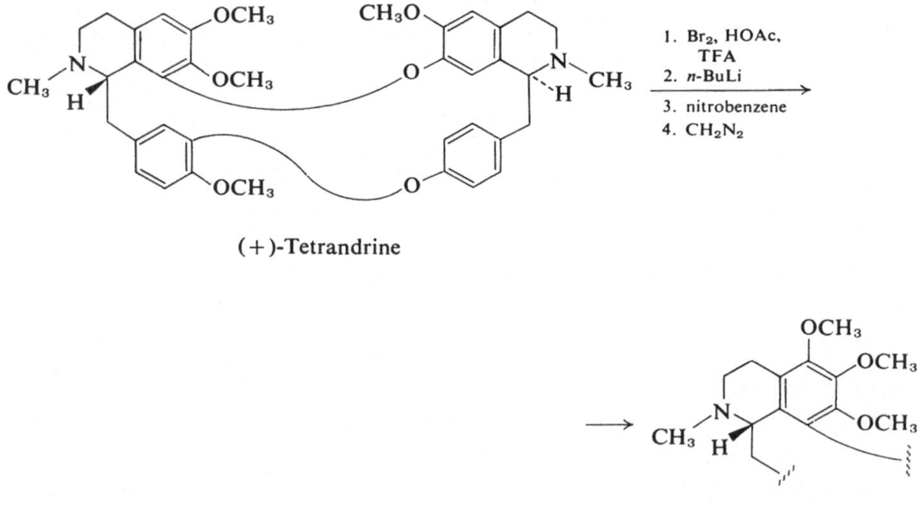

1. Br$_2$, HOAc, TFA
2. n-BuLi
3. nitrobenzene
4. CH$_2$N$_2$

(+)-Tetrandrine

(+)-Hernandezine

5.5. Some General Methods for Functional Group Modification of Isoquinoline Alkaloids

5.5.1. Selective Removal of an Aromatic Methylenedioxy Group

Boron trichloride in methylene chloride will selectively hydrolyze a methylenedioxy function, methoxyl substituents remaining unaffected. The resulting catechol can be converted to an unsubstituted benzene ring through the Musliner–Gates procedure.[16]

5.5.2. Deoxygenation of a Phenol

An efficient method for the deoxygenation of a phenol, predating the Musliner–Gates procedure, consists of the reaction of the phenol with diethyl phosphorochloridate. The resulting aryl diethyl phosphate ester is then hydrogenolyzed with sodium in liquid ammonia.[17]

$$\text{ArOH} + (\text{C}_2\text{H}_5\text{O})\text{POCl} \xrightarrow{\text{NaOH}} \text{ArOPO}(\text{OC}_2\text{H}_5) \xrightarrow{\text{Na/NH}_3} \text{ArH}$$

A more recent method for accomplishing the same goal converts the phenol into the potassium aryl sulfate by reaction with chlorosulfonic acid in the presence of a tertiary amine, followed by treatment with aqueous potassium hydroxide. The resulting sulfate ester is then catalytically hydrogenolyzed.[18]

$$\text{ArOH} \xrightarrow[\text{2. KOH, H}_2\text{O}]{\text{1. ClSO}_4\text{H, py., CS}_2} \text{ArOSO}_3^{\ominus} \text{K}^{\oplus} \xrightarrow[\text{or H}_2, \text{Pd/C}]{\text{Ni(R), OH}^{\ominus}} \text{ArH}$$

5.5.3. N-Demethylation of Tertiary Amines

An improved procedure for the *N*-demethylation of tertiary amines, which was applied in the morphine series, utilizes phenyl chloroformate to obtain an intermediate carbamate which is then easily cleaved with a 1:1 mixture of 64 and 95% hydrazine.[19] *p*-Nitrophenyl chloroformate[20] or the milder vinyl chloroformate[21] are sometimes preferred over phenyl chloroformate.

$$\text{>N---CH}_3 \xrightarrow[\text{KHCO}_3, \Delta]{\text{ClCOOPh,}} \text{>N---C---OPh} \xrightarrow{\text{1:1 64 and 95\% N}_2\text{H}_4} \text{>NH}$$

Alternate procedures include the use of silver nitrite in DMF to generate the *N*-nitroso derivative;[22] or of trichloroethyl chloroformate to obtain a urethan which is cleaved reductively with zinc in acetic acid.[23]

$$\text{>N—CH}_3 \xrightarrow[\text{DMF, 70°}]{\text{AgNO}_2,} \text{>N—N=O}$$

$$\Big\downarrow \quad \text{Cl}_3\text{CCH}_2\text{—O—C}\overset{O}{\underset{Cl}{<}}$$

$$\text{>N—C—O—CH}_2\text{CCl}_3 \xrightarrow{\text{Zn, HOAc}} \text{>NH}$$
$$\overset{\|}{O}$$

Where facile oxidation is precluded, as in the case of substituted *N*-methylanilines, the nor derivative may be generated by treatment with mercuric acetate.[23a]

A modified Polonovski reaction is another efficient method for *N*-demethylation. The amine *N*-oxide is treated with trifluoroacetic anhydride instead of the more usual acetic anhydride, and the resulting immonium salt is then hydrolyzed.[24]

5.5.4. N-Demethylation and N-Debenzylation of Quaternary Ammonium Salts

N-Methyl quaternary pyridinium salts are easily demethylated with triphenylphosphine in DMF.[25] Cuprous phenylmercaptide,[26] lithium *t*-butylmercaptide,[27] and lithium triethylborohydride,[28] are all powerful reagents for the *N*-demethylation of aliphatic *N*-methyl quaternary salts. Other potentially useful nucleophilic reagents for this reaction are triphenyl phosphine, thiourea, sodium thiosulfate, and sodium azide.[29] The *N*-debenzylation of ammonium salts has been performed at near 0° using lithium *n*-propylmercaptide in HMPA.[30]

5.5.5. O-Demethylation and O-Debenzylation of Aromatic Ethers

Nucleophilic scission of methyl aryl ethers can be achieved using alkyl sodium mercaptide reagents in DMF,[31] or through the related procedure utilizing sodium *p*-cresylmercaptide in toluene with a limited amount of HMPA.[32] Another useful reagent in this respect is pyridine hydrochloride in quinoline[33] or in pyridine.[16] Sodium benzylselenolate in refluxing DMF is a superior reagent for the *O*-demethylation of aryl methyl ethers, which also shows a high degree of regioselectivity.[33a] Refluxing a benzyl ether with sodium bis(2-methoxyethoxy)aluminum hydride (Redal) in xylene results in cleavage of the benzylic bond.[34]

Iodotrimethylsilane has recently been introduced as a reagent for cleaving alkyl–aryl ethers under neutral conditions, and is now available commercially.[34a]

5.5.6. Methylenation of Catechols

A new procedure for the conversion of a catechol to a methylenedioxy function utilizes phase-transfer catalysis. The catalyst is Adogen 464, a methyl trialkylammonium chloride, and the alkylating agent is methylene bromide. Usual yields are around 80%.[35] Another efficient method utilizes potassium fluoride and dibromomethane in DMF.[35a]

5.5.7. O-Methylation and Protection of Phenols

Weakly acidic phenols that do not react with diazomethane can be methylated with sodium hydride and methyl iodide in THF at room temperature.[36] The methoxymethyl ether moiety can be used to protect phenols. It is stable to alkali, Grignard reagents, lithium aluminum hydride, and catalytic hydrogenation, and is readily removed by mineral acid. Dimethoxymethane can be used in lieu of the carcinogenic chloromethyl methyl ether for this purpose.[37] Alternatively, phenols may be protected as methyl thiomethyl ethers.[37a] The O-acetylation of phenols in the presence of primary and secondary amines can be carried out with acetyl bromide and TFA.[37b]

5.5.8. Aromatic Hydroxylation or Methoxylation

For the introduction of a phenolic or methoxyl group into an aromatic ring, see Sec. 5.4.[15]

The Gibbs' color test, claimed to be positive for phenols possessing a free para position, is unreliable in the isoquinoline series, and supplies no meaningful structural information.

5.6. Biogenesis

The callus tissues derived from the tuber of *Stephania cepharantha* Hayata (Menispermaceae) can synthesize bisbenzylisoquinolines, but these dimers have been found to be somewhat different from those in the original plant. Only two compounds were formed in the callus tissues, namely aromoline and berbamine, whereas the whole plant produces cepharanthine, homoaromoline, cepharanoline, isotetrandrine, berbamine, and cycleanine. With the exception of cycleanine, the alkaloids of the whole plant can be considered to be derived

from either aromoline or from berbamine by *O*-methylation or through conversion of an *o*-methoxyphenol to a methylenedioxy group. Thus, the callus
tissues lack the specific enzymes necessary for *O*-methylation and methylenedioxy group formation.[38]

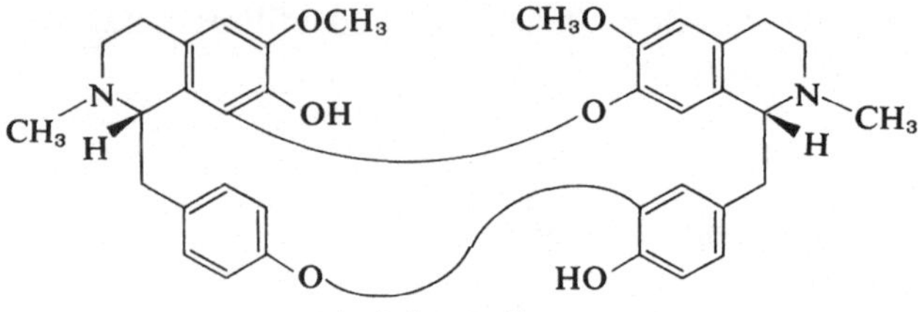

(+)-Aromoline

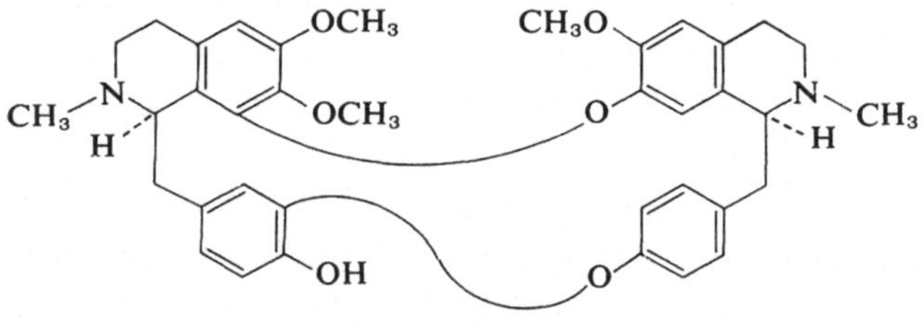

(+)-Berbamine

Alkaloids of type 6,7,8*,12-6,7*,12(11-11), e.g., funiferine and tiliageine,
have been found in the vine *Tiliacora dinklagei* (Menispermaceae) together
with dimers of type 6,7*,8†,12-6*,7†,12(11-11), such as tiliacorinine and nor-
tiliacorinine-A.[39] It is likely, therefore, that bases belonging to the former
type act as precursors for alkaloids of the latter variety.

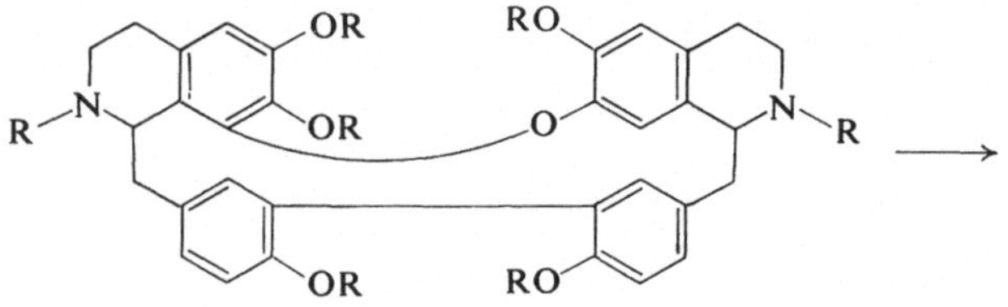

Type 6,7,8*,12-6,7*,12(11-11)

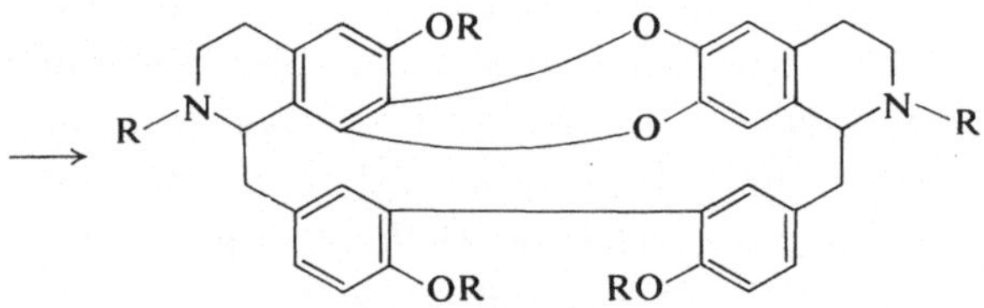

6,7*,8†,12-6*,7†,12(11-11)

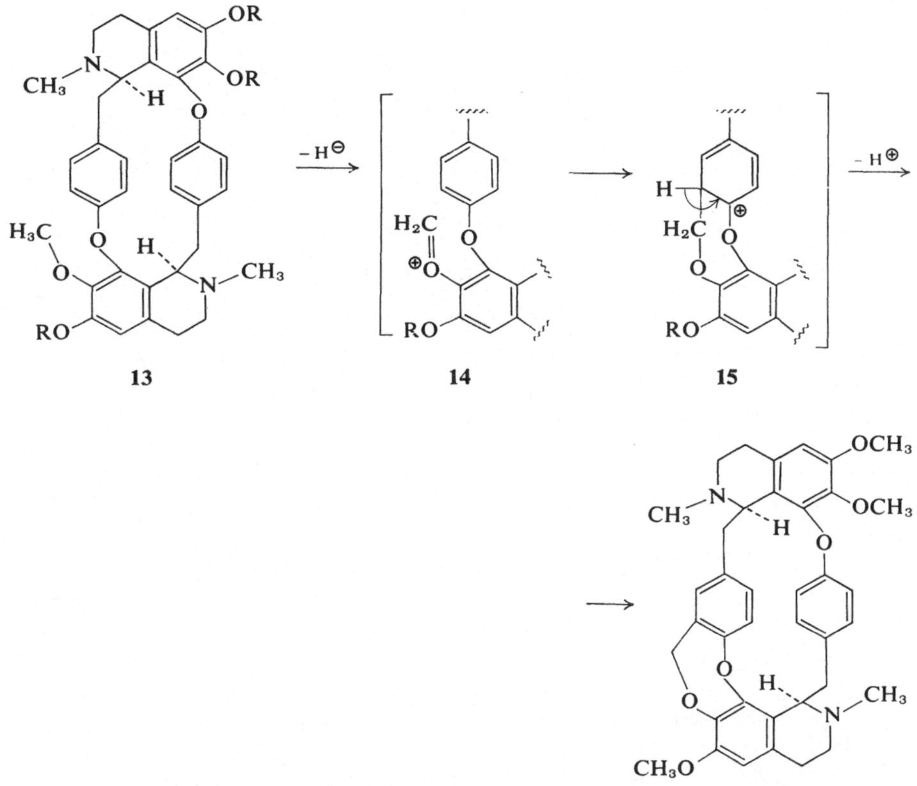

13   **14**   **15**

Insularine

The formation of the methylenoxy bridge has been discussed.[40] Assuming an ionic mechanism, insularine must be formed via the oxonium ion **14** derived from the cycleanine-type dimer **13**.

Furthermore, the identical intermediate **15**, or one of its close analogs, could undergo a 1,2-alkyl shift to afford **16**. Subsequent hydride addition and bond cleavage, as shown below, would result in eventual formation of the alkaloid cissampareine.[40]

A further bisbenzylisoquinoline with a methylenoxy bridge is repanduline which also incorporates an α-ketol function. Its probable biogenetic precursors must be represented by structures **17** and **18**, where **17** corresponds to the accompanying alkaloid nortenuipine.[40,41]

(+)-Tetrandrine (6,7,8*,11†,12-6,7*,12†) can be selectively N-demethylated at N-2′ using *Streptomyces griseus* or a wide variety of other microorganisms. However, *Cunninghamella blakesleeana* is unusual in that it effects the loss of the methyl group at N-2.[42]

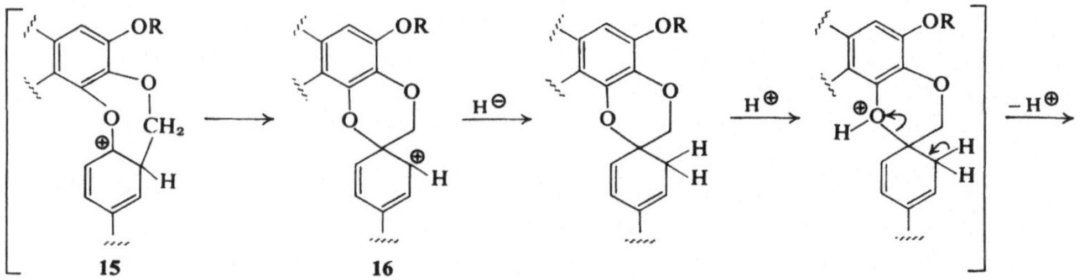

15
(or analog)

16

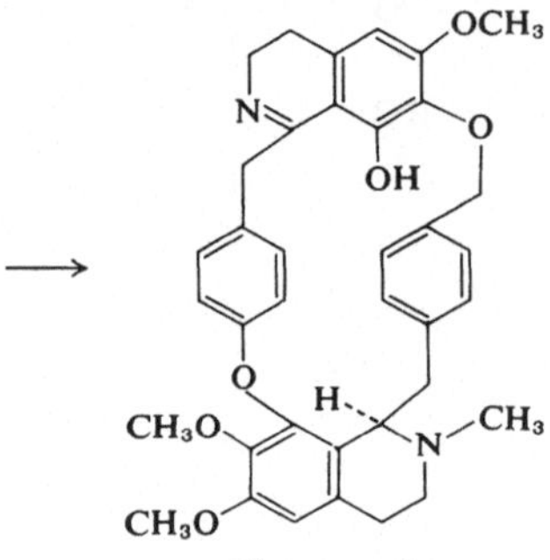

Cissampareine

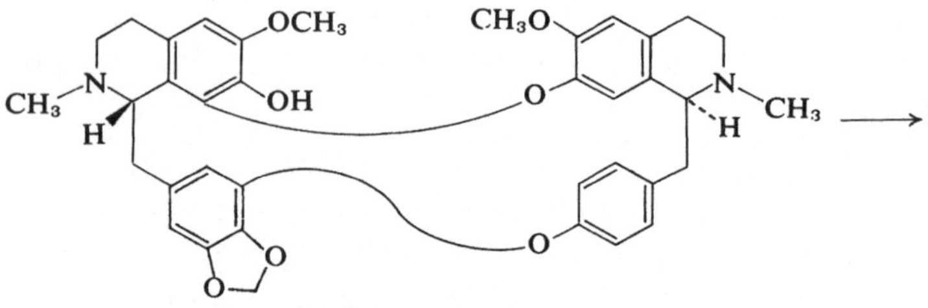

Nortenuipine, 17

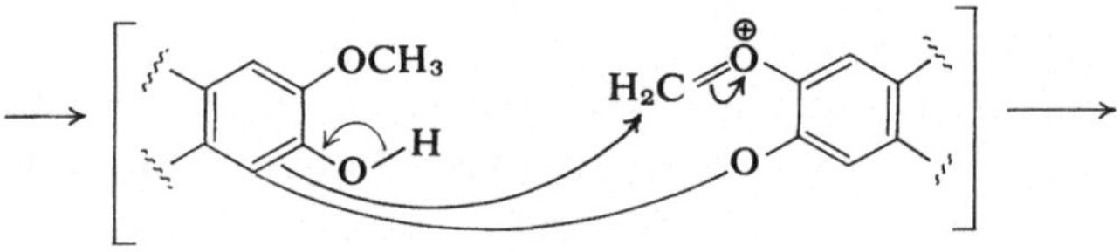

18

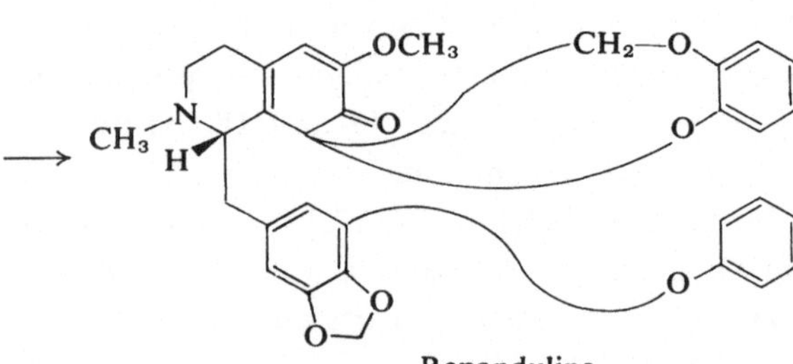

Repanduline

5.7. Pharmacology

The methiodide salts of the crude alkaloids from *Tiliacora racemosa* Colebr. (Menispermaceae),[4] and the alkaloids obamegine, thalrugosine, *O*-methylthalicberine, obaberine, aromoline,[42a] and thalistyline[4a] have shown hypotensive activity. Obamegine, thalidasine, thalrugosine, and thalrugosidine,[43] as well as aromoline,[43] thalistyline,[4a] thalibrine,[4a] hernandezine, and thalidezine,[43] exhibit some antimycobacterial activity.

Since bisbenzylisoquinolines such as cissampareine, thalidasine, and tetrandrine show some antitumor activity, an attempt has been made to define the structural requirements for such action. A macrocyclic ring is not required, and neither are *N*-methylation or specific stereochemistry at C-1 or C-1'.[44]

(+)-Isotubocurarine has been prepared, and its curarimimetic action was found to be about twice that of (+)-tubocurarine. It follows that the dimethyl-ammonium moiety should be adjacent to an asymmetric center of the *S*-configuration for maximum activity.[45] This result fits with the finding that for quaternary tetrahydropapaverine salts, monomers of the *S*-configuration are significantly more potent neuromuscular blocking agents than their analogs in the *R*-series.[46]

The quaternary dimers **19** and **20** of tetrahydropapaverine were found to be more potent neuromuscular blocking agents than (+)-tubocurarine, but less effective than *N,O,O*-trimethyl-(+)-tubocurarine.[47]

19, R = CH₃ (Laudexinium salt)
20, R = CH₂CH₃

5.8. Spectral Measurements and CD Curves

Baldas, Bick, and co-workers have made a systematic study of the mass spectral fragmentation of several bisbenzylisoquinolines. Modes of cleavage were partly supported by deuteration studies.[48]

A careful investigation of the 100-MHz PMR spectrum of *N*-acetyltiliam-osine (type 5,6,7*,8†,12-6*,7†,12(11-11)) has been carried out, and the assignments were confirmed by appropriate NOE (nuclear Overhauser effect) enhancements. The chemical shifts obtained are indicated below. A similar study was done on *N*-acetylnortiliacorinine-A (type 6,7*,8†,12-6*,7†,12(11-11)).[4]

N-Acetyltiliamosine

The PMR spectra of a series of alkaloids related to thalmine (6,7*,11†,12-5*,6,7,12†) have been interpreted and tabulated.[48]

Caution should be exercised in the interpretation of the PMR spectra of some bisbenzylisoquinolines. For example, the spectrum of thalsimine at room temperature indicated a 1:1 mixture of isomers, showing ten methoxyls and two *N*-methyls. The two isomers are stable conformers, but on heating to 95° the spectrum showed only the expected five methoxyl groups.[49] A similar phenomenon had previously been observed in the case of a related synthetic imino base.[9]

(+)-Thalsimine

The CD curves of a number of bisbenzylisoquinolines, including thalsimi-dine, thalsimine, hernandezine, thalisopine, thalmine, *O*-methylthalicberine, and thalfoetidine, have been published.[50]

5.9. X-Ray Diffraction

The first complete X-ray structural determination of a bisbenzylisoquinoline was carried out by Sobell and co-workers. N,O,O-Trimethyl-(+)-tubocurarine, a potent neuromuscular blocking agent, was studied in the form of its diiodide salt, which is monoclinic, space group $P2_1$, $a = 15.22$, $b = 18.36$, $c = 15.44$ Å, $\beta = 91.2°$.[51] The conformation of the molecule is such that it has a convex hydrophilic side containing the six ether oxygen atoms. The concave side is almost entirely hydrophobic. The molecule has a tight compact structure showing little flexibility, and the interquaternary nitrogen distance is rigidly fixed at 10.7 Å.

N,O,O-Trimethyl-(+)-tubocucarine

A second X-ray study was carried out on (+)-tubocurarine chloride.[52] In this instance the molecule assumes a folded conformation with the two tetrahydroisoquinoline rings turned towards the center of the molecule, so that the distance between the nitrogen atoms is only 8.97 Å. Protruding from the compact bulk of the molecule is ring C in which the phenolic proton forms a hydrogen bond with the chloride ion. This structure is more flexible than that of N,O,O-trimethyl-(+)-tubocurarine.

Thirdly, the crystal structure of the tumor-inhibitory (+)-tetrandrine has also been determined by direct X-ray analysis.[53] The crystals are orthorhombic,

(+)-Tetrandrine

space group $P2_12_12_1$ with $a = 38.368$, $b = 7.230$, $c = 12.05$ Å, and $Z = 4$. The molecule is shaped as a rough equilateral triangle. The N-2′ methyl group occupies a pseudoequatorial site with unrestricted access to the lone pair, while the N-2 methyl is in a pseudoaxial site.[53,54] It should be noted, however, that in spite of this conformational difference, N-2′ is not necessarily a generally more reactive center than N-2, since *N*-demethylation of the alkaloid using methyl chloroformate is nonregiospecific.[42]

References and Notes

1. For a review on the bisbenzylisoquinolines, see M. P. Cava, K. T. Buck, and K. L. Stuart, *Alkaloids*, **16**, 249 (1977).
2. For a review on the synthesis of bisbenzylisoquinolines see M. Shamma and V. St. Georgiev, *Alkaloids*, **16**, 319 (1977).
3. M. Shamma and J. L. Moniot, *Heterocycles*, **4**, 1817 (1976).
4. K. P. Guha, P. C. Das, B. Mukherjee, R. Mukherjee, G. P. Juneau, and N. S. Bhacca, *Tetrahedron Lett.*, 4241 (1976).
4a. W.-N. Wu, J. L. Beal, and R. W. Doskotch, *Tetrahedron Lett.*, 3687 (1976); and W.-N. Wu, J. L. Beal, R. P. Leu, and R. W. Doskotch, *Lloydia*, **40**, 281 (1977).
5. B. D. Gupta, S. K. Banerjee, and K. L. Handa, *Phytochemistry*, **15**, 576 (1976).
6. T. Ibuka, T. Konoshima, and Y. Inubushi, *Chem. Pharm. Bull.*, *Tokyo*, **23**, 114 (1975).
7. T. Ibuka, T. Konoshima, and Y. Inubushi, *Chem. Pharm. Bull.*, *Tokyo*, **23**, 133 (1975).
8. M. P. Cava and A. Afzali, *J. Org. Chem.*, **40**, 1553 (1975).
8a. L. D. Yakhontova, O. N. Tolkachev, D. A. Fesenko, M. E. Perelson, and N. F. Proskurnina, *Khim. Prir. Soedin.*, 234 (1977); through *Current Abstr. Chem.*, **66**, No. 7, item 259610 (1977).
9. Y. Inubushi, Y. Masaki, S. Matsumoto, and F. Takami, *J. Chem. Soc.*, *C*, 1547 (1969).
10. Y. Inubushi, Y. Ito, Y. Masaki, and T. Ibuka, *Tetrahedron Lett.*, 2857 (1976); and Y. Inubushi, Y. Ito, Y. Masaki, and T. Ibuka, *Chem. Pharm. Bull.*, *Tokyo*, **25**, 1636 (1977).
11. Y. Inubushi and M. Kozuka, *Chem. Pharm. Bull.*, *Tokyo*, **2**, 215 (1954); and M. Tomita and H. Furukawa, *J. Pharm. Soc. Japan*, **83**, 676 (1963).
12. D. C. Smith and F. D. Popp, *J. Heterocycl. Chem.*, **13**, 573 (1976).
13. M. Shamma and J. E. Foy, *Tetrahedron Lett.*, 2249 (1975).
14. M. Shamma, J. E. Foy, T. R. Govindachari, and N. Viswanathan, *J. Org. Chem.*, **41**, 1293 (1976).
15. P. Wiriyachitra and M. P. Cava, *J. Org. Chem.*, **42**, 2274 (1977).
16. S. Teitel and J. P. O'Brien, *J. Org. Chem.*, **41**, 1657 (1976). See also S. Teitel and A. Brossi, *Heterocycles*, **1**, 73 (1973); and S. Teitel, J. O'Brien, and A. Brossi, *J. Org. Chem.*, **37**, 1879 and 3368 (1972); and S. Teitel and J. P. O'Brien, *Heterocycles*, **5**, 85 (1976).
17. G. W. Kenner and N. R. Williams, *J. Chem. Soc.*, 522 (1955). For a later modification, see R. A. Rossi and J. F. Bunnett, *J. Org. Chem.*, **38**, 2314 (1974).
18. W. Lonsky, H. Traitler, and K. Kratzl, *J. Chem. Soc. Perkin I*, 169 (1975).
19. K. C. Rice, *J. Org. Chem.*, **40**, 1850 (1975). See also G. A. Brine, K. G. Boldt, C. K. Hart, and F. I. Carroll, *Org. Prep. Proc. Int.*, **8**, 103 (1976); and K. C. Rice and E. L. May, *J. Heterocycl. Chem.*, **14**, 665 (1977).

20. P. Pfäffli and H. Hauth, *Helv. Chim. Acta*, **56**, 347 (1973).
21. R. A. Olofson, R. C. Schnur, L. Bunes, and J. P. Pepe, *Tetrahedron Lett.*, 1567 (1977); S. W. Baldwin, P. W. Jeffs, and S. Natarajan, *Synthetic Comm.*, **7**, 79 (1977).
22. L. Bernardi and G. Bosisio, *Chem. Commun.*, 690 (1974).
23. T. A. Montzka, J. D. Matiskella, and R. A. Partyka, *Tetrahedron Lett.*, 1325 (1974).
23a. S. Kano, T. Yokomatsu, T. Ebate, and S. Shibuya, *Chem. Pharm. Bull.*, Tokyo, **25**, 1456 (1977).
24. R. Hohlbrugger and W. Klötzer, *Chem. Ber.*, **107**, 3457 (1974); A. Cavé, C. Kan-Fan, P. Potier, and J. Le Men, *Tetrahedron*, **23**, 4681 (1967). See also K. W. Bentley and A. W. Murray, *J. Chem. Soc.*, 2497 (1963).
25. U. Berg, R. Gallo, and J. Metzger, *J. Org. Chem.*, **41**, 2621 (1976).
26. G. H. Posner and J.-S. Ting, *Synth. Commun.*, **4**, 355 (1974).
27. S. M. Hecht and J. W. Kozarich, *Chem. Commun.*, 387 (1973).
28. M. P. Cooke and R. M. Parlman, *J. Org. Chem.*, **40**, 531 (1975).
29. T.-L. Ho, *Synth. Commun.*, **3**, 99 (1973). See also S. Gerzberg, R. T. Gaona, H. Lopez, and J. Comin, *Tetrahedron Lett.*, 1269 (1973).
30. J. P. Kutney, G. B. Fuller, R. Greenhouse, and I. Itoh, *Synth. Commun.*, **4**, 183 (1974).
31. G. I. Feutrill and R. N. Mirrington, *Aust. J. Chem.*, **25**, 1719 (1972); J. A. Lawson and J. I. DeGraw, *J. Med. Chem.*, **20**, 165 (1977).
32. C. Hansson and B. Wickberg, *Synthesis*, 191 (1976).
33. L. René, J. P. Buisson, and R. Royer, *Bull. Soc. Chim. France, 2° Partie*, 2763 (1975).
33a. R. Ahmad, J. M. Saá, and M. P. Cava, *J. Org. Chem.*, **42**, 1228 (1977).
34. T. Kametani, S.-P. Huang, M. Ihara, and K. Fukumoto, *J. Org. Chem.*, **41**, 2545 (1976).
34a. T.-L. Ho and G. A. Olah, *Angew. Chem. Int. Ed. Engl.*, **15**, 774 (1976).
35. A. P. Bashall and J. F. Collins, *Tetrahedron Lett.*, 3489 (1975).
35a. J. H. Clark, H. L. Holland, and J. M. Miller, *Tetrahedron Lett.*, 3361 (1976).
36. B. A. Stoochnoff and N. L. Benoiton, *Tetrahedron Lett.*, 21 (1973).
37. J. P. Yardley and H. Fletcher, 3rd, *Synthesis*, 244 (1976).
37a. R. A. Holton and R. G. Davis, *Tetrahedron Lett.*, 533 (1977).
37b. R. J. Borgman, R. V. Smith, and J. E. Keiser, *Synthesis*, 249 (1975).
38. M. Akasu, H. Itokawa, and M. Fujita, *Phytochemistry*, **15**, 471 (1976).
39. A. N. Tackie, D. Dwuma-Badu, J. S. K. Ayim, T. T. Dabra, J. E. Knapp, D. J. Slatkin, and P. L. Schiff, Jr., *Lloydia*, **38**, 210 (1975).
40. M. Shamma and J. L. Moniot, *Heterocycles*, **3**, 297 (1975).
41. J. Harley-Mason, A. S. Howard, W. I. Taylor, M. J. Vernengo, I. R. C. Bick, and P. S. Clezy, *J. Chem. Soc. C*, 1948 (1967).
42. P. J. Davis and J. P. Rosazza, *J. Org. Chem.*, **41**, 2548 (1976); and P. J. Davis, D. R. Wiese and J. P. Rosazza, *Lloydia*, **40**, 239 (1977).
42a. W.-N. Wu, J. L. Beal, L. A. Mitscher, K. N. Salman, and P. Patil, *Lloydia*, **39**, 204 (1976).
43. L. A. Mitscher, *Recent Adv. Phytochem.*, **9**, 264 (1975). See also L. A. Mitscher, W. Wu, R. W. Doskotch, and J. L. Beal, *Lloydia*, **35**, 167 (1972); and L. A. Mitscher, W. Wu, and J. L. Beal, *Experientia*, **28**, 500 (1972); and W.-N. Wu, J. L. Beal and R. W. Doskotch. *Lloydia*, **39**, 378 (1976); and W.-N. Wu, J. L. Beal, R.-P. Leu and R. W. Doskotch, *Lloydia*, **40**, 384 (1977).
44. S. M. Kupchan and H. W. Altland, *J. Med. Chem.*, **16**, 913 (1973). For a further discussion of this complex subject see H. Kuroda, S. Nakazawa, K. Katagiri, O. Shiratori, M. Kozuka, K. Fujitani, and M. Tomita, *Chem. Pharm. Bull.*, Tokyo, **24**, 2413 (1976).
45. T. O. Soine and J. Naghaway, *J. Pharm. Sci.*, **63**, 1643 (1974).
46. J. B. Stenlake, W. D. Williams, N. C. Dhar, and I. G. Marshall, *Eur. J. Med. Chem.-Chim. Ther.*, **9**, 233 (1974).

47. J. B. Stenlake, W. D. Williams, N. C. Dhar, and I. G. Marshall, *Eur. J. Med. Chem.-Chim. Ther.*, **9**, 239 (1974); through *Chem. Abstr.*, **82**, 57976v (1975).
48. J. Baldas, I. R. C. Bick, T. Ibuka, R. S. Kapil, and Q. N. Porter, *J. Chem. Soc. Perkin I*, 592, 599 (1972); J. Baldas, I. R. C. Bick, M. R. Falco, J. X. de Vries, and Q. N. Porter, *J. Chem. Soc. Perkin I*, 597 (1972); and J. Baldas, Q. N. Porter, I. R. C. Bick, G. K. Douglas, M. R. Falco, J. X. de Vries, and S. Yu. Yunosov, *Tetrahedron Lett.*, 6315 (1968).
49. J. M. Saá, M. V. Lakshmikantham, M. J. Mitchell, M. P. Cava, and J. L. Beal, *Tetrahedron Lett.*, 513 (1976).
50. G. P. Moisseeva, Z. F. Ismailov, and S. Yu. Yunusov, *Khim. Prir. Soedin.*, 705 (1970); through *Chem. Nat. Compds.*, 715 (1973).
51. H. M. Sobell, T. D. Sakore, S. S. Tavale, F. G. Canepa, P. Pauling, and T. J. Petcher, *Proc. Natl. Acad. Sci. USA*, **69**, 2212 (1972).
52. P. W. Codding and M. N. G. James, *Chem. Commun.*, 1174 (1972).
53. C. J. Gilmore, R. F. Bryan, and S. M. Kupchan, *J. Am. Chem. Soc.*, **98**, 1947 (1976).
54. S. M. Kupchan, A. J. Liepa, R. L. Baxter, and H. P. J. Hintz, *J. Org. Chem.*, **38**, 1846 (1973).

BALUCHISTANAMINE:
AN ISOQUINOLONE–BENZYLISOQUINOLINE
DIMER

Occurrence: Berberidaceae
Structure:

Baluchistanamine

6.1. Structure Elucidation and Synthesis

Baluchistanamine, $C_{37}H_{38}N_2O_8$, a new dimeric base isolated from *Berberis baluchistanica* Ahrendt, is the first member of a new group of alkaloids, the isoquinolone–benzylisoquinolines.[1] The IR spectrum of the optically active base shows maxima at 1640 (6.10 μ) (conjugated tertiary δ-lactam) and 1720 cm^{-1} (5.81 μ) (conjugated aldehyde), while a bathochromic shift in the UV spectrum upon addition of base is indicative of a phenolic function. The mass spectrum of baluchistanamine shows a weak molecular ion at m/e 638, and a base peak at m/e 411. Consistent with the presence of one nonbasic amidic nitrogen, no doubly charged ion derived from the base peak m/e 411 is present. Other important fragments are m/e 365, 227, 206, 204, and 120.

The PMR spectrum of baluchistanamine exhibits signals for an amine N-methyl, an amide N-methyl, three methoxyl groups, ten aromatic protons, and one aldehydic proton. On the supposition that baluchistanamine is derived biogenetically from oxyacanthine, also found in *B. baluchistanica*, the latter base was oxidized with permanganate in acetone[2] to afford, in low (5%) yield, a product identical with baluchistanamine.

m/e 411, $C_{23}H_{27}N_2O_5$

m/e 365, $C_{21}H_{21}N_2O_4$

m/e 227, $C_{14}H_{11}O_3$

m/e 206, $C_{12}H_{16}NO_2$
m/e 204, $C_{12}H_{14}NO_2$

m/e 120, C_7H_4O

It has been indicated that isoquinolone alkaloids originate in plants from oxidation of simple benzylisoquinolines. A parallel assumption is that hernandaline is formed by oxidation of a thalicarpine-type aporphine–benzylisoquinoline. It is self-evident from the structure of baluchistanamine that *in vivo* oxidation of an oxyacanthine type alkaloid to an isoquinolone–benzylisoquinoline dimer, as well as of a simple monomeric benzylisoquinoline to an isoquinolone, is an intrinsic feature of the alkaloidal catabolic process within *B. baluchistanica*.

An isoquinoline

(+)-Hernandaline, R = H
(+)-Thaliadine, R = OCH_3

A relevant and very recent development is the isolation and characterization by Beal, Doskotch, and co-workers, of the new alkaloid (+)-thaliadine from *Thalictrum minus* L. race B (Ranunculaceae). This alkaloid is accompanied in the plant by such aporphine–benzylisoquinoline dimers as (+)-adiantifoline, (+)-desmethyladiantifoline, and (+)-thaliadanine (Sec. 12.1), and must be formed from their oxidative degradation.[3]

6.2. PMR Spectroscopy

The protons on the aldehyde-bearing ring of baluchistanamine form a complex ABX pattern centered about $\delta 7.38$. The methoxyl singlets are at $\delta 3.62$, 3.85, and 3.92. The PMR chemical shifts for baluchistanamine are indicated below.

6.3. UV Spectroscopy and Circular Dichroism

Baluchistanamine: λ_{max}^{EtOH} 224, 260, 270, 282 sh, 294 sh, and 305 sh nm (4.57, 4.05, 4.06, 3.97, 3.90, and 3.80).

CD curve (MeOH) $[\theta]_{290} = 0$, $[\theta]_{263} = +2560$, $[\theta]_{253} = 0$, $[\theta]_{231} = -14,000$ and $[\theta]_{220} = 0$.

O-Methylbaluchistanamine: λ_{max}^{EtOH} 226, 262, 270, 283 sh, 292 sh, and 305 sh nm (4.58, 4.12, 4.10, 3.99, 3.85, and 3.77).

CD curve (MeOH) $[\theta]_{290} = 0$, $[\theta]_{263} = +2020$, $[\theta]_{253} = 0$, $[\theta]_{231} = -14,900$ and $[\theta]_{220} = 0$.

References

1. M. Shamma, J. E. Foy, and G. A. Miana, *J. Am. Chem. Soc.*, **96**, 7809 (1974).
2. M. Shamma and J. E. Foy, *Tetrahedron Lett.*, 2249 (1975).
3. W.-t. Liao, J. L. Beal, W.-N. Wu, and R. W. Doskotch, *Lloydia*, in press.

THE CULARINES

Structures:

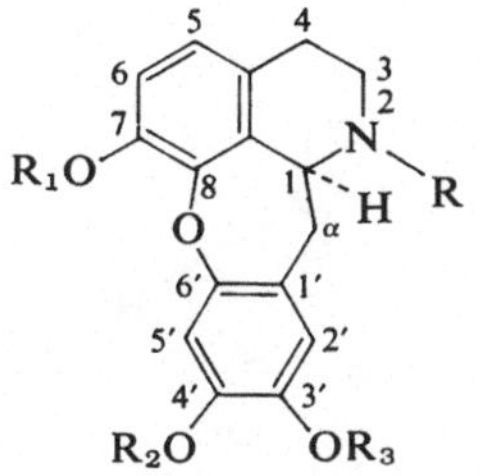

(+)-Cularine, R = R$_1$ = R$_2$ = R$_3$ = CH$_3$
(+)-Cularimine, R$_1$ = R$_2$ = R$_3$ = CH$_3$, R = H
(+)-Cularidine, R = R$_2$ = R$_3$ = CH$_3$, R$_1$ = H
(+)-Cularicine, R = CH$_3$, R$_1$ = H, R$_2$ + R$_3$ = CH$_2$

7.1. Synthesis

The synthetic approaches to the cularine bases involve (a) formation of the diaryl ether linkage in the initial stages of the synthesis and subsequent elaboration of rings B and C, or (b) construction of an appropriately substituted benzylisoquinoline, with diaryl ether formation via Ullmann condensation or phenolic oxidative coupling at a late stage in the synthetic scheme. All four of the naturally occurring alkaloids in this group, cularine,[1] cularimine,[1] cularicine,[2] and cularidine,[3] as well as the unnatural positional isomer isocularine[4] have been synthesized in the classical fashion by the Ullmann route.[5]

The full details of the phenolic coupling method for the preparation of cularine have been reported[6] and the details of a previous study,[7] attempting the rearrangements of spirodienones to cularine analogs, have indicated that the opening of the spirodienones to the phenolic benzylisoquinolines isolated probably proceeds via reaction with solvent as shown in Scheme 7.1.[8,9]

The phenolic base cularicine has been prepared by a synthesis which closely parallels the original synthesis of cularine.[10] However, variations in the scheme were introduced to accommodate the sensitivity of the required phenol protecting group to Lewis acid.[11]

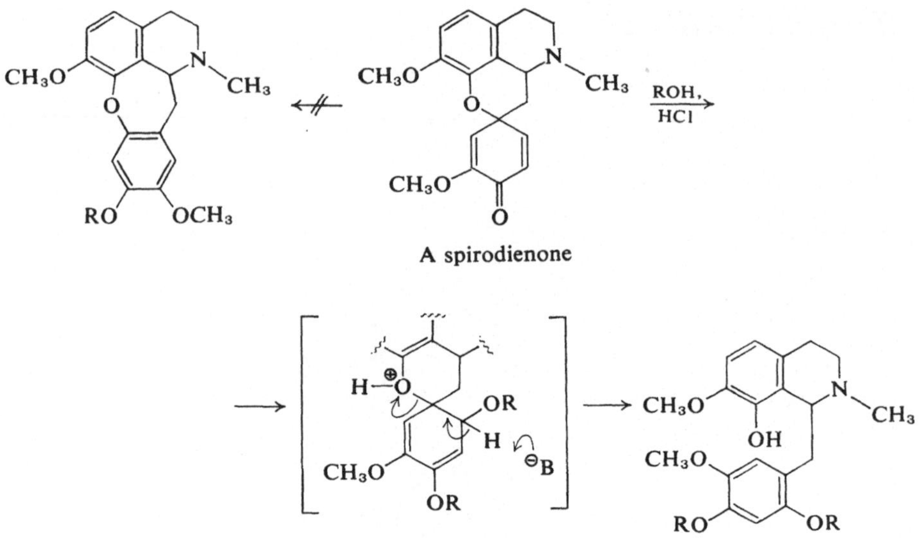

A spirodienone

SCHEME 7.1

The diphenyl ether **1** was converted to the morpholinamide **2**, which after cyclization to the enamine **3** and hydrolysis furnished the tricyclic ketone **4**. Interestingly, in the modified Pomeranz-Fritsch cyclization step, one of a pair of ethoxyl rather than hydroxyl racemates was isolated. Catalytic hydrogenolysis and reductive N-methylation completed the synthesis shown below:

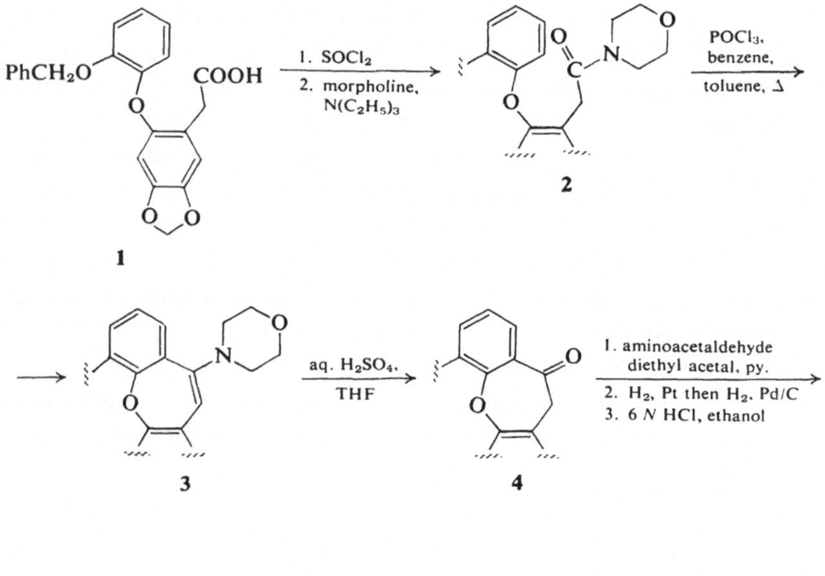

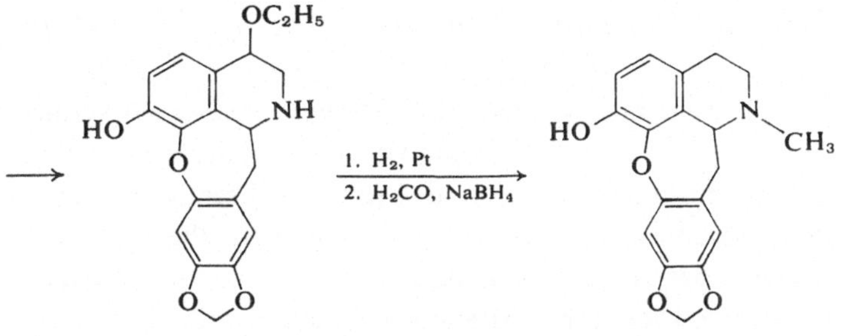

Cularicine

Oxidation of isocularine derivatives with a variety of oxidants to obtain 1,2-dehydro analogs led instead to ring-B aromatic α-oxo products.[12]

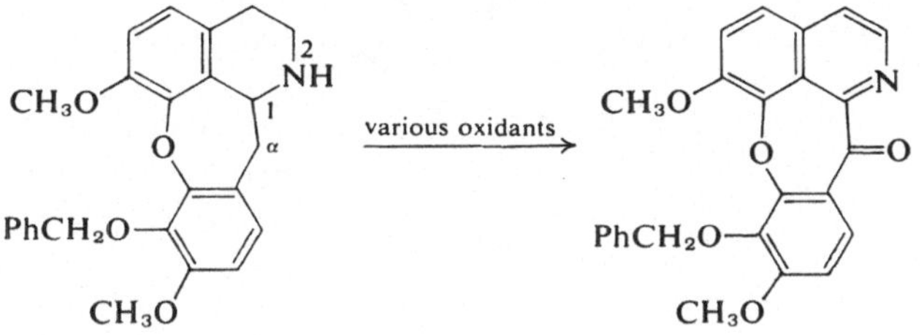

Isocularine analog

7.2. Absolute Configuration

Application of the anomalous dispersion method in the X-ray crystallographic analysis of cularine methiodide has confirmed that the asymmetric center of the alkaloid has the *S*-configuration. In the crystal lattice, the dihydro-oxepine ring has the twist-boat shape, the oxygen atom being the bow. The dihedral angles between rings A and D in the two molecules in the asymmetric unit were determined to be slightly different, namely 59.3 and 54.6°, respectively.[13]

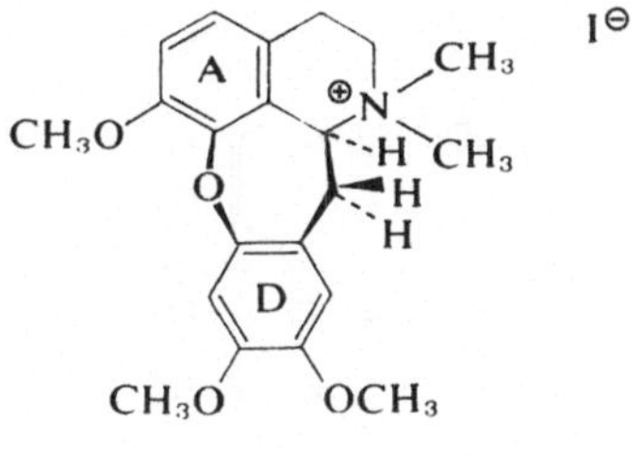

7.3. An Approach to the Synthesis of Cancentrine

Cancentrine, isolated from *Dicentra canadensis* (Goldie) Walp. (*Fumariaceae*), is a dimeric base composed of a cularine unit joined to a morphinane unit by an unusual spiro bridge (darkened section).[14] Two naturally occurring dehydrocancentrines are also known, namely, the yellow dehydrocancentrine-A, which possesses a C-8,14 double bond, and dehydrocancentrine-B, which is unsaturated between C-31 and -32. Catalytic reduction of either of these dehydro bases using Adams catalyst yields cancentrine.[14]

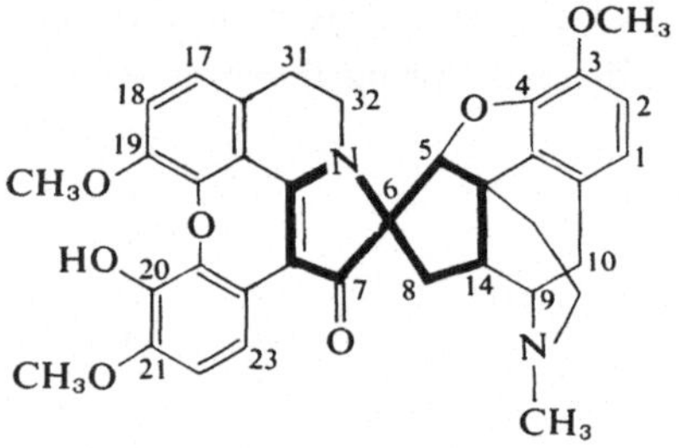

Cancentrine

In a model study, suggested by a proposed biogenetic scheme for cancentrine,[14] 3,4-dihydropapaverine was condensed with 1,2-cyclohexanedione in the presence of Triton B and pyridine to afford the spiro compound **5** in good yield[15]:

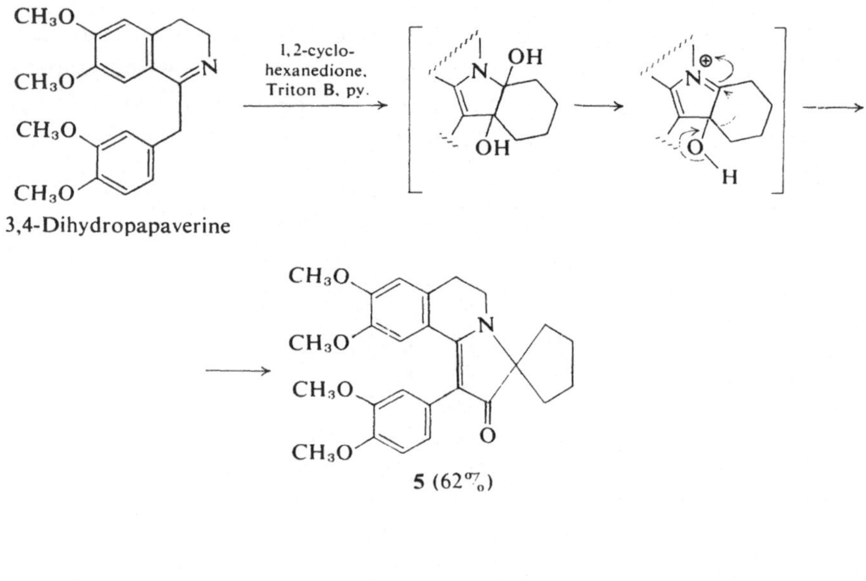

3,4-Dihydropapaverine

1,2-cyclo-hexanedione,
Triton B, py.

5 (62%)

7.4. CMR Spectroscopy

The CMR chemical shifts for cularine are indicated in the structure below.[16]

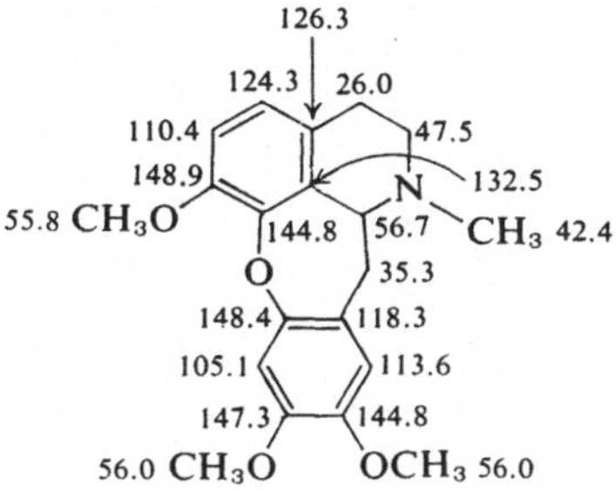

References

1. H. C. Hsu, T. Kikuchi, S. Aoyagi, and H. Iida, *J. Pharm. Soc. Japan*, **92**, 1030 (1972).
2. H. Iida, H. C. Hsu, T. Kikuchi, and K. Kawano, *J. Pharm. Soc. Japan*, **92**, 1242 (1972).
3. H. Iida, H. C. Hsu, and T. Kikuchi, *Chem. Pharm. Bull.*, *Tokyo*, **21**, 1001 (1973).
4. T. Kametani, K. Fukumoto, and M. Fujihara, *Chem. Pharm. Bull.*, *Tokyo*, **20**, 1800 (1972).
5. M. Shamma, *The Isoquinoline Alkaloids*, Academic Press, New York (1972), p. 153 ff.
6. A. H. Jackson, G. W. Stewart, G. A. Charnock, and J. A. Martin, *J. Chem. Soc. Perkin I*, 1911 (1974).
7. A. H. Jackson and G. W. Stewart, *Chem. Commun.*, 149 (1971).
8. A. J. Birch, A. H. Jackson, P. V. R. Shannon, and G. W. Stewart, *J. Chem. Soc. Perkin I*, 2492 (1975).
9. A. M. Choudhury, *J. Chem. Soc. Perkin I*, 132 (1974).
10. T. Kametani and K. Fukumoto, *J. Chem. Soc.*, *London*, 4289 (1963).
11. I. Noguchi and D. B. MacLean, *Can. J. Chem.*, **53**, 125 (1975).
12. T. Kametani, K. Fukumoto, Y. Kato, and M. Fujihara, *J. Pharm. Soc. Japan*, **93**, 1094 (1973).
13. T. Kametani, T. Honda, H. Shimanouchi, and Y. Sasada, *Chem. Commun.*, 1072 (1972).
14. R. Rodrigo, R. H. F. Manske, D. B. MacLean, L. Baczynskyj, and J. K. Saunders, *Can. J. Chem.*, **50**, 853 (1972); D. B. MacLean, L. Baczynskyj, R. Rodrigo, and R. H. F. Manske, *Can. J. Chem.*, **50**, 862 (1972).
15. S. Ruchirawat and V. Somchitman, *Tetrahedron Lett.*, 4159 (1976).
16. E. Wenkert, B. L. Buckwalter, I. R. Burfitt, M. J. Gasic, H. E. Gottlieb, E. W. Hagaman, F. M. Schell, and P. M. Wovkulich, in *Topics in Carbon-13 NMR Spectroscopy*, G. C. Levy, ed., Vol. 2, Wiley-Interscience, New York (1976), p. 106.

THE DIBENZOPYRROCOLINES

8.1. Synthesis

8.1.1. Benzyne Intermediates

Recent developments in the synthesis of dibenzopyrrocolines have centered mainly on variations of the benzyne approach.[1] Potassium amide in liquid ammonia treatment of the bromobenzylisoquinoline 1 affords the dibenzopyrrocoline 2 along with the corresponding aporphine 3.[2-4]

In the absence of a C-7 phenol in the cyclization precursor, the reaction proceeds further and styrenes such as 4 are isolated, presumably by oxidation of the Hofmann elimination product.[4]

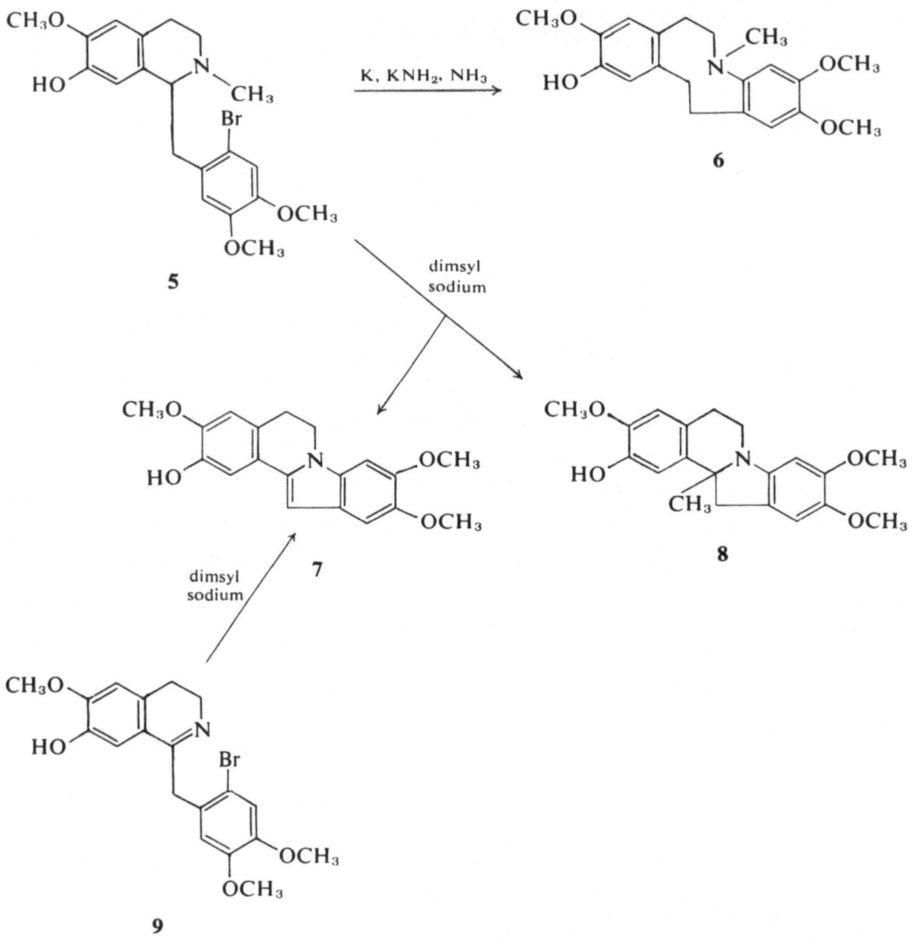

SCHEME 8.1

Reaction of the tetrahydrobenzylisoquinoline **5** in the presence of potassium amide and metallic potassium led to the benzazonine **6**.[4] On the other hand, when the nucleophilic base dimsyl sodium was employed to effect the cyclization of **5**, the products were the indole **7** and the dibenzopyrrocoline tertiary base **8**.[5] The migration of the *N*-methyl group occurs possibly by a Stevens rearrangement. Base **7** was also produced by a similar cyclization of the 3,4-dihydrobenzylisoquinoline **9** (see Scheme 8.1).[6]

Displacement of the initially formed methiodide was observed when dimsyl sodium was used to cyclize the C-6 phenolic and the nonphenolic bromobenzylisoquinolines **10** and **11** (Scheme 8.2).[7]

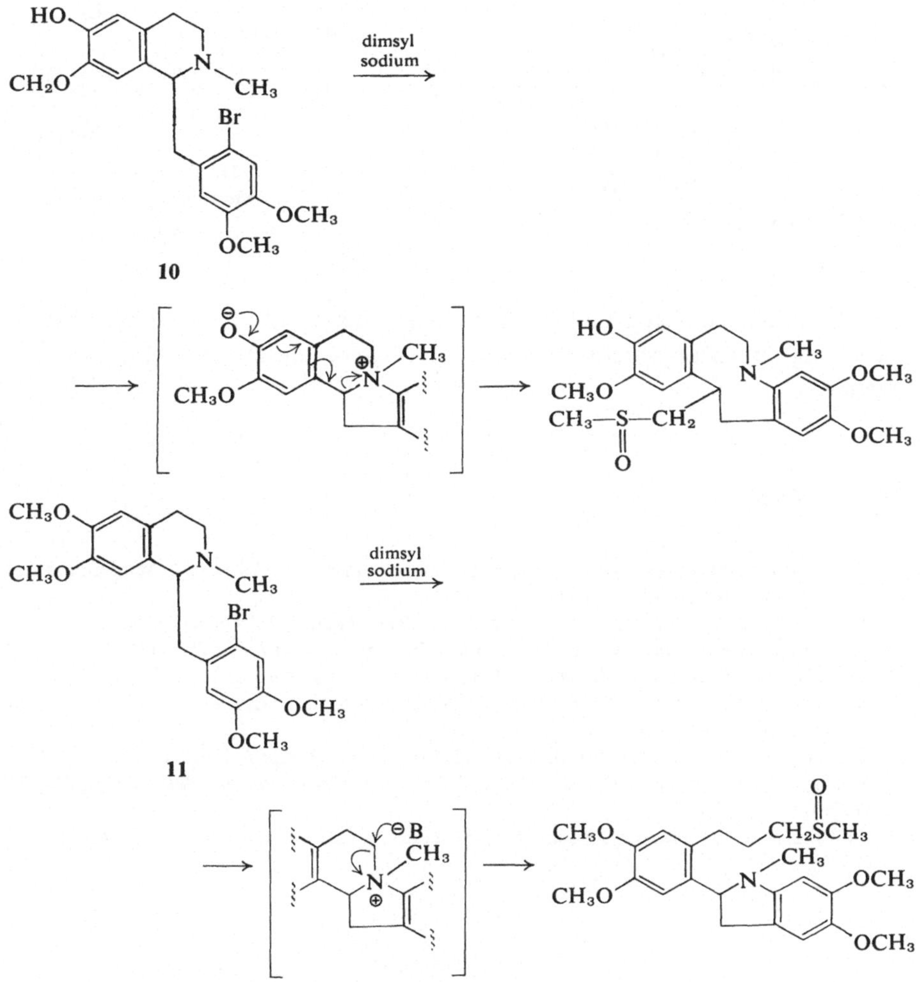

SCHEME 8.2

8.1.2. Enzymic Processes

In an interesting synthesis of O-methylcryptaustoline iodide, Brossi's group has shown that enzymic oxidation of alkaloids can be a useful preparative method.[8] Optically active laudanosoline was oxidized with horseradish peroxidase and hydrogen peroxide to furnish the tetraphenol **12** in 81% yield, with complete retention of configuration. Subsequent O-methylation afforded O-methylcryptaustoline. For a parallel enzymic conversion to an aporphine see Sec. 10.2.8.

(+)-Laudanosoline

1. horseradish peroxidase, H_2O_2, pH 5
2. CH_2N_2

12, R = H
(−)-*O*-Methylcryptaustoline,
R = CH_3

It should be noted that to date there remain only two known naturally occurring dibenzopyrrocolines, namely, (−)-cryptaustoline and (−)-cryptowoline.[1]

References

1. For leading references, see M. Shamma, *The Isoquinoline Alkaloids*, Academic Press, Inc., New York (1972), p. 169.
2. T. Kametani, K. Fukumoto, and T. Nakano, *Tetrahedron*, **28**, 4667 (1972).
3. S. V. Kessar, R. Randhawa, and S. S. Gandhi, *Tetrahedron Lett.*, 2923 (1973).
4. S. V. Kessar, P. S. Pawanjit, and S. S. Gandhi, *Indian J. Chem.*, **13**, 1116 (1975).
5. S. Kano, T. Yokomatsu, N. Yamada, K. Matsumoto, S. Tokita, and S. Shibuya, *Chem. Pharm. Bull., Tokyo*, **22**, 1607 (1974).
6. T. Kametani, S. Shibuya, and S. Kano, *J. Chem. Soc. Perkin I*, 1212 (1973).
7. S. Kano, E. Komiyama, K. Nawa, and S. Shibuya, *Chem. Pharm. Bull., Tokyo*, **24**, 310 (1976).
8. A. Brossi, A. Ramel, J. O'Brien, and S. Teitel, *Chem. Pharm. Bull., Tokyo*, **21**, 1839 (1973).

THE PROAPORPHINES

9.1. Introduction

Proaporphine alkaloids have now also been found in members of the plant families Annonaceae[1] and Liliaceae.[2] Some recently isolated proaporphines are shown below.

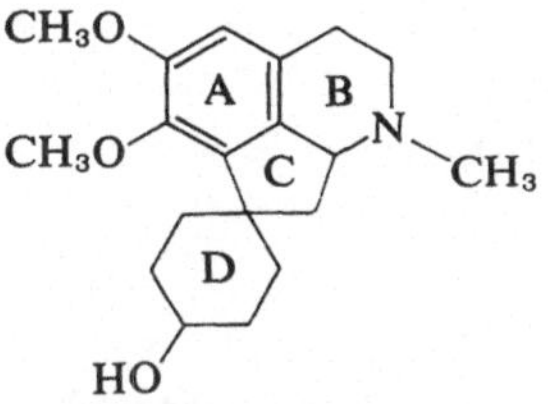

(+)-N-Methyllitsericine[2]

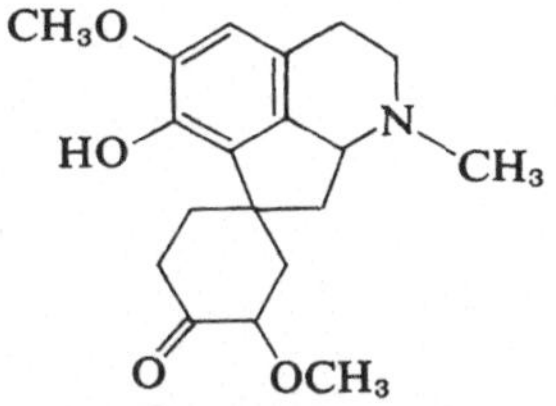

(−)-Roehybrine[3]

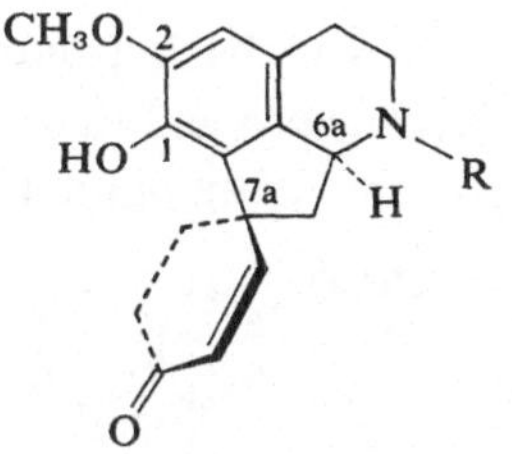

(+)-Isocrotsparinine
(R = H)[4,5]
(+)-N-Methylisocrots-
parinine[4] (R = CH₃)

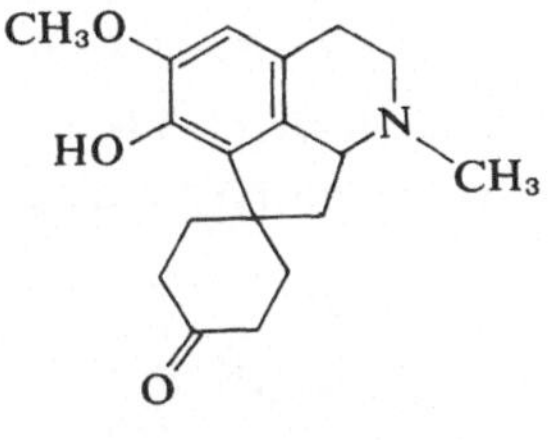

(+)-Tetrahydroglaziovine[4]

9.2. Revision of Stereochemistry of the Reduced Proaporphines

An X-ray study of (±)-11,12-dihydroglaziovine (1) hydrochloride revealed the relative configuration to be such that H-6a and the olefinic double bond in ring D are on opposite sides of the molecule.[4] The free base 1, upon O-methylation with diazomethane, led to (±)-amuronine (2).[4] These findings are in direct

opposition to the conclusions based on circular dichroism data for (+)-lineari-
sine (2-O-demethylamuronine) in which the olefinic bond in ring D was placed
on the same side as H-6a.[6]

Likewise, the structure of the 1,2-methylenedioxy analog of amuronine and
linearisine, namely roemeronine, should also be reconsidered,[4] since its stereochemical assignment was based on optical rotatory dispersion arguments[7] similar to those originally used in connection with amuronine and linearisine.

1, R = H
2, R = CH₃

9.3. Synthesis

Synthetic efforts in the proaporphine series have centered mainly around
efficient ways to obtain (+)-glaziovine which has a promising potential as a
tranquilizer.[8]

Irradiation of the phenolic diazonium salt 3 in basic solution produced
glaziovine in 45% yield. This is an appreciably higher yield than that obtained
by irradiation of the corresponding phenolic bromo compound 4.[9]

3, R = N₂⊕
4, R = Br

Glaziovine

A convenient synthesis of glaziovine proceeds from the dihydroproaporphine
amuronine which is available by the Bernauer tricyclic ketone approach. Thus
when amuronine was refluxed with 20–36% hydrochloric acid, O-demethylation
occurred preferentially at C-1 to yield 11,12-dihydroglaziovine which was
O-acetylated, and then converted to glaziovine through bromination and de-
hydrobromination. It should be mentioned here that preferential O-demethylation cannot be applied directly to a proaporphine since under such strongly

acidic conditions dienone–phenol rearrangement occurs with formation of an aporphine.[10]

Amuronine 2 $\xrightarrow[\Delta,\ 15\text{–}24\ \text{hr}]{20\text{–}36\%\ \text{HCl},}$ 11,12-Dihydroglaziovine 1 $\xrightarrow[\substack{3.\ \text{base, }(-\text{HBr})\\4.\ \text{hydrolysis}}]{\substack{1.\ \text{Ac}_2\text{O}\\2.\ \text{Br}_2}}$ Glaziovine

An alternate route utilizes a modified Bernauer conversion of the tricyclic ketone **5** to the aldehyde **6**; this was then condensed with methyl vinyl ketone to furnish the enone **7** which was converted to glaziovine.[4]

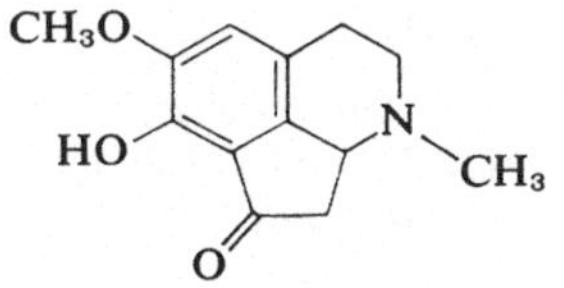

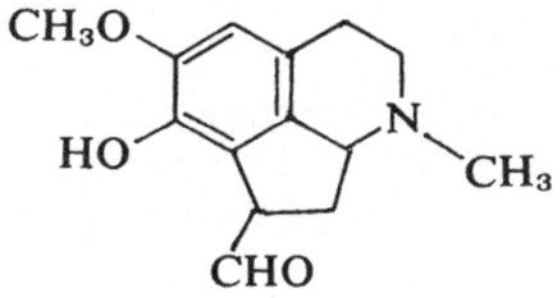

9.4. Pharmacology

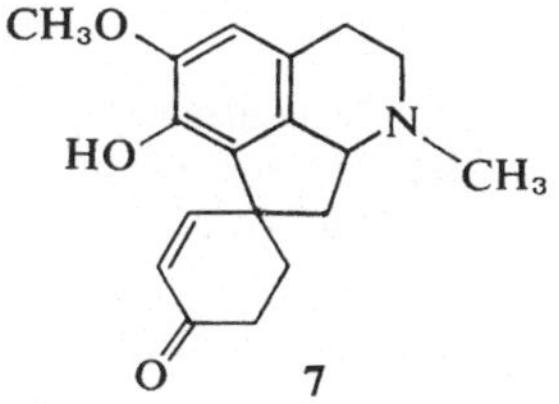

Beside the potential use of glaziovine as a tranquilizer referred to above,[8] a claim has been made (no data) that proaporphines with a dienone system possess analgesic and hypotensive activity.[11] *N*-Methylcrotsparine does show some hypotensive activity.[12] The two proaporphine alcohols obtained by sodium borohydride reduction of the ketonic functions of (±)-8,9-dihydroglaziovine and of (±)-11,12-dihydroglaziovine had sedative activity when tested on rabbits.[12a]

9.5. Spectral Studies

The PMR spectra of several reduced proaporphines have been recorded both in deuteriochloroform and in pyridine-d$_5$. PMR chemical shifts for 8,9-dihydroglaziovine in deuteriochloroform are shown below.[5] The CMR spectra of fifteen reduced proaporphines have been recorded.[12b]

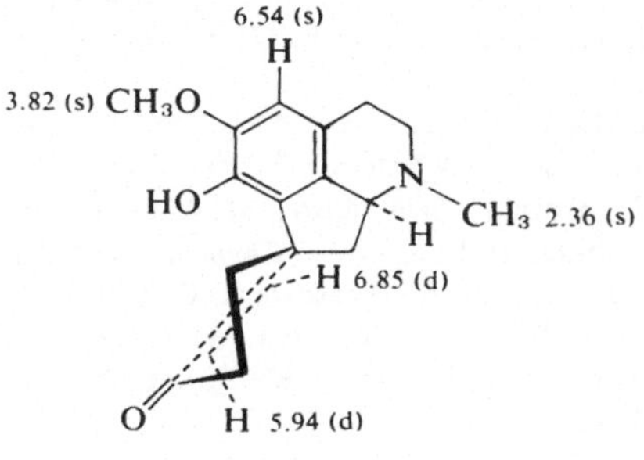

The UV and IR spectra of alkaloids with a cyclohexadienone or cyclohexenone ring, including proaporphines, homoproaporphines, and morphinandienones have been discussed and compared. Some of the available UV data are given below.[13]

Stepharine
λ_{max}^{ROH} 232 and 280 nm
(4.4 and 3.5)

Mecambrine
λ_{max}^{ROH} 230 and 291 nm
(4.5 and 3.7)

Amuronine
λ_{max}^{ROH} 227 and 277 sh nm
(4.4 and 3.3)

The fragmentation process in the mass spectra of proaporphines is strongly dependent on the character of the spiranoid ring D, as well as on the substitution pattern in the aromatic ring A and the relative stereochemistry. Cleavage patterns should, therefore, be compared only among structurally close analogs.[14]

References and Notes

1. P. E. Sonnet and M. Jacobson, *J. Pharm. Sci.*, **60**, 1254 (1971).
2. S. T. Lu, T. L. Su, and E. C. Wang, *J. Chinese Chem. Soc.* (Taiwan), **22**, 349 (1975).
3. J. Slavík, L. Dolejš, and L. Slavíková, *Collect. Czech. Chem. Commun.*, **39**, 888 (1974).
4. C. Casagrande, L. Canonica, and G. Severini-Ricca, *J. Chem. Soc. Perkin I*, 1652 (1975); and A. Colombo, *J. Chem. Soc. Perkin II*, 1218 (1976).
5. For assignments of stereochemistry to jacularine and Base E, see C. Casagrande, L. Canonica, and G. Severini-Ricca, *J. Chem. Soc. Perkin I*, 1659 (1975); and Ref. 4 above.
6. G. Snatzke and G. Wollenberg, *J. Chem. Soc. C*, 1681 (1966).
7. J. Slavík, P. Sedmera, and K. Bláha, *Collect. Czech. Chem. Commun.*, **35**, 1558 (1970). For a separate discussion of the absolute configuration of the proaporphines crotsparine and N-methylcrotsparine, see D. S. Bhakuni, S. Satish, and M. M. Dhar, *Tetrahedron*, **28**, 4579 (1972).
8. C. Casagrande and L. Canonica, Ger. Offen. 2,363,529; through *Chem. Abstr.*, **81**, 105780e and 105783h (1974). G. Ferrari and C. Casagrande, *Farmaco, Ed. Sci.*, **25**, 449 (1970); and B. Buffa, G. Costa, and P. Ghirardi, *Current Therap. Res., Clin. Exp.*, **16**, 621 (1974).
9. C. Casagrande and L. Canonica, *J. Chem. Soc. Perkin I*, 1647 (1975). For related syntheses of proaporphines, see J. Warnant, A. Farcilli, I. Medici, and E. Toromanoff, Ger. Offen. 2,409,007; through *Chem. Abstr.*, **82**, 4449u (1975); and Z. Horii, S. Uchida, Y. Nakashita, E. Tsuchida, and C. Iwata, *Chem. Pharm. Bull., Tokyo*, **22**, 583 (1974); and T. Kametani, K. Takahashi, K. Ogasawara, C. V. Loc, and K. Fukumoto, *Collect. Czech. Chem. Commun.*, **40**, 712 (1975). Also, W. V. Curran, *J. Heterocycl. Chem.*, **10**, 307 (1973); and T. Kametani, F. Satoh, K. Fukumoto, H. Sugi, and K. Kigasawa, *Heterocycles*, **1**, 47 (1973).

10. J. S. Bindra and A. Grodski, *J. Org. Chem.*, **42**, 910 (1977); and S. A. Siphar, Ger. Offen. 2,534,410; through *Chem. Abstr.*, **85**, 33245s (1976).
11. S. Ishiwatari, K. Itakura, and K. Misawa, Jap. Pat. 73 26,015; through *Chem. Abstr.*, **80**, 3688t (1974).
12. M. P. Dubey, R. C. Srimal, and B. N. Bhawan, *Indian J. Pharmacol.*, **7**, 73 (1969).
12a. S. A. Siphar, French Demande 2,302,737 through *Chem. Abstr.*, **86**, 190318f (1977).
12b. G. S. Ricca, *Org. Magn. Reson.*, **9**, 8 (1977).
13. S. Dvořáčková, L. Hruban, V. Preininger, and F. Šantavý, *Heterocycles*, **3**, 575 (1975). The numerical values given here are estimated from graphs.
14. L. Dolejš, *Collect. Czech. Chem. Commun.*, **39**, 571 (1974).

THE APORPHINES

10.1. Introduction[1]

Some new aporphines of interest are shown in Scheme 10.1.

Oliveridine is the first C-7 methoxylated aporphine isolated.[4] Several N-acetylated noraporphines are now known, all of them originating from *Liriodendron* species.[7,8]

10.2. Synthesis[1]

The following discussion describes new, high-yield, syntheses of aporphines.

10.2.1. Photolysis

Oxidative photocyclization of the α-hydroxylated benzylisoquinoline **1**, followed by O-acetylation, provided the oxoaporphine **2** which could be isolated

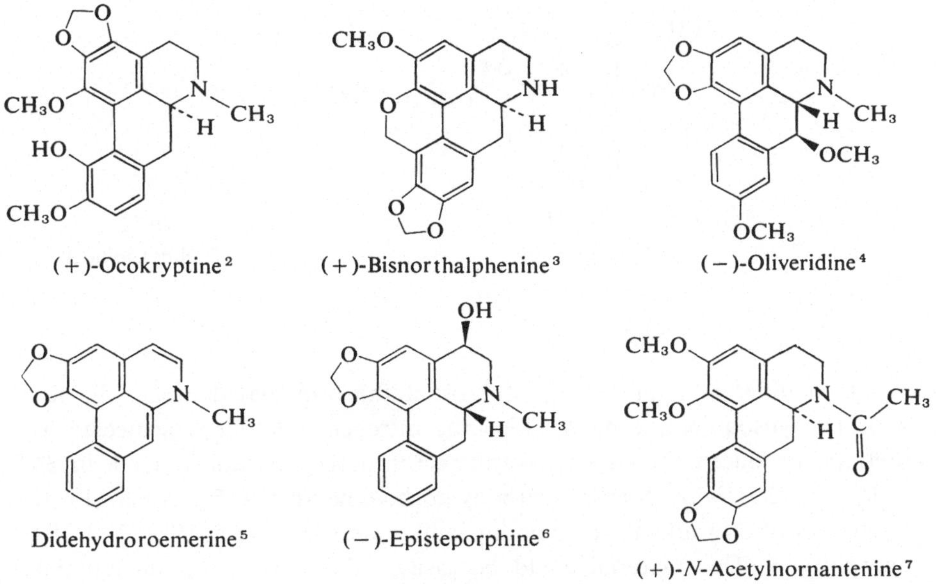

(+)-Ocokryptine[2]   (+)-Bisnorthalphenine[3]   (−)-Oliveridine[4]

Didehydroroemerine[5]   (−)-Episteporphine[6]   (+)-N-Acetylnornantenine[7]

SCHEME 10.1

123

and readily converted to corunnine and subsequently to thaliporphine (3), as shown below[9]:

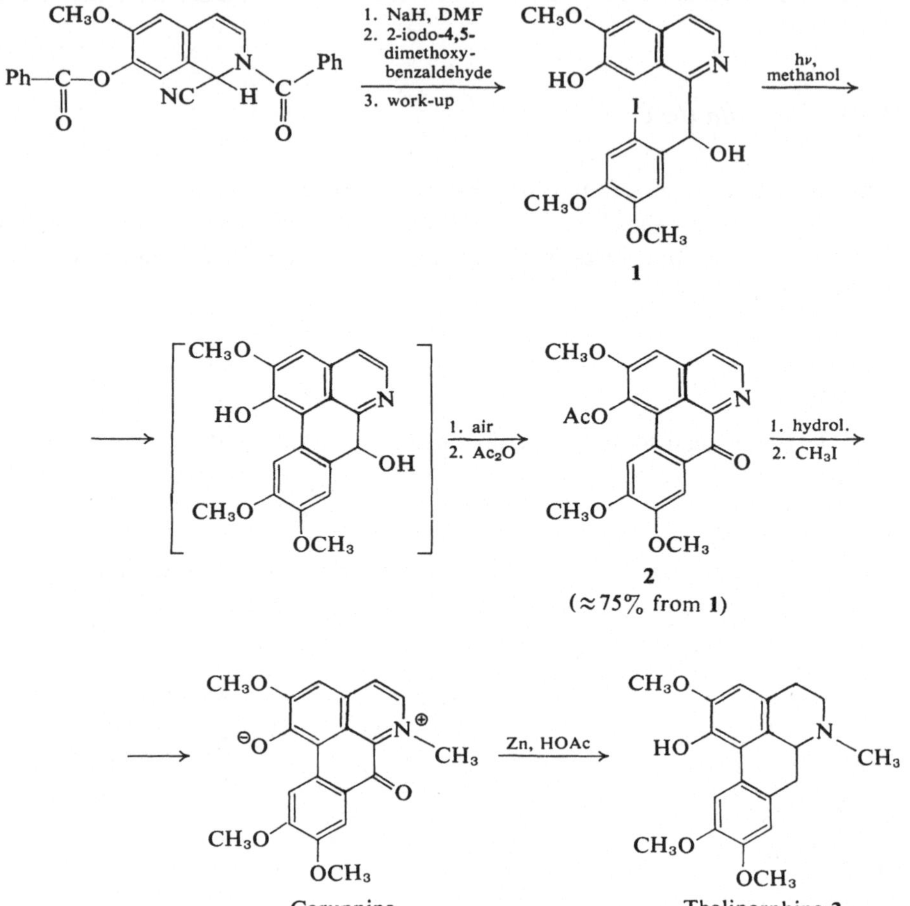

1

2
(≈75% from **1**)

Corunnine   Thaliporphine **3**

In a different approach, photolysis of the brominated diphenolic tetra-hydrobenzylisoquinoline **4**, in which the nitrogen function is protected as a urethan, produced the neoproaporphine (or proerythrinadienone) **5** in 34% yield, together with *N*-ethoxycarbonylnorboldine (**6**) in 5% yield. Further photolysis of **5** in ethanol containing sodium acetate gave a 44% yield of the aporphine **6**. This material could be easily reduced with lithium aluminum hydride to boldine (**7**), thus providing the first total synthesis of this alkaloid.[10]

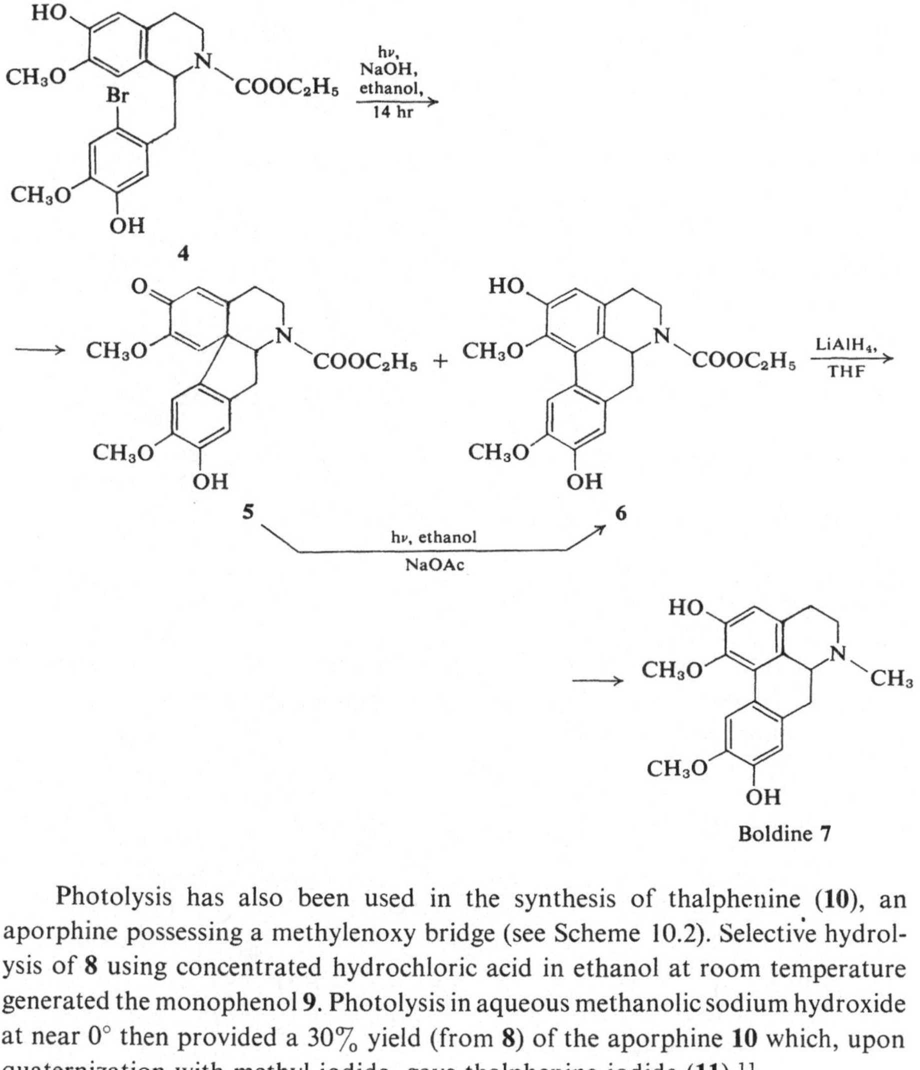

Photolysis has also been used in the synthesis of thalphenine (**10**), an aporphine possessing a methylenoxy bridge (see Scheme 10.2). Selective hydrolysis of **8** using concentrated hydrochloric acid in ethanol at room temperature generated the monophenol **9**. Photolysis in aqueous methanolic sodium hydroxide at near 0° then provided a 30% yield (from **8**) of the aporphine **10** which, upon quaternization with methyl iodide, gave thalphenine iodide (**11**).[11]

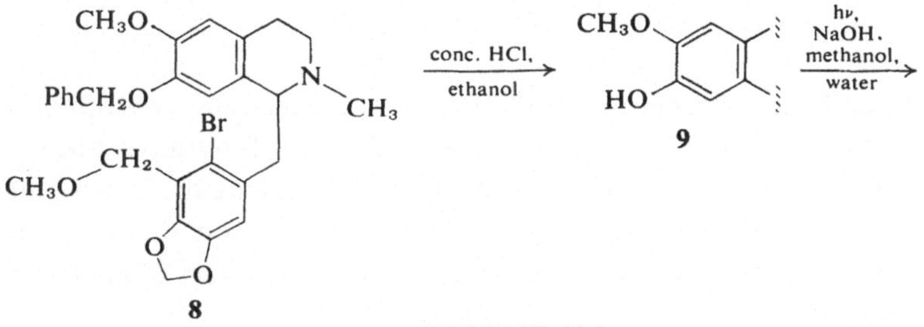

SCHEME 10.2

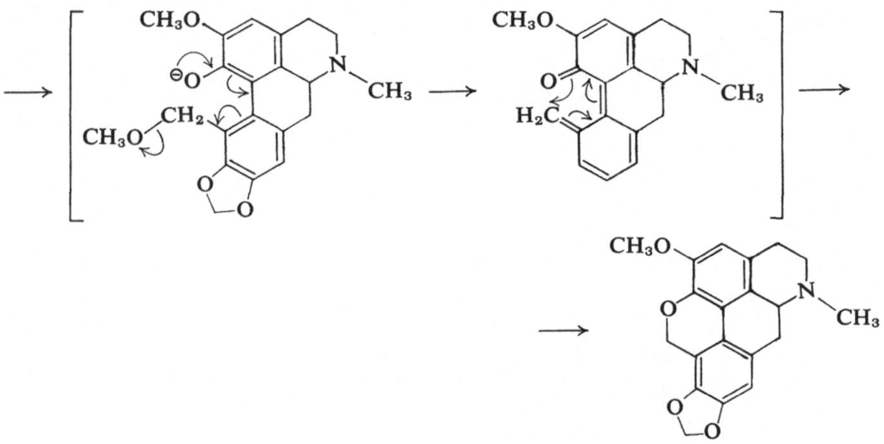

SCHEME 10.2 (*Continued*)

10.2.2. The Use of VOCl₃

Vanadium oxytrichloride is a reagent of general utility in the oxidative coupling of phenolic tetrahydrobenzylisoquinolines. Oxidation of reticuline (**12**) perchlorate with vanadium oxytrichloride provided isoboldine (**13**) in 53% yield.[12]

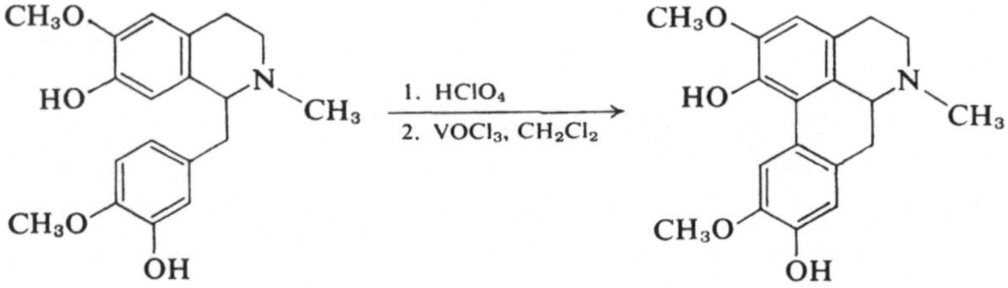

Reticuline **12**   Isoboldine **13**

10.2.3. The Use of VOF₃–TFA

Kupchan and his colleagues were the first to use vanadium oxytrifluoride in trifluoroacetic acid (VOF₃–TFA) for the oxidation of tetrahydrobenzylisoquinolines in the synthesis of aporphines and related alkaloids. This significant development has allowed the ready preparation of a variety of aporphines in high yield, and has also provided an insight into possible modes of aporphine biogenesis in plants.[13]

Oxidation with VOF$_3$–TFA of the nonphenolic *N*-formylnorlaudanosine provides in 55% yield the neospirene **15** through the intermediacy of the morphinandienone **14** which undergoes loss of the original C-6 methyl group.

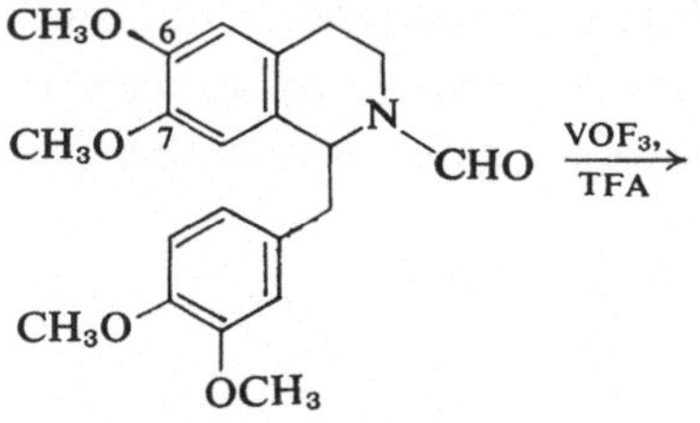

N-Formylnorlaudanosine

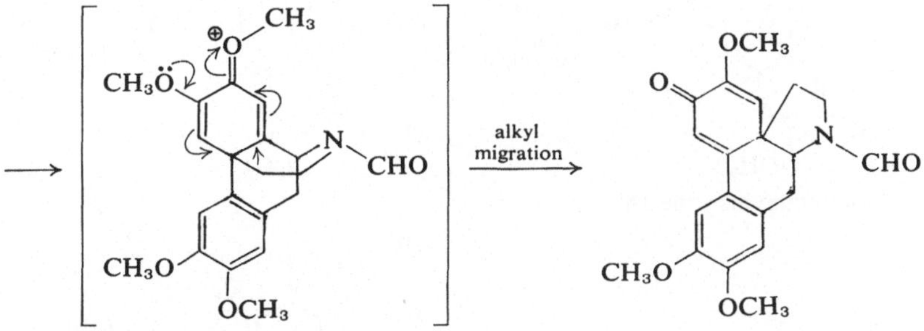

14　　　　　　　　　　**15**

With laudanosine (**16**) itself, the reaction takes a different course, and glaucine (**17**) is produced. This result can best be interpreted in terms of an

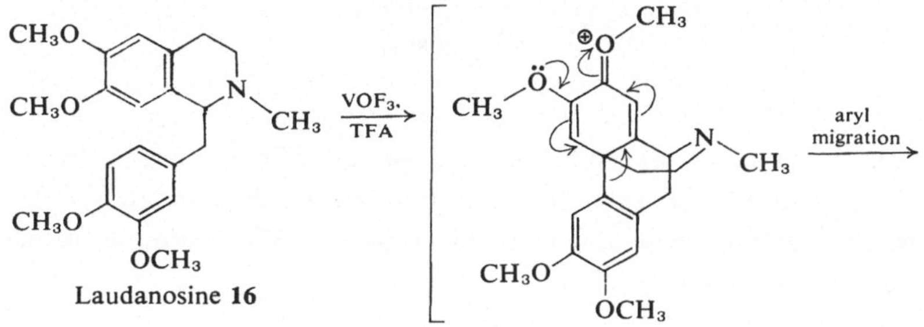

Laudanosine **16**

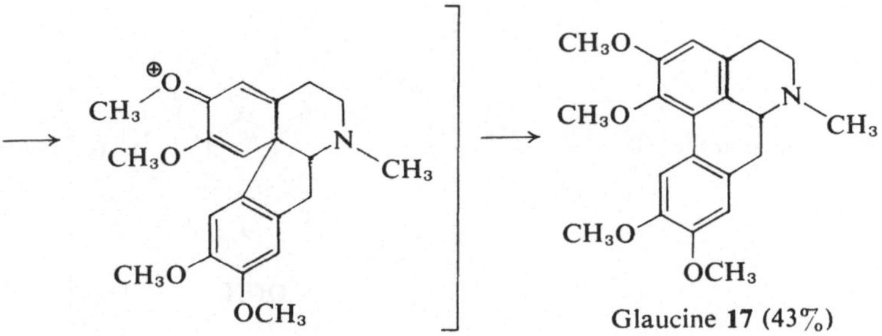

Glaucine **17** (43%)

intermediate morphinandienone which rearranges to a neoproaporphine leading to the product isolated.[14]

Since morphinandienones were postulated as intermediates in these transformations, it was only fitting that their acid-catalyzed rearrangement also be investigated. Indeed, it was established that they too lead to aporphines via neoproaporphines as shown below in the case of *O*-methylflavinantine (**18**), the catalyst being concentrated hydrochloric acid.[14]

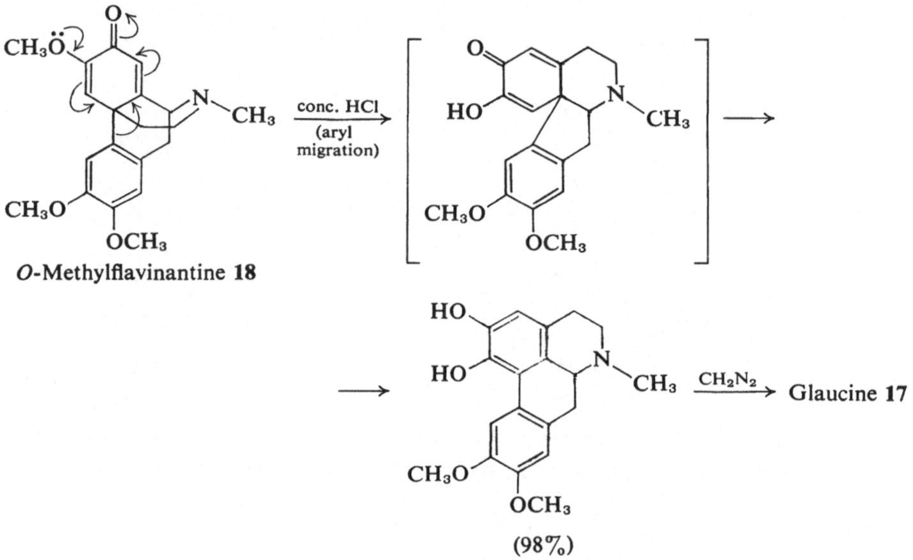

O-Methylflavinantine **18**

Glaucine **17**

(98%)

If, on the other hand, BF$_3$–ether is used instead of concentrated hydrochloric acid, alkyl shift rather than aryl migration occurs, the ultimate product after catalytic reduction being the alkaloid erybidine (**19**).[15] (For a further discussion, see Sec. 18.3.3.)

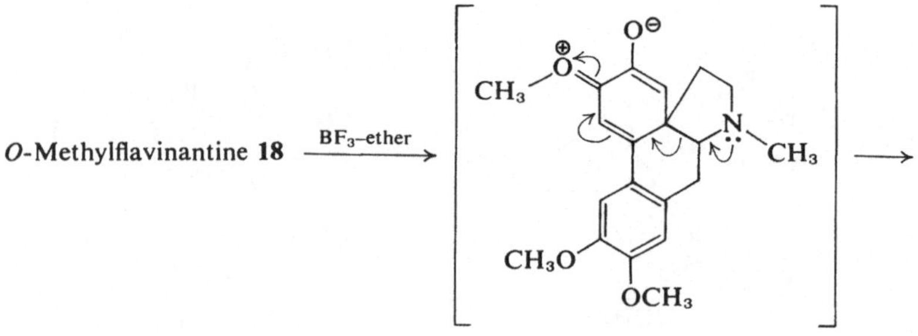

O-Methylflavinantine **18** $\xrightarrow{\text{BF}_3\text{–ether}}$

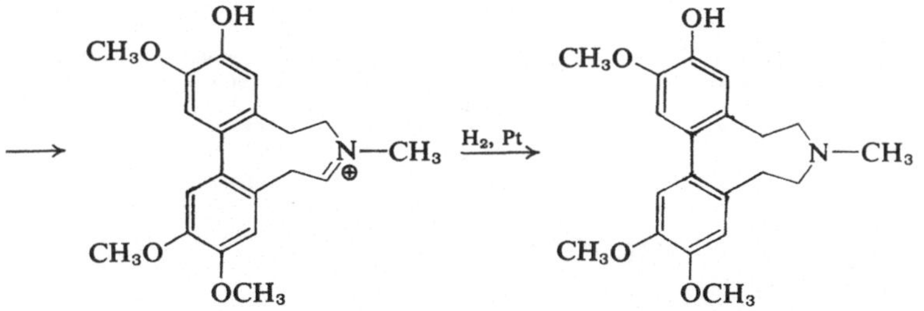

Erybidine 19
(85% overall yield)

A neoproaporphine can also rearrange either to an aporphine in acid solution or to a dibenzazonine in a basic medium, as proven through treatment of the neoproaporphine–borane complex **20** (prepared as shown) with concentrated hydrochloric acid to furnish the aporphine predicentrine (**21**), or alternatively with alkali followed by borohydride reduction to generate erybidine (**19**)[16]:

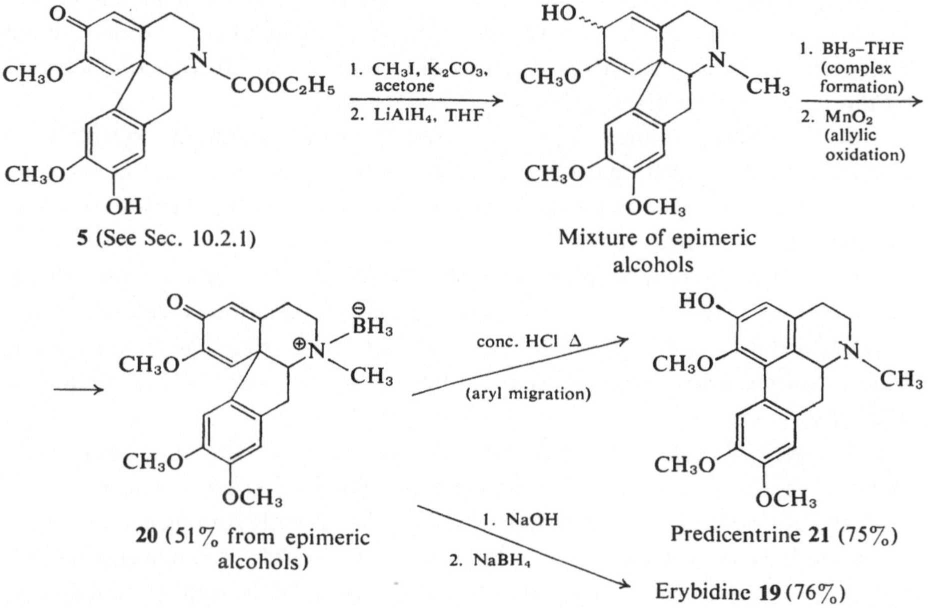

5 (See Sec. 10.2.1)

Mixture of epimeric alcohols

20 (51% from epimeric alcohols)

Predicentrine 21 (75%)

Erybidine 19 (76%)

In contrast, an amine neospirene will rearrange only to a dibenzazonine, regardless of whether acid or base is used as the catalyst, as shown in the sequence below where again judicious use was made of borane complexes.[16]

23, an isomer of
erybidine **19**

To summarize, the synthetically useful transformations are:

a. Morphinandienone amines or amides can be prepared from tetrahydro-benzylisoquinolines by electrolytic oxidation (see Sec. 2.3.8).

b. Morphinandienone amines rearrange in protonic acids by aryl migration to furnish neoproaporphines which subsequently can be transformed into aporphines.

c. Morphinandienone amines as well as amides upon treatment with strong Lewis acids undergo alkyl migration to furnish neospirenes. The neospirene amides can be isolated, whereas the neospirene amines are further transformed by the Lewis acid to dibenzazonines.

d. Neospirene amides can be prepared from the vanadium oxytrifluoride oxidation of *N*-acyltetrahydrobenzylisoquinolines. The amides are convertible to neospirene–borane complexes which, on treatment with either a Lewis acid or base, yield dibenzazonines.

e. *N*-Acylneoproaporphines (see Sec. 10.2.1), on reduction with lithium aluminum hydride, borane complexation, and reoxidation with manganese dioxide, furnish neoproaporphine–boranes. The latter rearrange to aporphines using protonic acids at room temperature, or alternatively to dibenzazonines with base.

The general picture becomes even more intricate at elevated temperatures. When *O*-methylflavinantine (**18**) was treated with BF$_3$–ether in refluxing benzene and the resulting mixture immediately reduced with Adams catalyst, four compounds were isolated, namely thaliporphine (**3**) (28%), predicentrine (**21**) (8%), 2,9,10-trimethoxy-3-hydroxyaporphine (**24**) (36%), and erybidine (**19**) (8%), for a total overall yield of 80%.[16] The synthesis of the four products can thus be written as in Scheme 10.3.[16]

The formation of these products can best be rationalized in terms of an equilibrium, under the experimental conditions used, between morphinandienone,

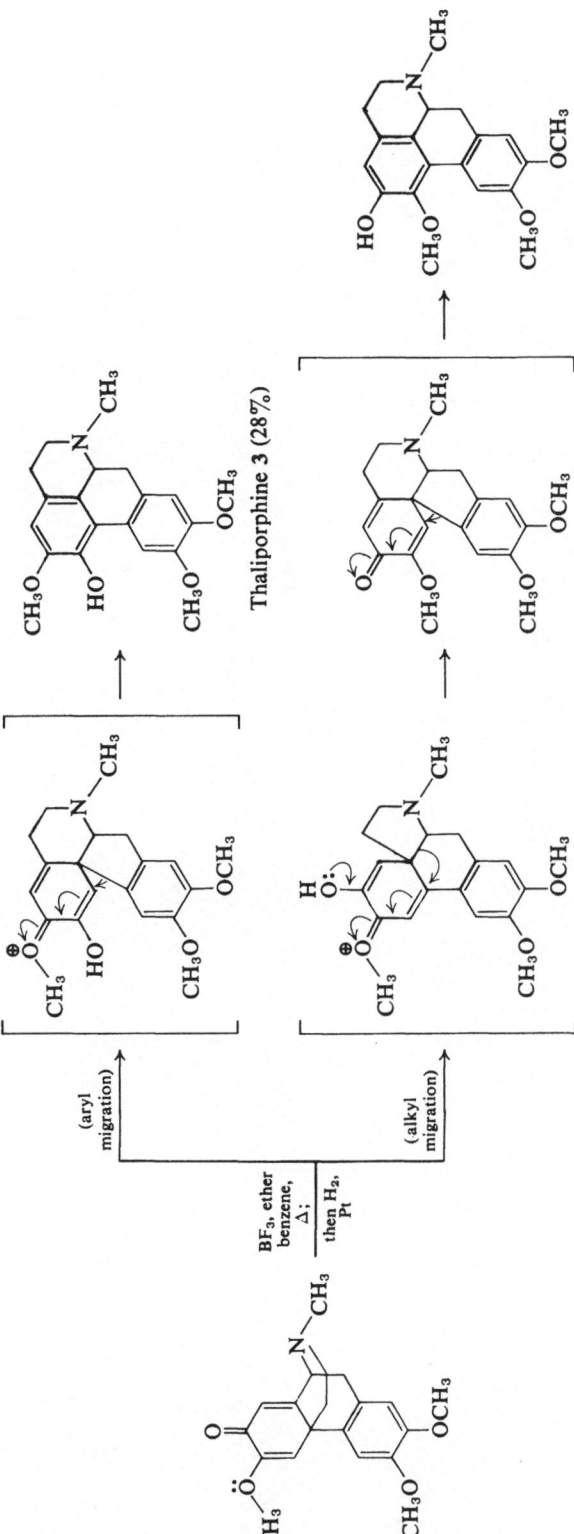

SCHEME 10.3 (Continued overleaf)

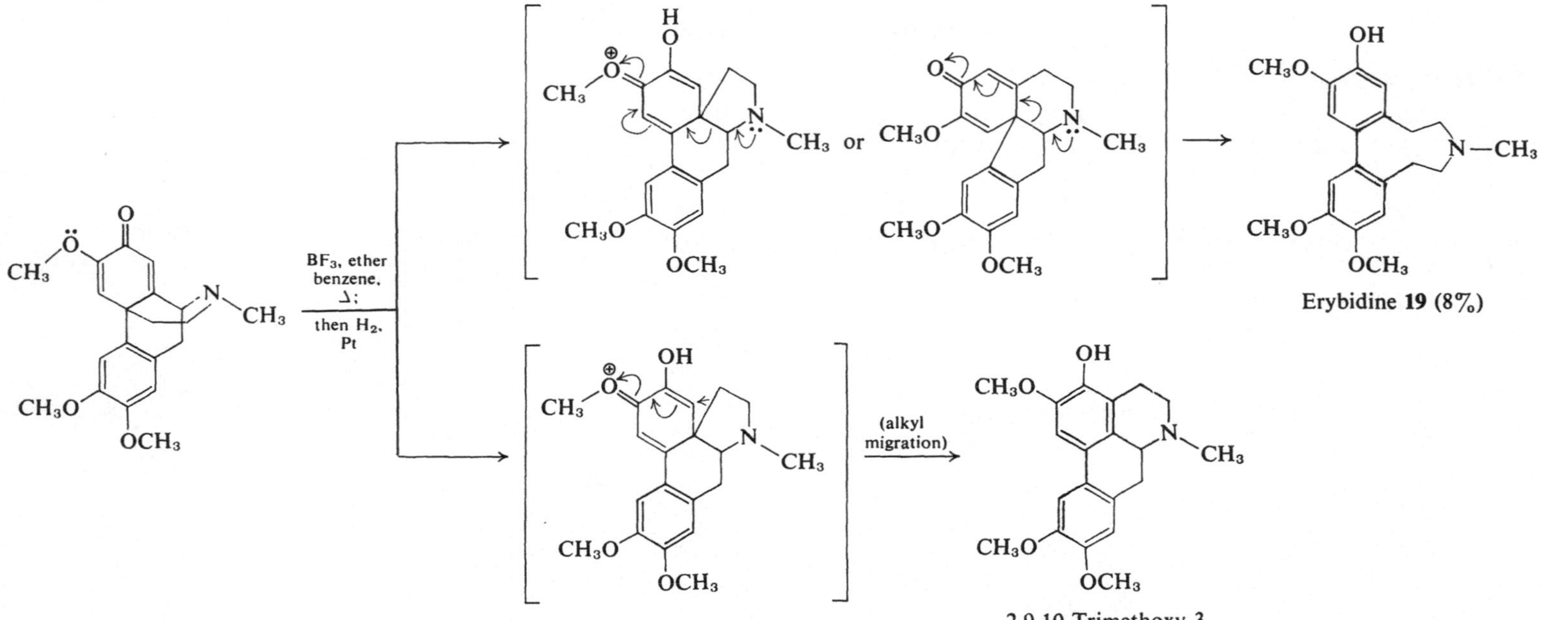

Erybidine **19** (8%)

2,9,10-Trimethoxy-3-hydroxyaporphine **24** (36%)

(alkyl migration)

BF$_3$, ether benzene, Δ; then H$_2$-Pt

SCHEME 10.3 (*Continued*)

neospirene, and neoproaporphine[16]:

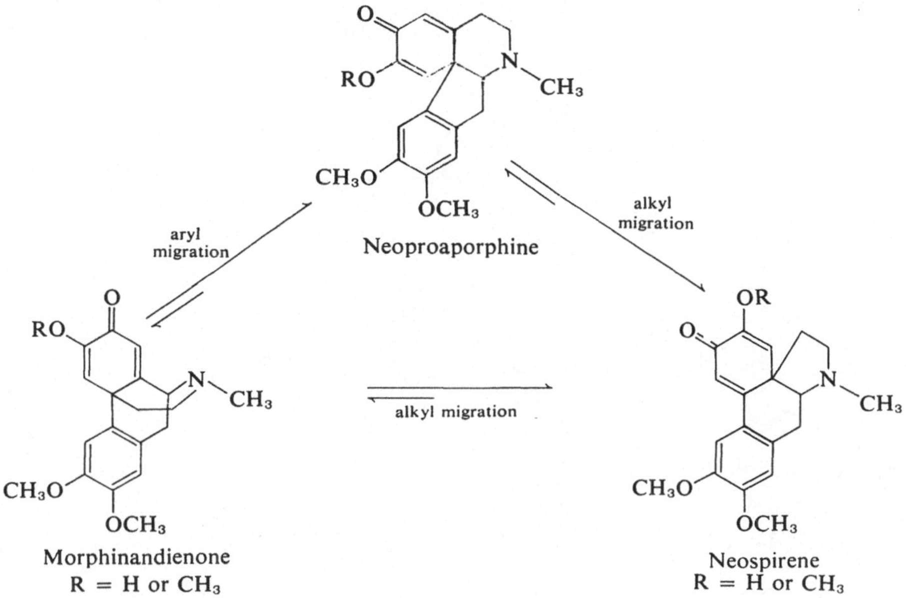

Neoproaporphine

Morphinandienone
R = H or CH₃

Neospirene
R = H or CH₃

One can also start with a neoproaporphine or a neospirene and obtain results very similar to the above since treatment of the neoproaporphine–borane complex **20** with BF₃ at elevated temperatures, followed by catalytic reduction, yielded predicentrine (**21**), erybidine (**19**), and 2,9,10-trimethoxy-3-hydroxyaporphine (**24**). The neospirene–borane complex **22**, when subjected to the same conditions, gave rise to thaliporphine (**3**), the erybidine isomer **23**, and another nonnatural aporphine, 2-hydroxy-3,9,10-trimethoxyaporphine (**25**)[16]; both pathways are shown below:

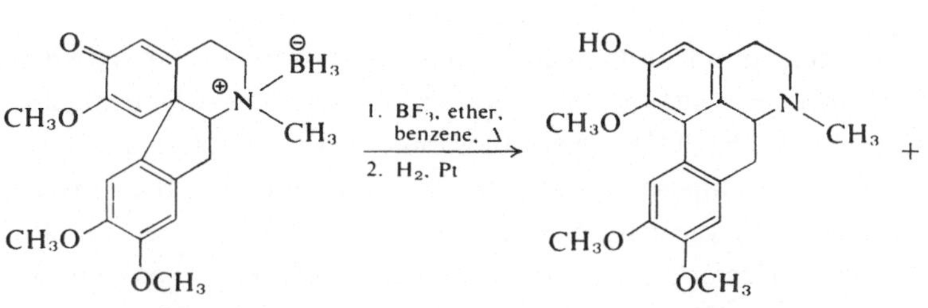

20

21 (47%)

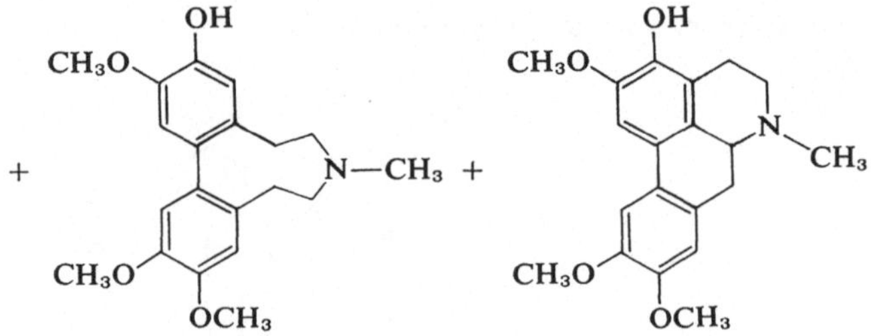

19 (24%)

24 (8%)

and

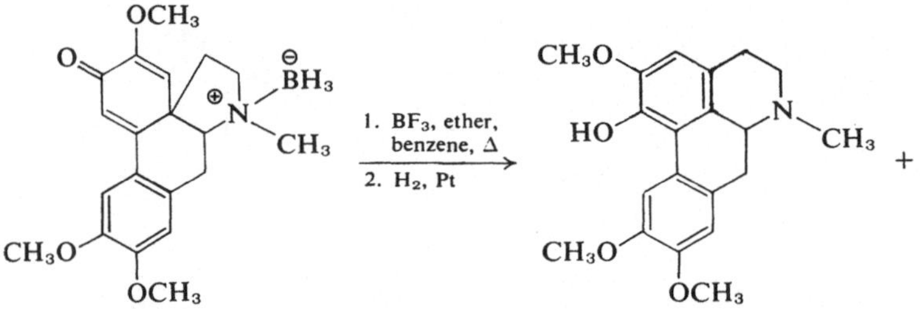

22

1. BF₃, ether, benzene, Δ
2. H₂, Pt

3 (8%)

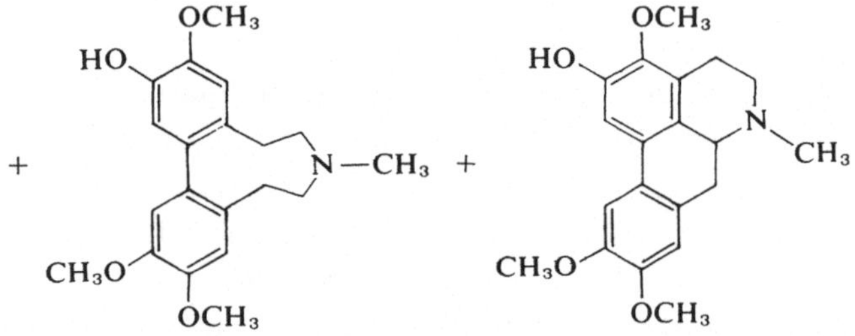

23 (35%)

25 (25%)

Finally, it has been found only recently that the intramolecular oxidative coupling of appropriately substituted monophenolic tetrahydrobenzyliso-quinolines induced by VOF₃–TFA constitutes one of the simplest and most efficient routes to the aporphines. The only limitation here is that the nitrogen atom must be protected as the trifluoroacetyl amide or the borane complex, as illustrated by the two examples shown in Scheme 10.4. The yields are appreciably lower if the amine function is unprotected.[17]

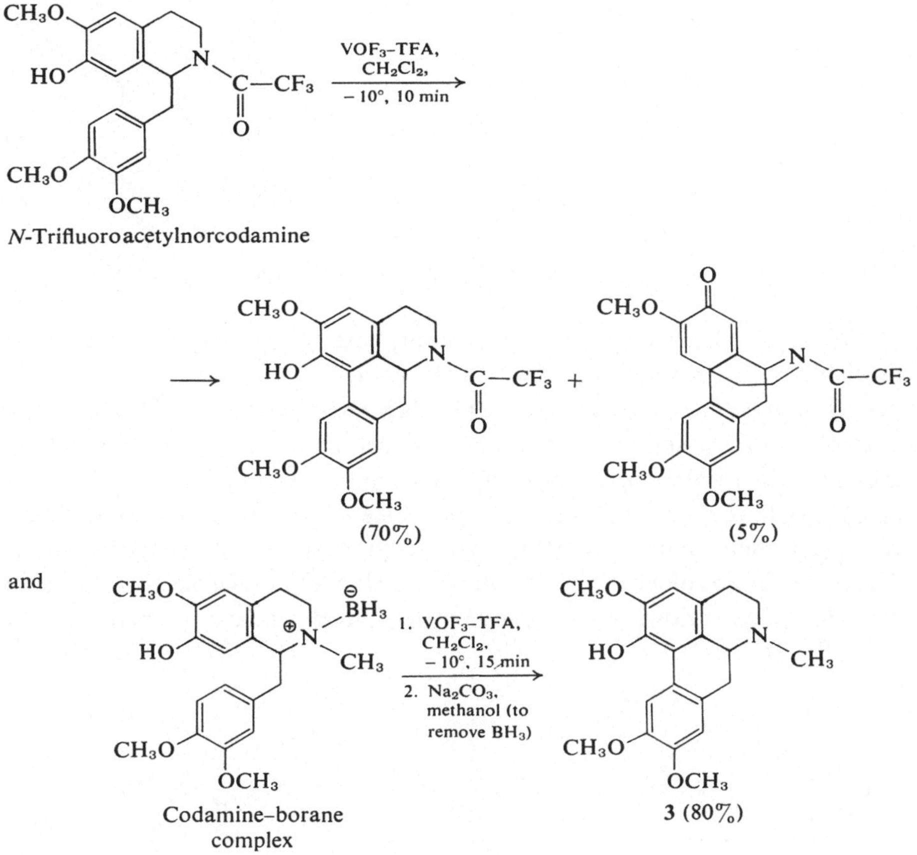

CH$_3$O, HO ... N—C—CF$_3$, O, CH$_3$O, OCH$_3$

VOF$_3$–TFA,
CH$_2$Cl$_2$,
$-10°$, 10 min

N-Trifluoroacetylnorcodamine

CH$_3$O, HO ... N—C—CF$_3$, O, CH$_3$O, OCH$_3$
(70%)

+

CH$_3$O, O, N—C—CF$_3$, O, CH$_3$O, OCH$_3$
(5%)

and

CH$_3$O, HO, N$^{\oplus}$ $^{\ominus}$BH$_3$, CH$_3$, CH$_3$O, OCH$_3$
Codamine–borane
complex

1. VOF$_3$–TFA,
   CH$_2$Cl$_2$,
   $-10°$, 15 min
2. Na$_2$CO$_3$,
   methanol (to
   remove BH$_3$)

CH$_3$O, HO, N—CH$_3$, CH$_3$O, OCH$_3$
3 (80%)

SCHEME 10.4

10.2.4. Electrolytic Oxidation

Electrolytic oxidation of laudanosine (**16**) in TFA leads to a 17% yield of glaucine (**17**), the presumed intermediate again being a morphinandienone.[15]

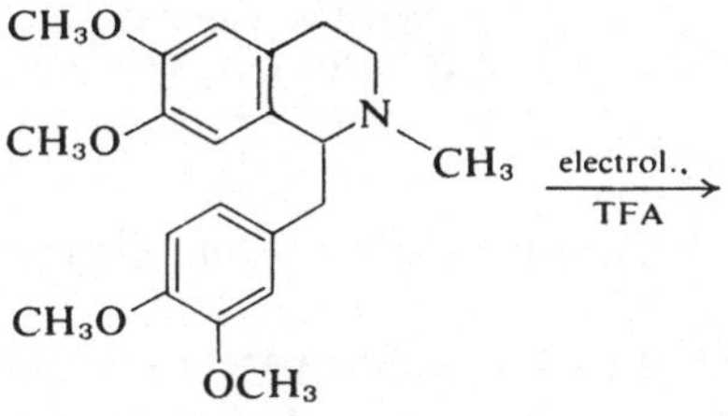

CH$_3$O, CH$_3$O, N, CH$_3$, CH$_3$O, OCH$_3$
16

electrol.,
TFA

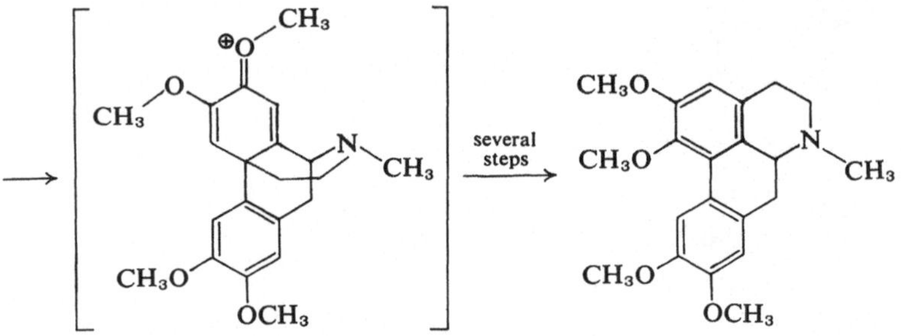

17

A novel preparative synthesis of aporphines which should prove of appreciable utility in the future involves the cathodic cyclization of a 1-(*o*-iodobenzyl)isoquinoline methiodide salt. Gottlieb and Neumeyer have shown that electrolysis of 1-(*o*-iodobenzyl)isoquinoline methiodide in dry acetonitrile containing tetraethylammonium bromide, and using a mercury cathode, furnished an 86% yield of the yellow didehydroaporphine which was reduced over Adams catalyst in methanolic hydrochloric acid to produce aporphine hydrochloride. The formation of didehydroaporphine proceeds via two one-electron reduction steps as shown below. 10,11-Dimethoxyaporphine was also prepared by this route.[18]

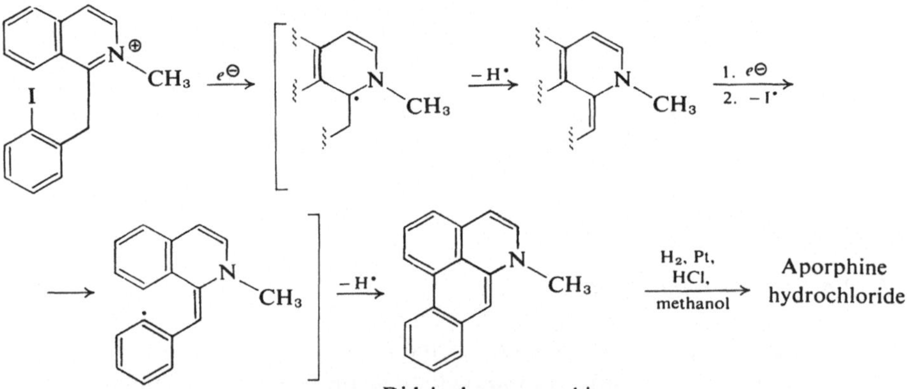

Didehydroapoporphine
λ_{max}^{MeOH} 228, 275, 300, 339,
352, 404, and 427 nm
(4.46, 4.20, 3.41, 4.03, 4.09,
3.77, and 3.75)

10.2.5. Oxidation with Cuprous Chloride and Oxygen

Phenolic oxidation of (+)-reticuline perchlorate with cuprous chloride and oxygen in pyridine supplied (+)-corytuberine, (+)-isoboldine, and a small amount of the morphinandienone pallidine. This mode of oxidation, investigated by Kametani, encourages ortho–ortho coupling so that more corytuberine than isoboldine was formed. Homoaporphines may also be obtained by such a procedure, starting with the appropriate diphenolic phenethylisoquinoline.[19]

(+)-Reticuline perchlorate

$\xrightarrow{\text{Cu}_2\text{Cl}_2, \text{O}_2, \text{py.}}{\text{room temp.}}$

(+)-Corytuberine (28%)   (+)-Isoboldine (8%)

10.2.6. The Use of Lead Tetraacetate; Hydroxylation at C-4

Oxidation of codamine with lead tetraacetate and subsequent treatment with acetic anhydride and concentrated sulfuric acid is known to generate 4-acetoxy-*O*-acetylthaliporphine and *O*-acetylthaliporphine.[20] It has now been authoritatively reported that lead tetraacetate oxidation of thaliporphine (**3**) itself affords 4β-acetoxythaliporphine whose hydrolysis and *O*-methylation produces cataline in 87% overall yield.[21] Alternatively, reaction of 4β-acetoxy-thaliporphine with methanol leads to 4β-methoxythaliporphine, probably through the intermediacy of a quinone methide.[21]

Thaliporphine 3 $\xrightarrow{\text{Pb(OAc)}_4,\ \text{HOAc}}$

4β-Acetoxythaliporphine   1. hydrol.   Cataline
                    2. CH₂N₂

methanol, room temp.

4β-Methoxythaliporphine

In a new variation of this series of reactions, treatment of the initial quinol acetate, obtained from lead tetraacetate oxidation of codamine, with TFA resulted in a substantial increase in the yield of the corresponding aporphine, namely thaliporphine (**3**), while no C-4 oxygenated aporphine could be detected.[22]

Protonated quinol acetate   TFA   → **3** (96%)

10.2.7. Benzyne Intermediates

Immonium salts can be used in the preparation of aporphines via benzyne intermediates. Treatment of salt **26** with dimsyl sodium in DMSO provided a 25% yield of dehydrodomesticine.[23]

26   DMSO, NaH   Dehydrodomesticine

In the benzyne approach, the nitrogen must not be in the form of a urethan since this function can become involved, and benzodiazepines will be obtained.[24]

A urethan

A benzodiazepine

10.2.8. Enzymatic Oxidative Coupling

In certain cases, enzymatic coupling of tetrahydrobenzylisoquinolines can be as efficient as purely chemical methods in the preparation of aporphines. Horseradish peroxidase coupling of (+)-laudanosoline methiodide (27) at neutral pH yielded a 60% yield of the corresponding aporphine. The reaction proceeds with retention of configuration.[25]

27

Incubation of (±)-reticuline with rat liver homogenate in the presence of such cofactors as NAD, NADP, or NADPH results in formation of isoboldine together with the morphinandienone pallidine, and the tetrahydroprotoberberines coreximine and scoulerine.[25a]

10.2.9. Reissert Intermediates

Reissert compounds are useful in the preparation of the benzylisoquinolines required for aporphine synthesis, and the field has been reviewed.[26] Reissert intermediates have also been used in the preparation of 7-hydroxylated aporphines by the oxazolinone method.[27]

10.2.10. Improved Pschorr Cyclizations

Pschorr cyclization of 7-hydroxylated tetrahydrobenzylisoquinolines proceeds in high yield. For example, cyclization of the aminophenol **28** by diazotization followed by treatment with copper powder gave an 88% yield of the aporphine **29**.[28]

Nuciferine (1,2-dimethoxyaporphine) has been prepared in good yield by a modification of the Pschorr cyclization step in which freshly prepared copper powder was employed as the catalyst. The intermediate *N*-benzylnornuciferine was thus obtained in 60% yield. This represents one of the highest yields for a classical Pschorr cyclization reported to date.[29]

10.2.11. Coupling with Thallium Tristrifluoroacetate (TTFA)

E. C. Taylor and co-workers have shown that nonphenolic oxidative coupling of *N*-methyltetrahydrobenzylisoquinolines can be performed in good to excellent yield using TTFA. Thus treatment of the tetrahydrobenzylisoquinoline **29a** with TTFA at −40° led to a 46% yield of ocoteine.[29a]

Attempted coupling of **29a** with thallium(III) acetate at 0° in the same solvent mixture surprisingly led to 8-acetoxyocoteine.[29a]

10.3. Reactions of Aporphines

10.3.1. Oxidation and Formation of Dehydroaporphines and Oxoaporphines

An efficient method for the conversion of aporphines to dehydroaporphines involves catalytic dehydrogenation using 10% Pd/C in refluxing acetonitrile. This procedure does not apply to noraporphines or phenolic aporphines.[30]

Oxidation of nonphenolic aporphines by iodine affords the corresponding dehydroaporphine. In contrast, oxidation of nonphenolic noraporphines proceeds all the way to the oxoaporphine stage.[31] Oxoaporphines can also be obtained from nonphenolic aporphines using lead tetraacetate.[32]

The phenolic aporphines N-methylnandigerine and bulbocapnine are converted to a blue 10,11-o-quinone using mercuric chloride.[31] This same oxidation product has also been found as a natural product in *Corydalis cava* Schweigg et Korte (Fumariaceae).[33]

(+)-N-Methylnandigerine

A 10,11-orthoquinone   (+)-Bulbocapnine

The nonphenolic dehydroaporphines dehydronuciferine (1,2-dimethoxy-dehydroaporphine) and dehydrodicentrine (1,2-methylenedioxy-9,10-dimethoxy-dehydroaporphine) are oxidized by oxygen at pH 6 to give the corresponding oxoaporphines. The transformation of dehydronuciferine to the oxoaporphine can also be achieved using peracetic acid or benzoyl peroxide.[31]

The eosin sensitized photooxidation of glaucine, dehydroglaucine, or norglaucine, in ethanol yields the corresponding oxoaporphine O-methylatheroline in high yield.[33a]

10.3.2. Selective Cleavage of the Methylenedioxy Substituent

Treatment of (+)-bulbocapnine methyl ether (**30**) with excess boron tri-chloride in methylene chloride containing 0.3% ethanol as stabilizer gives the expected catechol **31** together with the novel aporphine **32**. Similarly, treat-ment of **30** with boron tribromide affords the diphenol **33**, so that with either reagent the original C-1,2-methylenedioxy ring opens and subsequently cyclizes on the C-11 oxygen, forming a seven-membered ring. Both compounds **32** and **33** can be methylated to the same dimethyl ether **34**, and prolonged hydrolysis of the latter with dilute hydrochloric acid leads to the known aporphine (+)-corytuberine (see Scheme 10.5).[34] Deoxygenation of the dextrorotatory catechol **31** at C-1 and C-2 by the Musliner–Gates procedure also led to racemization of the isolated 10,11-dimethoxyaporphine.[35]

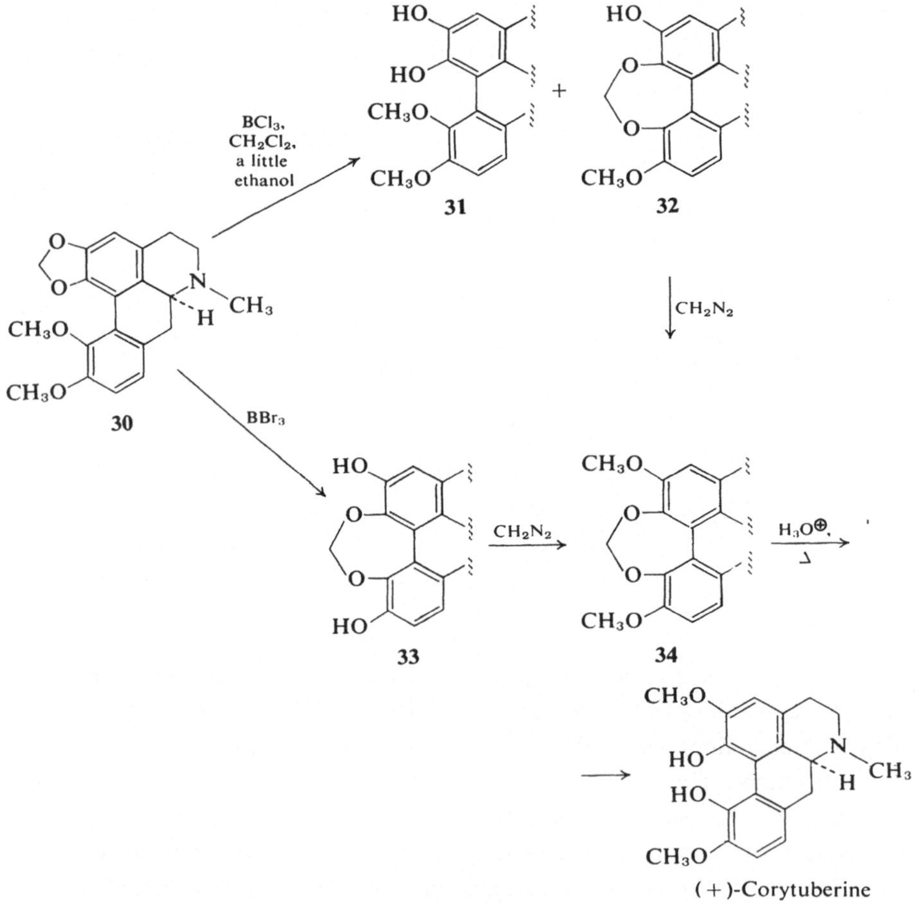

SCHEME 10.5

In ethanol-free methylene chloride, bulbocapnine methyl ether (**30**) with boron tribromide affords a mixture of diphenol **33** and phenanthrene **35**.

The presence of a little ethanol in the solvent leads, upon the addition of boron tribromide, to some hydrobromic acid resulting in formation of the hydrobromide salt of the initial aporphine. No phenanthrene is ultimately obtained under these conditions. Bonding of boron tribromide to the basic nitrogen seems to be the initial step in the opening of ring B. This is supported by the observation that reaction of (+)-bulbocapnine methyl ether (**30**) with boron tribromide in the presence of a strong base such as 1,8-bis(dimethylamino)-naphthalene furnishes mainly the phenanthrene derivative **35**. On the other hand, the hydrobromide salt of **30** when treated with boron tribromide produces only the diphenol **33**.[34]

Natural (+)-bulbocapnine can easily be converted into its unnatural (−)-enantiomer by oxidation to dehydrobulbocapnine, reduction with zinc in hydrochloric acid, and resolution with L-tartaric acid. Sequential treatment of (−)-bulbocapnine first with boron trichloride and then with boron tribromide affords (−)-1,2,10,11-tetrahydroxyaporphine.[34]

10.3.3. Dimerization

In the first authenticated dimerization of an aporphine, oxidation of (+)-bulbocapnine methyl ether (**30**) with iodine provided the dimer **36**. Reduction of **36** with zinc in dilute sulfuric acid led to racemic **30**, whereas reduction with lithium aluminum hydride supplied a pair of diastereomeric dimers **37**.[36]

Treatment of dehydroglaucine with mercuric acetate in aqueous solution, followed by reduction with sodium borohydride similarly gave a (C-7)–(C-7) glaucine dimer.[36a]

10.3.4. Hofmann and Emde Degradations

The size of the base used can affect the direction of elimination of quaternary aporphine salts. Hofmann elimination of (−)-nuciferine methiodide using sodium ethoxide led to the phenanthrene derivative **38**, but use of the bulkier base potassium triethylcarbinolate afforded the chiral dihydrophenanthrene **39**. The optically inactive dihydrophenanthrene **40** was obtained through Emde reduction of the starting methiodide salt (see Scheme 10.6).[29]

10.3.5. O-Acetylation

A mild O-acetylation of phenolic aporphines, avoiding phenanthrene formation, utilizes acetyl bromide in trifluoroacetic acid. The diacetate of apomorphine was thus obtained in high yield.[37]

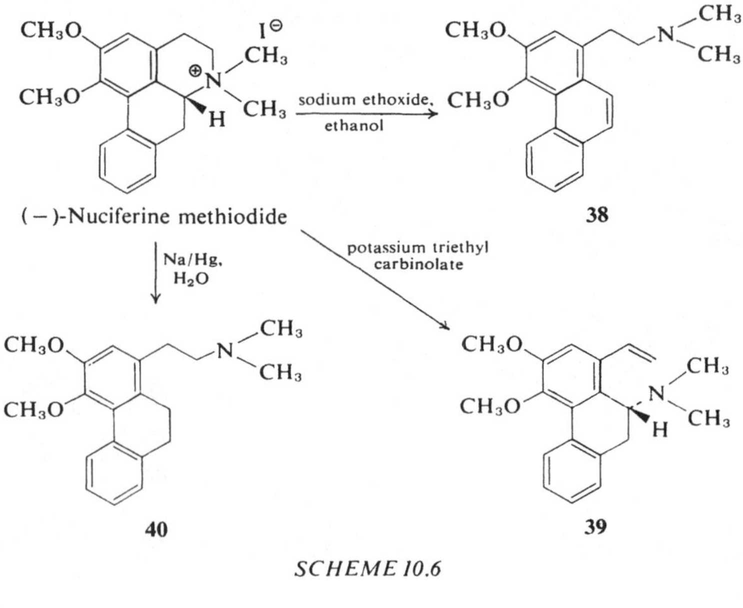

SCHEME 10.6

10.3.6. Hydroxylation

Taking advantage of the observation that phenol is formed from nitrobenzene and phenyllithium at low temperature, Wiriyachitra and Cava have succeeded in converting nuciferine into 3-hydroxynuciferine. The sequence involves bromination of the aporphine in a strongly acidic medium, reaction with *n*-butyllithium to form the aromatic lithio compound, and finally oxidation with nitrobenzene.[38]

Nuciferine   3-Hydroxynuciferine

10.3.7. Selective O-Demethylation and N-Demethylation

Good yields of monophenolic bases are obtained owing to regioselective O-demethylation by sodium benzylselenolate in refluxing DMF. The affected positions are C-1, C-8, and C-11, while methylenedioxy groups survive the reaction. Two examples are cited below.[38a]

Apomorphine dimethyl ether   Apocodeine

Ocopodine   Leucoxine

Apomorphine (10,11-dihydroxyaporphine) was *N*-demethylated by treatment with ethyl chloroformate followed by hydrazine cleavage of the crude urethan to give an 87% yield of norapomorphine.[38aa]

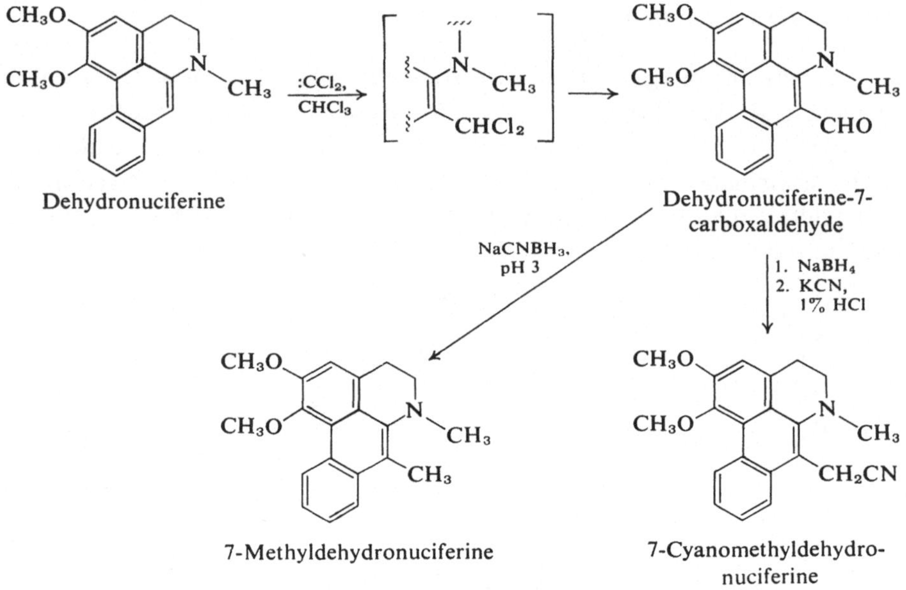

SCHEME 10.7

10.3.8. Protonation and Formylation of Dehydroaporphines

Dehydroaporphines initially undergo both *N* protonation and C-7 protonation in trifluoroacetic acid; but the C-7 protonated immonium salts are formed almost completely under equilibration conditions, thus indicating a degree of enamine-type character to dehydroaporphines.[38b]

Reaction of dichlorocarbene with dehydronuciferine yields dehydronuciferine-7-carboxaldehyde. Reduction of this aldehyde with sodium cyanoborohydride at pH 3 affords in good yield 7-methyldehydronuciferine. Alternatively, sodium borohydride reduction of the aldehyde succeeded by treatment with hydrogen cyanide affords 7-cyanomethyldehydronuciferine.[38c] (See Scheme 10.7.)

10.4. Absolute Configuration

Substituents are related to absolute configuration by the following simple rule: Aporphines substituted at both C-10 and C-11, or at both C-9 and C-10, are usually dextrorotatory and belong to the *L(S)*-series. Aporphines unsubstituted or monosubstituted in ring D can belong either to the *L(S)*- or the *D(R)*-series.

$(+)$-$L(S)$ $(-)$-$D(R)$

The CD curves of several aporphines have been measured, and these have been correlated with the absolute configurations.[7,39]

10.5. Biogenesis

The work of Barton and Battersby[39a] proved conclusively that the biogenesis of aporphines, unsubstituted or only monosubstituted in ring D, proceeds through the intermediacy of proaporphines. However, the exact pathway or pathways by which tetrasubstituted 1,2,9,10- and 1,2,10,11-aporphines are formed is still under active investigation.

Brochmann-Hanssen[40] and Blaschke[41] have supported the thesis that direct phenolic coupling of tetrahydrobenzylisoquinolines can lead to tetrasubstituted aporphines. Along this line, it has been shown that $(\pm)$-reticuline specifically labeled with ^{14}C either at C-3 or on the N-methyl group, when administered to opium poppies (*Papaver somniferum* L., Papaveraceae) was incorporated into isoboldine to the extent of 0.08% (see Scheme 10.7). But no incorporation of reticuline was observed in magnoflorine. It was, therefore, concluded that 1,2,9,10-tetrasubstituted aporphines such as isoboldine can be biosynthesized by direct phenolic coupling, while steric factors preclude such an avenue for 1,2,10,11 aporphines.[40] Further support for direct oxidative coupling comes from the observation that reticuline is present in *Corydalis cava* Schweigg et Korte (Fumariaceae).[42] This is of some importance since it had previously been demonstrated that labeled $(\pm)$-reticuline is converted to $(+)$-bulbocapnine by the same plant (see Scheme 10.8).[41]

Furthermore, evidence has been advanced for the sequence norreticuline → reticuline → isoboldine → boldine in *Litsea glutinosa* (Lauraceae). $(\pm)$-[3-^{14}C]-Norreticuline as well as $(\pm)$-[2',6',8-[3]H$_3$]-reticuline were found to be efficient precursors of $(+)$-boldine, while $(+)$-[8-^{3}H]-isoboldine was also readily incorporated into $(+)$-boldine (7).[43] The plan can easily convert norboldine into boldine.[43]

On the other hand, it has been shown by Battersby and co-workers that

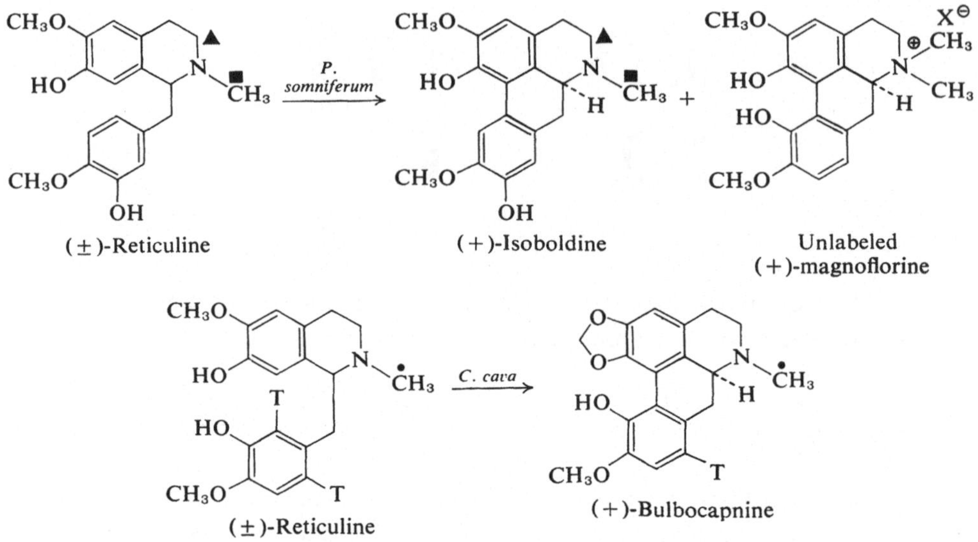

(±)-Reticuline   (+)-Isoboldine   Unlabeled
                                                (+)-magnoflorine

(±)-Reticuline   (+)-Bulbocapnine

SCHEME 10.8

in *Dicentra eximia* (Ler) Torr. (Fumariaceae), the 1,2,10,11-tetrasubstituted
aporphine (+)-corydine, and the 1,2,9,10-substituted analogs (+)-boldine,
(+)-dicentrine, and (+)-glaucine are formed not from reticuline or orientaline,
but rather from 4'-O-methylnorlaudanosoline and norprotosinomenine via the
hypothetical intermediacy of reactive neoproaporphines (see Scheme 10.9).[44]

In view of Kupchan's *in vitro* synthesis of aporphines via the rearrangement of morphinandienones, it is rational to conceive of a Battersby–Kupchan

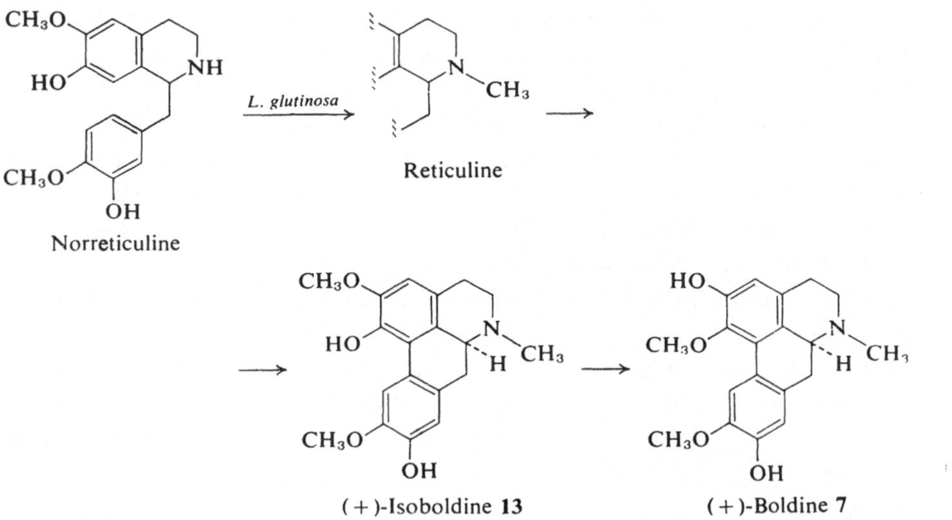

Norreticuline   Reticuline

(+)-Isoboldine **13**   (+)-Boldine **7**

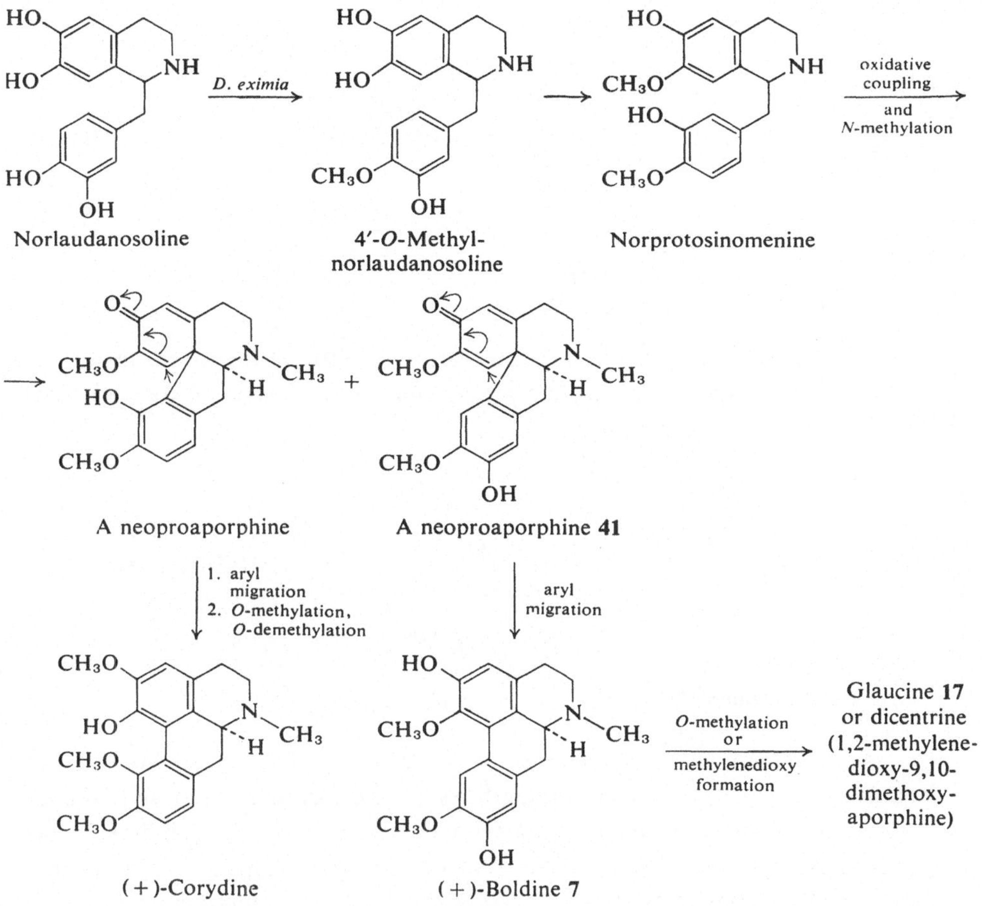

Norlaudanosoline → 4'-O-Methyl-norlaudanosoline → Norprotosinomenine

A neoproaporphine + A neoproaporphine **41**

(+)-Corydine   (+)-Boldine **7**

Glaucine **17** or dicentrine (1,2-methylenedioxy-9,10-dimethoxyaporphine)

SCHEME 10.9

biogenetic mechanism which would involve the sequence: tetrahydrobenzyl-isoquinoline → morphinandienone → neoproaporphine → aporphine.[14-16] The formation of (+)-boldine (**7**) would then entail the following transformations, assuming that *O*-demethylation of the morphinandienone intermediate is slower and less favored than aryl migration, to furnish the neoproaporphine **41** which in turn would then rearrange to (+)-boldine (**7**).

It is clear that additional work with labeled precursors is required before a complete picture of the biogenesis of aporphines can be formulated.

The possible origin of the methylenoxy bridge in thalphenine has been discussed. (+)-Nantenine methochloride has been found to accompany thal-phenine (**11**) in *Thalictrum polygamum* Muhl. (Ranunculaceae). Assuming an

Norproto-sinomenine $\xrightarrow{[O]}$ [A morphinandienone structure] $\xrightarrow{\text{aryl migration}}$ **41** $\xrightarrow{\text{aryl migration}}$ (+)-Boldine **7**

A morphinandienone

ionic mechanism, the most likely biogenetic mode of formation for the methylen-oxy bridge appears to be through an oxonium ion of type **42** derived from nantenine or its *N*-metho salt.[45]

(+)-Nantenine **41a** → **42** → (+)-Thalphenine **10**

A *Streptomyces* species can convert 10,11-dimethoxyaporphine into an equal mixture of apocodeine (10-methoxy-11-hydroxyaporphine) and isoapo-codeine (10-hydroxy-11-methoxyaporphine). Isoapocodeine is, however, the predominant product when cultures of *Cunninghamella* spp. are used, so that selective *O*-demethylation may be achieved with different microorganisms.[46]

The microbial transformation of glaucine using *Streptomyces griseus* produces norglaucine and predicentrine (1,9,10-trimethoxy-2-hydroxyaporphine). On the other hand, *Fusarium solani* gives didehydroglaucine. This dehydrogenation occurs with (+)-glaucine, but not with its enantiomer.[46]

10.6. Pharmacology

Pharmacological studies on aporphines have centered mainly around (−)-apomorphine ((−)-10,11-dihydroxyaporphine),[47] and its close analogs[48] because (−)-apomorphine stimulates the dopaminergic system, and has anti-Parkinson activity.[49,50] It also has a hypotensive effect,[49,50] and can decrease serum prolactin levels in patients with hyperprolactinemia.[51]

Apocodeine, which shows only about one-fourth the CNS (central nervous system) activity of apomorphine in rats is converted *in vivo* into apomorphine and norapomorphine, thus explaining its limited activity.[52] On the other hand, incubation of apomorphine with catechol *O*-methyltransferase produces a mixture of the two possible isomeric monomethyl ethers, but with the 10-methylated isomer greatly predominating.[53]

11-Hydroxyaporphine can lower the blood pressure of cats.[49] On the other hand, 8-hydroxy-*N*-*n*-propylnoraporphine shows no dopamine agonist activity.[54a] In fact, the C-10,11 diphenolic system does not seem to be a requirement for dopaminergic activity since 10-hydroxy- and 11-hydroxy-*N*-*n*-propylnoraporphine show appreciable activity in rats.[50,54] *N*-Ethyl and *N*-*n*-propylnorapomorphine are potent emetics in dogs. The hypotensive effect of apomorphine in the anesthetized cat was also observed with *N*-*n*-propylnorapomorphine.[55]

The diphenolic tertiary amine **43**, structurally related to *N*-*n*-propylnorapomorphine ranks as a strong dopamine receptor agonist, and may thus find use in the treatment of Parkinson's disease.[56] The dopamine agonist activity of apomorphine has been related to its shape and conformation.[56a]

43

Anonaine (1,2-methylenedioxynoraporphine) hydrochloride has shown activity as an antimicrobial agent.[57]

Very recently, it has been observed that oliveroline shows antiparkinsonian activity similar to apomorphine. Oliveridine and oliverine caused a dose-dependent hypotension in normal rats, which was followed by a secondary hypertension. In the isolated rabbit ear, oliveridine had a vasodilating effect comparable to that of papaverine. All three aporphines were obtained from West African Annonaceae.[57a]

(−)-Oliveroline   (−)-Oliveridine   (−)-Oliverine

10.7. NMR Spectroscopy

Several studies on the CMR spectra of aporphines in CDCl$_3$ have appeared. Table 10.1 lists the chemical shifts for nine aporphines.[21,58,59] The methyl signal of a C-1 methoxyl group appears at lower field than that of other methoxyl groups because of increased steric hindrance.[58] Any aromatic carbon appearing downfield from 140 ppm is ipso to an oxygenated function (OH, OCH$_3$, OCH$_2$O).[59] O-Methylation of a phenolic aporphine will shift the signal of the ipso aromatic carbon downfield by about 3.5 ppm. It has also been noted that O-acetylation of a phenolic aporphine will shift the ipso carbon downfield by about 7 ppm.[59]

The PMR spectra of aporphines incorporating a methylenedioxy function can yield useful information concerning the location of this function. The presence of a C-1,2 methylenedioxy group is evidenced by an upfield shift of the H-11 which appears between δ7.47 and 7.86; whereas when a hydroxyl or a methoxyl is at C-1, the H-11 signal is located further downfield, between δ7.80 and 8.21. Moreover, the ics for the two protons of the methylenedioxy group is large (4–12 Hz) when the group is located at C-1,2, and small (2–4 Hz) when at C-2,3. No splitting of the methylenedioxy protons is observed when the group is at C-9,10. If the methylenedioxy is at C-10,11, the PMR spectrum is characterized by the absence of an H-11 downfield aromatic signal, and by a large ics (≈8 Hz) of the two methylene protons.[60]

The PMR spectral values for the N-acetyl amide of the new noraporphine (−)-elmerrillicine, present in the South-East Asian tree *Elmerrillia papuana* (Schltr.) Dandy (Magnoliaceae), are given below.[61]

Table 10.1. CMR Spectra of Aporphines in CDCl₃ (δ)

Carbon atom	Nuciferine[a] (40a)	Glaucine[a] (17)	Nantenine[a] (41a)	Iso-corydine[a]	Domesticine[b]	Thaliporphine[c] (3)	Predicentrine[c] (21)	Boldine[c] (7)	Dicentrine[c]
1	144.6	143.9	144.0	141.7	140.73	140.7	142.3	142.0	141.7
1a	126.3	126.5	126.4	135.4	119.47*	119.5	126.3	126.8	116.6
1b	128.1	128.6	128.2	129.8	127.22*	127.2	125.9	125.9	126.4
2	151.4	151.5	151.4	150.8	145.83	145.8	148.2	148.1	146.6
3	110.9	110.1	110.3	110.8	109.73	108.7	113.5	113.3	106.1
3a	127.5	127.0	127.0	128.8	123.64	123.9	129.6	129.9	126.6
4	28.9	29.1	29.0	29.1	28.81	29.0	28.7	28.9	29.2
5	52.8	53.1	52.9	52.4	53.31	53.5	53.3	53.4	53.6
6a	61.9	62.3	62.1	62.6	62.45	62.7	62.5	62.6	62.4
7	34.8	34.4	34.9	35.6	34.94	34.5	34.2	34.2	34.3
7a	135.9	129.1	130.4	129.6	130.20[d]	128.9	129.2	130.2	128.3
8	127.7	110.6	107.8	118.6	108.15[d]	110.9	110.7	114.2	110.5
9	126.7	147.7	146	110.7	145.83	147.6	148.1	145.1	148.2
10	126.4	147.1	145.9	149.0	145.83	147.1	147.6	145.6	147.6
11	127.3	111.4	108.4	143.6	108.74[d]	112.0	110.0	110.1	111.2
11a	131.6	124.2	125.1	119.8	125.76[d]	124.8	124.1	123.6	123.4
NMe	43.5	43.4	43.6	43.6	43.91	44.0	43.8	44.0	44.0
C-1 OMe	59.7	59.8	59.8	61.7	—	—	60.3	60.2	—
Other OMe	59.7	55.5	55.4	55.5	56.03	55.9	55.8	56.1	55.9
	55.3	55.5		55.8		56.0	56.0		56.1
	55.7	55.7							

[a] Reference 58.   [b] Reference 23.   [c] Reference 59.
[d] The values for C-1a and C-1b, for C-7a and C-11a, and for C-8 and C-11, have been exchanged, and are therefore different from those given in the original paper.

References and Notes

1. For a complete listing of naturally occurring aporphines, with accompanying spectral data, see H. Guinaudeau, M. Leboeuf, and A. Cavé, *Lloydia*, **38**, 275 (1975).
2. M. P. Cava, Y. Watanabe, K. Bessho, and M. J. Mitchell, *Tetrahedron Lett.*, 2437 (1968).
3. M. Shamma and J. L. Moniot, *Heterocycles*, **2**, 427 (1974).
4. M. Hamonnière, M. Leboeuf, and A. Cavé, *C. R. Acad. Sci. Ser. C*, **278**, 921 (1974); and *Phytochemistry*, **16**, 1029 (1977); and M. Nieto, A. Cavé, and M. Leboeuf, *Lloydia*, **39**, 350 (1976).
5. V. Preininger and V. Tosnarova, *Planta Med.*, **23**, 233 (1973).
6. H. Guinaudeau, M. Leboeuf, M. M. Debray, A. Cavé, and R. R. Paris, *Planta Med.*, **27**, 304 (1975).
7. C. D. Hufford and M. J. Funderburk, *J. Pharm. Sci.*, **63**, 1338 (1974).
8. C.-L. Chen, H. Chang, and E. B. Cowling, *Phytochemistry*, **15**, 547 (1976).
9. S. M. Kupchan and P. F. O'Brien, *Chem. Commun.*, 915 (1973).
10. S. M. Kupchan, C. Kim, and K. Miyano, *Chem. Commun.*, 91 (1976).
11. M. Shamma and D.-Y. Hwang, *Tetrahedron*, **30**, 2279 (1974).
12. M. A. Schwartz, *Synth. Commun.*, **3**, 33 (1973). See also B. Frank and V. Teetz, *Angew. Chem. Int. Ed. Engl.*, **10**, 411 (1971).
13. For early papers on this subject, see S. M. Kupchan and A. J. Liepa, *J. Am. Chem. Soc.*, **95**, 4062 (1973); and S. M. Kupchan, A. J. Liepa, V. Kameswaran, and R. F. Bryan, *J. Am. Chem. Soc.*, **95**, 6861 (1973).
14. S. M. Kupchan, V. Kameswaran, J. T. Lynn, D. K. Williams, and A. J. Liepa, *J. Am. Chem. Soc.*, **97**, 5622 (1975).
15. S. M. Kupchan and C.-K. Kim, *J. Am. Chem. Soc.*, **97**, 5623 (1975).
16. S. M. Kupchan and C.-K. Kim, *J. Org. Chem.*, **41**, 3210 (1976).
17. S. M. Kupchan, O. P. Dhingra, and C.-K. Kim, *J. Org. Chem.*, **41**, 4049 (1976).
18. R. Gottlieb and J. L. Neumeyer, *J. Am. Chem. Soc.*, **98**, 7108 (1976).
19. T. Kametani, Y. Satoh, M. Takemura, Y. Ohta, M. Ihara, and K. Fukumoto, *Heterocycles*, **5**, 175 (1976).
20. O. Hoshino, T. Toshioka, and B. Umezawa, *Chem. Pharm. Bull., Tokyo*, **22**, 1302 (1974).
21. O. Hoshino, H. Hara, M. Ogawa, and B. Umezawa, *Chem. Pharm. Bull., Tokyo*, **23**, 2578 (1975); and *Heterocycles*, **5**, 207 (1976).
22. H. Hara, O. Hoshino, and B. Umezawa, *Chem. Pharm. Bull., Tokyo*, **24**, 262 (1976). If an 8-chlorotetrahydrobenzylisoquinoline is oxidized with lead tetraacetate, low yields of the corresponding aporphine, morphinanedienone, and isopavine are obtained, see H. Hara, O. Hoshino, and B. Umezawa, *Heterocycles*, **5**, 213 (1976).
23. S. Kano, Y. Takahagi, E. Komiyama, T. Yokomatsu, and S. Shibuya, *Heterocycles*, **4**, 1013 (1976). For related cyclizations involving benzyne intermediates, see S. V. Kessar, R. Randhawa, U. K. Nadir, and S. S. Gandhi, *Indian J. Chem.*, **13**, 1113 (1975); S. V. Kessar, S. Batra, U. K. Nadir, and S. S. Gandhi, *Indian J. Chem.*, **13**, 1109 (1975); and Z. Horii, S. Uchida, Y. Nakashita, E. Tsuchida, and C. Iwata, *Chem. Pharm. Bull., Tokyo*, **22**, 583 (1974).
24. R. J. Spangler, D. C. Boop, and J. H. Kim, *J. Org. Chem.*, **39**, 1368 (1974).
25. A. Brossi, A. Ramel, J. O'Brien, and S. Teitel, *Chem. Pharm. Bull., Tokyo*, **21**, 1839 (1973).
25a. T. Kametani, Y. Ohta, M. Takemura, M. Ihara, and K. Fukumoto, *Heterocycles*, **6**, 415 (1977).

26. F. D. Popp, *Heterocycles*, **1**, 165 (1973).
27. F. E. Granchelli and J. L. Neumeyer, *Tetrahedron*, **30**, 3701 (1974).
28. S. M. Kupchan, V. Kameswaran, and J. W. A. Findlay, *J. Org. Chem.*, **38**, 405 (1973).
29. J. G. Cannon, P. R. Khonje, and J. P. Long, *J. Med. Chem.*, **18**, 110 (1975). For the synthesis of ring B homoaporphines by the Pschorr cyclization, see D. Berney and K. Schuh, *Helv. Chim. Acta*, **59**, 2059 (1976).
29a. E. C. Taylor, J. G. Andrade, and A. McKillop, *Chem. Commun.*, 538 (1977).
30. M. P. Cava, D. L. Edie, and J. M. Saá, *J. Org. Chem.*, **40**, 3601 (1975).
31. M. P. Cava, A. Venkateswarlu, M. Srinivasan, and D. L. Edie, *Tetrahedron*, **28**, 4299 (1972).
32. L. Castedo, R. Suau, and A. Mouriño, *Heterocycles*, **3**, 449 (1975).
33. V. Preininger, R. S. Thakur, and F. Šantavý, *J. Pharm. Sci.*, **65**, 294 (1976).
33a. L. Castedo, R. Suau, and A. Mouriño, *An. Quim.*, **73**, 289 (1977).
34. M. Gerecke, R. Borer, and A. Brossi, *Helv. Chim. Acta*, **59**, 2551 (1976); and A. Brossi, private communication.
35. S. Teitel and J. P. O'Brien, *Heterocycles*, **5**, 85 (1976).
36. M. Gerecke, R. Borer, and A. Brossi, *Helv. Chim. Acta*, **58**, 185 (1975).
36a. L. Castedo, R. Riguera, J. M. Saá, and R. Suau, *Heterocycles*, **6**, 677 (1977).
37. R. J. Borgman, R. V. Smith, and J. E. Keiser, *Synthesis*, 249 (1975). A crown ether of apomorphine has also been prepared, see F. Vögtle and B. Jansen, *Tetrahedron Lett.*, 4895 (1976).
38. P. Wiriyachitra and M. P. Cava, *J. Org. Chem.*, **42**, 2274 (1977).
38a. R. Ahmad, J. M. Saá, and M. P. Cava, *J. Org. Chem.*, **42**, 1228 (1977).
38aa. J. C. Kim, *Org. Prep. Proced. Int.*, **9**, 1 (1977).
38b. M. P. Cava and A. Venkateswarlu, *Tetrahedron*, **32**, 2079 (1976).
38c. J. M. Saá and M. P. Cava, *J. Org. Chem.*, **42**, 347 (1977).
39. M. Tomita and H. Furukawa, *J. Pharm. Soc. Japan*, **82**, 1199 (1962).
39a. D. H. R. Barton and T. Cohen, *Festschrift Prof. Dr. Arthur Stoll Siebzigsten Geburstag 1957*, p. 117; and A. R. Battersby, R. T. Brown, J. H. Clements, and G. G. Iverach, *Chem. Commun.*, 230 (1965).
40. E. Brochmann-Hanssen, C.-C. Fu, and L. Y. Misconi, *J. Pharm. Sci.*, **60**, 1880 (1971). See also E. Brochmann–Hanssen, C. H. Chen, H.-C. Chiang, C.-C. Fu, and H. Nemoto, *J. Pharm. Sci.*, **62**, 1291 (1973).
41. G. Blaschke, *Arch. Pharm. (Weinheim)*, **301**, 432 (1968); and **303**, 358 (1970).
42. G. Blaschke, G. Waldheim, M. Schantz, and P. Peura, *Arch. Pharm. (Weinheim)*, **307**, 122 (1974).
43. S. Tewari, D. S. Bhakuni, and R. S. Kapil, *Chem. Commun.*, 940 (1974); and *J. Chem. Soc. Perkin I*, 706 (1977).
44. A. R. Battersby, J. L. McHugh, J. Staunton, and M. Todd, *Chem. Commun.*, 985 (1971).
45. M. Shamma and J. L. Moniot, *Heterocycles*, **3**, 297 (1975).
46. J. P. Rosazza, A. W. Stocklinski, M. A. Gustafson, J. Adrian, and R. V. Smith, *J. Med. Chem.*, **18**, 791 (1975); R. V. Smith and J. P. Rosazza, *J. Pharm. Sci.*, **64**, 1737 (1975); and P. J. Davis, D. Wiese, and J. P. Rosazza, *J. Chem. Soc. Perkin I*, 1 (1977).
47. A. Tagliamonte, R. Gessa, and G. L. Gessa, *Riv. Farmacol. Terapia*, **4**, 3 (1973).
48. For the synthesis and pharmacological testing of several aporphines, see F. Schneider, M. Gerold, and K. Bernauer, *Helv. Chim. Acta*, **56**, 759 (1973).
49. W. S. Saari, S. W. King, V. J. Lotti, and A. Scriabine, *J. Med. Chem.*, **17**, 1086 (1974).
50. J. L. Neumeyer, F. E. Granchelli, K. Fuxe, U. Ugerstedt, and H. Corrodi, *J. Med. Chem.*, **17**, 1090 (1974).

51. S. Lal, *Symposium on Central Dopamine Receptors*, Abstracts Los Angeles ACS Meeting, 1974. For a review on the pharmacology and biochemistry of apomorphine, see F. C. Colpaert, W. F. M. Van Bever, and J. E. M. F. Leysen, *Int. Rev. Neurobiol.*, **19**, 225 (1976).
52. R. V. Smith and M. R. Cook, *J. Pharm. Sci.*, **63**, 161 (1974).
53. J. G. Cannon, R. V. Smith, A. Modiri, S. P. Sood, R. J. Borgman, M. A. Aleem, and J. P. Long, *J. Med. Chem.*, **15**, 273 (1972).
54. J. L. Neumeyer, J. F. Reinhardt, W. P. Dafeldecker, J. Guarino, D. L. Kosersky, K. Fuxe, and L. Agnati, *J. Med. Chem.*, **19**, 25 (1976).
54a. J. L. Neumeyer, W. P. Dafeldecker, B. Costall, and R. J. Naylor, *J. Med. Chem.*, **20**, 190 (1977).
55. E. R. Atkinson, F. J. Bullock, F. E. Granchelli, S. Archer, F. J. Rosenberg, D. G. Teiger, and F. C. Nachod, *J. Med. Chem.*, **18**, 1000 (1975).
56. J. Z. Ginos, G. C. Cotzias, E. Tolosa, L. C. Tang, and A. LoMonte, *J. Med. Chem.*, **18**, 1194 (1975).
56a. H. Sheppard, C. R. Burghardt, and S. Teitel, *Mol. Pharmacol.*, **12**, 854 (1976).
57. R. C. Chen, J. L. Beal, R. W. Doskotch, L. A. Mitscher, and G. H. Svoboda, *Lloydia*, **37**, 493 (1974).
57a. A. Quevauviller and M. Hamonnière, *C. R. Acad. Sci. Ser. D*, 284 (1977).
58. E. Wenkert, B. L. Buckwalter, I. R. Burfitt, M. J. Gasic, H. E. Gottlieb, E. W. Hagaman, F. M. Schell, and P. M. Wovkulich, in *Topics in C-13 NMR Spectroscopy*, G. C. Levey, ed., Vol. 2, Wiley-Interscience, New York (1976), p. 105.
59. L. M. Jackman, J. C. Trewella, J. L. Moniot, and M. Shamma, in preparation. Also M. Shamma in *Specialist Periodical Reports, The Alkaloids*, Vol. 7, M. F. Grundon, Sr. Reporter, The Chemical Society, London (1977), p. 163.
60. M. Shamma and J. L. Moniot, *Experientia*, **32**, 282 (1976).
61. L. Cleaver, S. Nimgirawath, E. Ritchie, and W. C. Taylor, *Aust. J. Chem.*, **29**, 2003 (1976).

PAKISTANAMINE: A PROAPORPHINE-BENZYLISOQUINOLINE DIMER

Structure:

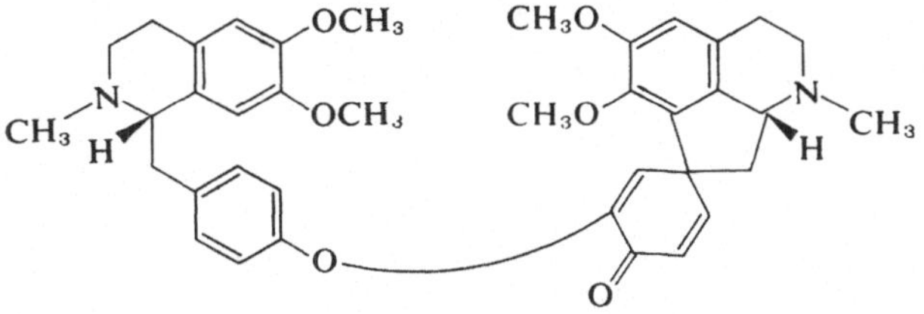

(+)-Pakistanamine
Berberis baluchistanica Ahrendt

11.1. Synthesis

Two approaches to the synthesis of racemic pakistanamine, the only known proaporphine–benzylisoquinoline dimer,[1] have been published, the full details of which are not available at time of publication.

Kametani's group subjected a mixture of berbamunines to ferricyanide oxidation to furnish the phenolic proaporphine–benzylisoquinolines 1. Diazomethane methylation then afforded a compound similar to pakistanamine.[2,3]

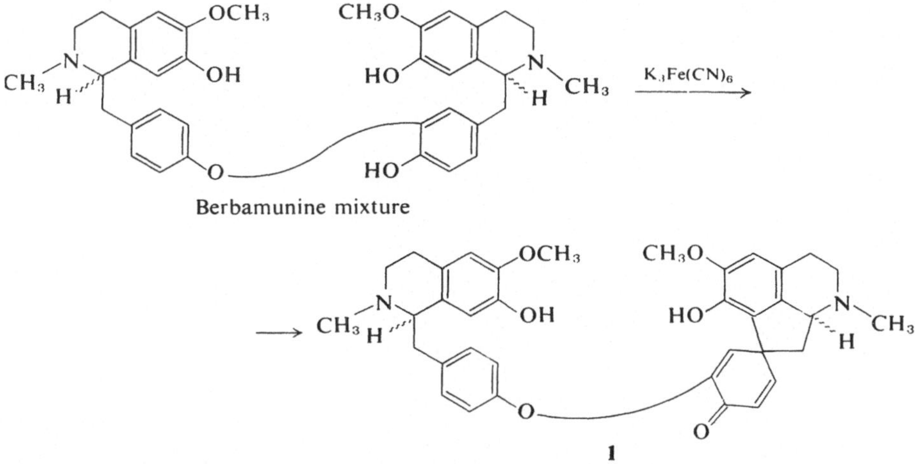

Berbamunine mixture

1

In the second approach, 8-bromodauricine (**2**), prepared by a six-step sequence from ($\pm$)-*N,O*-dimethylcoclaurine, was photolyzed in refluxing liquid ammonia containing potassium *t*-butoxide to produce directly a mixture of pakistanamines.[4]

2

References and Notes

1. M. Shamma, J. L. Moniot, S. Y. Yao, G. A. Miana, and M. Ikram, *J. Am. Chem. Soc.*, **95**, 5742 (1973).
2. T. Kametani, H. Teresawa, and F. Satoh, *Heterocycles*, **2**, 159 (1974).
3. It has been found, however, that enone systems such as compound **1** react with diazomethane by cycloaddition, thus complicating this approach (M. Shamma, J. L. Moniot, and R. W. Lagally, unpublished results).
4. S. Kokrady and R. E. Harmon, Abstract of 168th ACS Natl. Meeting, Atlantic City, N.J., Sept. 9–13 (1974).

THE APORPHINE–BENZYLISOQUINOLINE DIMERS

Some new aporphine–benzylisoquinoline dimers are shown below:

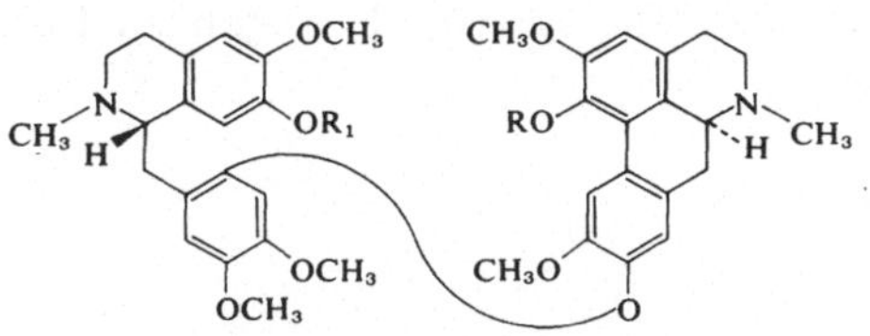

(+)-Thalictropine,[1] R = H, R$_1$ = CH$_3$
(+)-Thalictrogamine,[1] R = R$_1$ = H

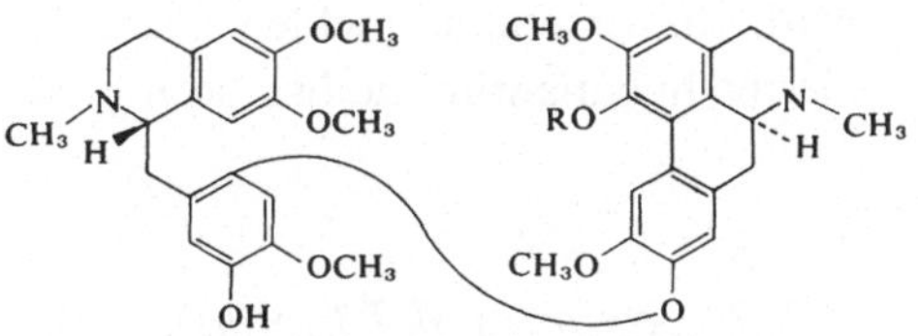

(+)-Pennsylvanine,[2,3] R = CH$_3$
(+)-Pennsylvanamine,[2] R = H

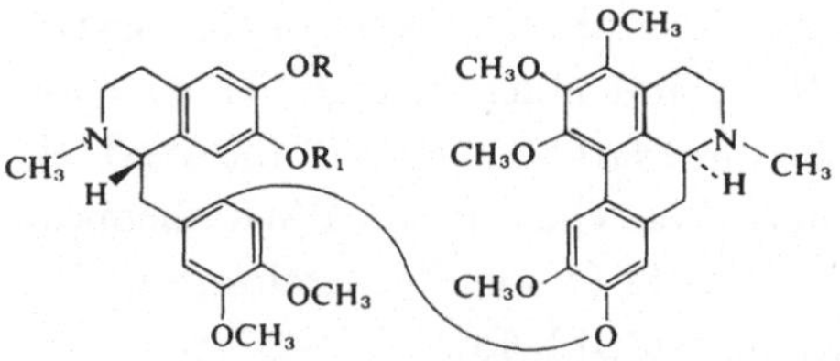

(+)-Desmethyladiantifoline,[5,5a]
R = H, R$_1$ = CH$_3$
(+)-Thaliadanine,[5a] R = CH$_3$, R$_1$ = H

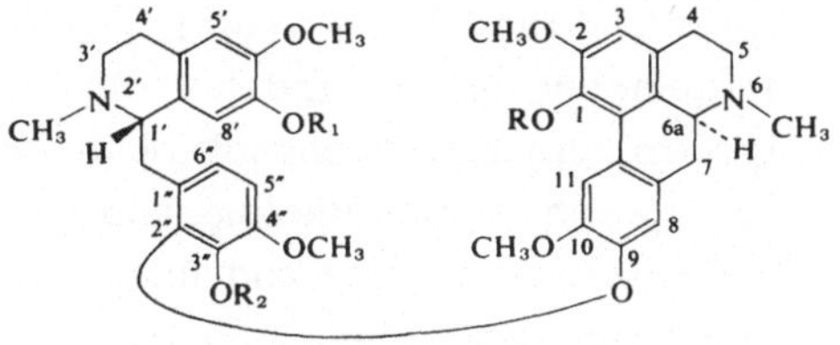

(+)-Fetidine,[4,6] R = H, R$_1$ = R$_2$ = CH$_3$
(+)-Revolutopine,[6a],
R = CH$_3$, R$_1$ = R$_2$ = H

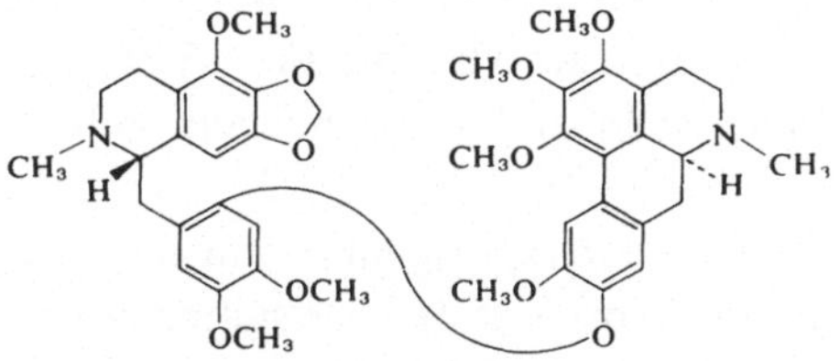

(+)-Thalmelatidine[5,7]

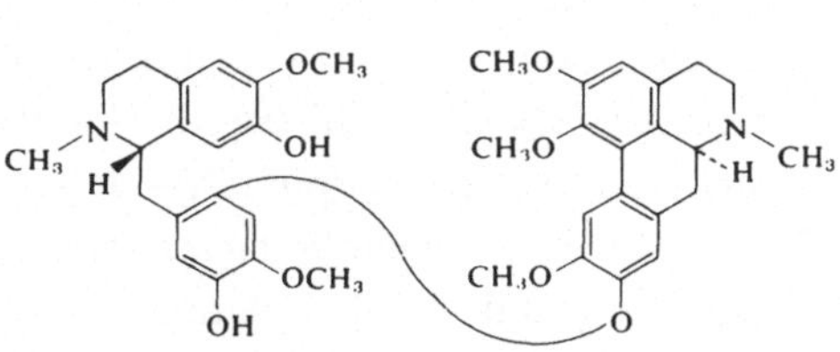

(+)-Thalipine[6a,8]

SCHEME 12.1

12.1. Introduction

The ten aporphine–benzylisoquinolines shown in Scheme 12.1 have all originated in the genus *Thalictrum* (Ranunculaceae). Of these, five dimers, namely, thalictropine, thalictrogamine, thalipine, pennsylvanine, and pennsylvanamine were chemically interrelated with thalicarpine, the first aporphine–benzylisoquinoline to be fully characterized.[9,10] The structure of thalmelatidine was elucidated by degradation,[7] whereas that originally proposed for fetidine has been corrected.[6] Revolutopine is the first known analog of fetidine, and its *O*-methylation with diazomethane yielded *O*-methylfetidine.[6a] The structural assignment for desmethyladiantifoline has been modified.[5a]

12.2. Structural Elucidation

Structural assignments for thalicarpine and its phenolic analogs are readily derived from a combination of the UV, mass spectral, and PMR data of the alkaloids and their acetate derivatives.[2] The major fragments in the mass spectra of these bases result from cleavage of the benzylic bond of the benzylisoquinoline moiety and homolysis of the diaryl ether linkage. Thus, mass spectroscopy detects phenols on any of the three large moieties of the alkaloid, i.e., aporphine, isoquinoline, and the benzyl ring. The relevant PMR chemical shifts are summarized and interpreted in Sec. 12.6; some generalizations for specifically locating phenolic functions are presented below.

1. The Aporphine Moiety:

a. A C-1 phenol gives rise to a strong bathochromic shift in the UV spectrum upon addition of base; the C-1 methoxyl signal near $\delta 3.71$ is absent in the PMR spectrum. Upon acetylation, the aporphine H-11 PMR signal shifts upfield to $\sim \delta 7.60$ and the acetate methyl signal appears at $\delta 2.34$.

b. The PMR spectrum of a C-2 phenolic base exhibits only one methoxyl signal below $\delta 3.85$ (assignable to C-10). Upon acetylation, the $\delta 3.71$ C-1 methoxyl signal is shifted to slightly higher field and no significant shift is observed for the aporphine H-11 signal.

c. A C-10 phenolic base exhibits only one methoxyl signal downfield from $\delta 3.85$, assignable to C-2. Upon acetylation, the aporphine H-11 is shifted downfield to $\sim \delta 8.30$ and the acetate methyl signal is present at $\delta 2.20$.

2. The Isoquinoline Moiety:

a. A C-6′ phenol is evidenced by the presence of a high-field C-7′ methoxyl signal at $\delta 3.58$ and the H-8′ signal at $\sim \delta 6.23$ in the PMR spectrum.

b. A C-7' phenolic base lacks the high-field δ3.58 methoxyl signal, and the H-8' signal is located downfield near δ6.40 rather than at δ6.2. Upon acetylation, the H-8' signal is not shifted while the acetate methyl signal appears at δ2.19.

3. The Benzyl Ring:

a. The C-5' phenol gives rise to a bathochromic shift in the UV spectrum on addition of base. In the PMR spectrum, the H-8' signal appears near δ6.23 and one aromatic proton singlet is at ~δ6.80. Upon acetylation, the lowest-field aromatic proton is at ~δ6.90; no upfield shifts of aromatic protons are observed and one methoxyl signal is shifted upfield to ~δ3.70.

Chemical interconversion to thalicarpine and application of these generalizations allow structural assignments for the phenolic analogs of thalicarpine.

A recent detailed 220-MHz PMR study[6] has led to the correction of the structure originally proposed for fetidine.[4] The most telling feature in that study was the presence of an AB quartet centered at δ6.75 (1H) and 6.81 (1H), J = 8.5 Hz. This pattern is indicative of the presence in fetidine of an ortho pair of aromatic protons. The revised structure of fetidine is shown in Scheme 12.1. Thus fetidine is the first aporphine–benzylisoquinoline alkaloid known to possess the 2'',3'',4''-oxygenation pattern on the benzyl ring.[6,10]

12.3. Synthesis

The two major hurdles in the synthesis of aporphine–benzylisoquinolines are the formations of the biphenyl system of the aporphine (see Sec. 10.2) and the diaryl ether linkage. In the past, two general routes for the synthesis of these dimeric alkaloids have been used successfully. In the synthesis of thalicarpine[11] the diaryl ether was formed prior to the heterocyclic systems; whereas for the synthesis of adiantifoline,[12] an Ullmann condensation was employed to join the resolved halves of the dimer. Condensations of this type generally suffer from poor yields; but a recent paper reports a significantly improved Ullmann ether synthesis.[13] It has been shown that the use of commercially available pentafluorophenyl copper (PFPC) in dry pyridine as the coupling reagent for Ullmann condensation increases yields greatly. As an example, an excess of cassythicine[14] was reacted with S-6'-bromolaudanosine and PFPC in dry pyridine to give the crystalline thalicarpine analog **1** in 51% yield. Related condensations to form bisbenzylisoquinoline and aporphine–benzylisoquinoline dimers proceeded in 42–54% yields, based on the bromo component.

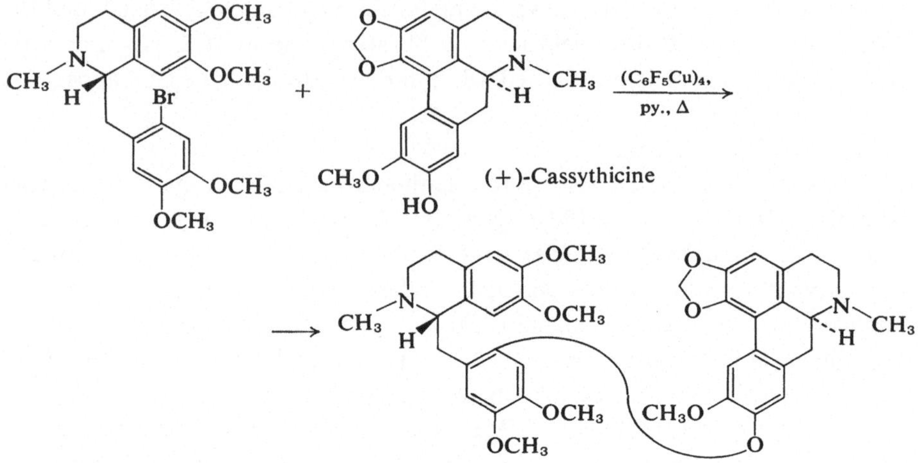

(+)-Cassythicine

1

(For another example of the use of PFPC, see Sec. 5.3.1.)

12.4. Reactions

Thalicarpine can be selectively *O*-demethylated using sodium benzyl-selenolate in refluxing DMF to give the diphenol **2**.[14a]

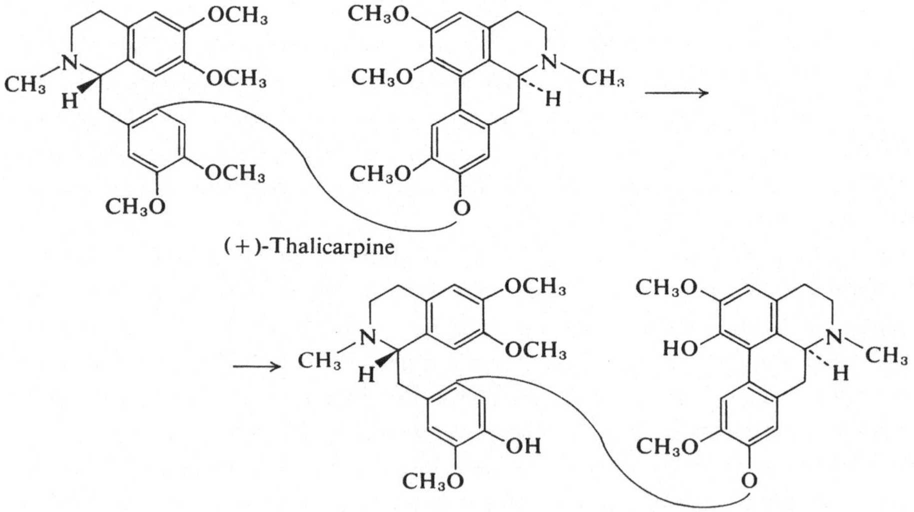

(+)-Thalicarpine

2

The microorganism *Streptomyces punipalus* can convert (+)-thalicarpine into (+)-hernandalinol.[14b]

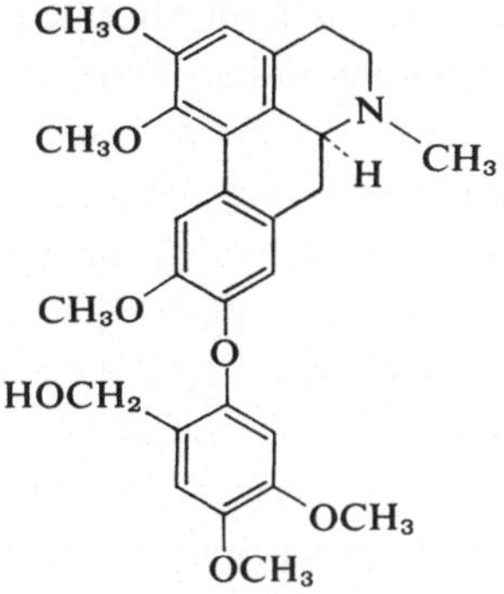

(+)-Hernandalinol

12.5. Biogenesis

A sufficiently large number of aporphine–benzylisoquinoline dimers have now been isolated to draw generalizations concerning their biogenesis. Thalicarpine and its phenolic analogs must be formed via phenolic oxidative coupling of a molecule of (+)-reticuline with a 1,2,9,10-tetraoxygenated aporphine, the latter formed by intramolecular oxidative coupling of a (+)-reticuline unit.[15] This is in contrast to pakistanine which is probably formed via the biscoclaurine (+)-berbamunine. The latter can undergo intramolecular oxidative coupling to a proaporphine–benzylisoquinoline, followed by rearrangement and selective methylation to generate pakistanine.[16]

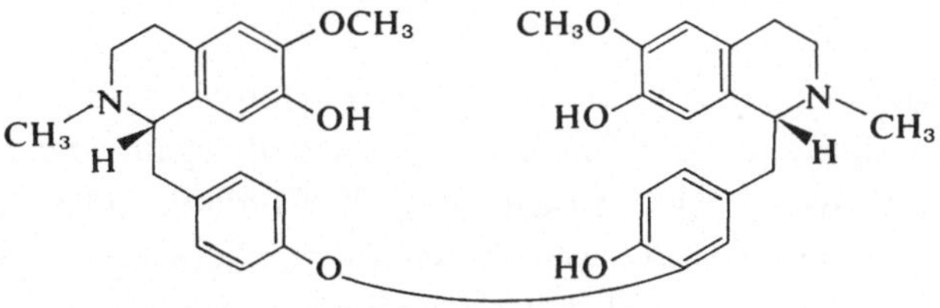

(+)-Berbamunine

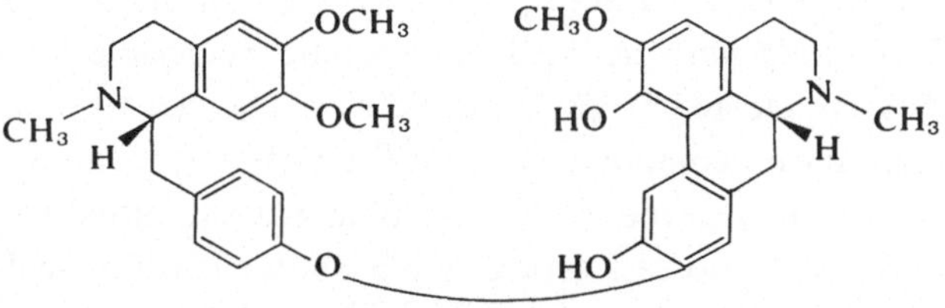

(+)-Pakistanine

It is worth pointing out that if thalicarpine or its analogs were to be formed through a bisbenzylisoquinoline, such an intermediate would have to originate through coupling of two reticuline units. None of the more than one hundred bisbenzylisoquinolines presently known are derived from condensation of a pair of reticuline units. This same biogenetic argument can be applied to the aporphine–benzylisoquinoline dimers which incorporate one reticuline and one 5-hydroxylaudanosoline unit[17] (adiantifoline), or two 5-hydroxylaudanosoline units (thalmineline or thalmelatidine), i.e., the aporphine subunit is preformed and later condenses with the benzylisoquinoline unit. It follows that the thalicarpine and pakistanine dimers are formed through radically different pathways, with only the pakistanine series proceeding via a dimeric proaporphine-benzylisoquinoline intermediate.

(+)-Reticuline   (+)-5-Hydroxylaudanosoline

12.6. Pharmacology

The alkaloid (+)-thalicarpine has shown cytotoxic activity against human KB cells in monolayer culture, and antitumor activity against the rat Walker 256 carcinosarcoma over a wide dosage range.[18] Cardiovascular effects such as hypotension and bradycardia were noted in monkeys treated with doses of 2.5 mg/kg and it was proposed that such effects were due to direct myocardial depression in combination with nonspecific vasodilation.[19] Thalicarpine was found to inhibit DNA, RNA, and protein synthesis in mouse L1210 leukemia cells,[20] and in a study with tritiated thalicarpine, reversible binding with calf thymus DNA was detected.[21] Thalicarpine was reported to bind to some unidentified human serum component *in vivo*,[21] and to inhibit aniline hydroxylase activity of rat liver microsomes.[22] Thalicarpine was submitted to clinical trials, but was soon found to possess serious CNS and/or cardiovascular toxicity at doses below those required for antitumor activity.[23,24,25]

Table 12.1. PMR Chemical Shifts of Some Aporphine-Benzylisoquinoline Alkaloids (δ Values)

Proton	Thalic-tropine	Thalic-trogamine	Pennsyl-vanine	Pennsyl-vanamine	Thalipine	Thalmela-tidine[a]
N—CH₃	2.47	2.49	2.46	2.47	2.48	2.45
	2.48	2.52	2.50	2.53	2.54	2.50
C-7′ OCH₃	3.58	—	3.58	3.60	—	—
C-1 OCH₃	—	—	3.71	—	3.70	3.62
C-10 OCH₃	3.97	3.95	3.92	3.94	3.96	3.96
Other methoxyl groups						
⎧3.78	3.78	3.79	3.79	3.80	3.76	3.78
C-2, C-6′, ⎪3.78	3.78	3.83	—	—	—	3.80(2)
C-3″, C-4″ ⎨3.82	3.82	3.83	3.82	3.83	3.76	3.90
⎩3.88	3.88	3.92	3.90	3.92	3.90	3.96
C-8′ H	6.20	6.40	6.22	6.25	6.35	5.86
C-11 H	8.18	8.18	8.18	8.20	8.09	8.04
Other aromatic protons						
⎧6.55	6.55	6.51	6.52	6.55	6.45	6.41
C-3, C-8, ⎪6.55	6.55	6.57	6.56	6.55	6.51	6.41
C-5, C-6″ ⎨6.55	6.55	6.57	6.59	6.58	6.55	6.50
⎪6.59	6.59	6.57	6.62	6.58	6.71	—
⎩6.67	6.67	6.78	6.76	6.80	6.76	—

[a] The methylenedioxy group signal of thalmelatidine appears at δ5.94.

12.7. PMR Spectroscopy

The relevant PMR chemical shifts of some of the more recently isolated aporphine–benzylisoquinoline alkaloids are given in Table 12.1.

References and Notes

1. M. Shamma and J. L. Moniot, Tetrahedron Lett., 775 (1973).
2. M. Shamma and J. L. Moniot, Tetrahedron Lett., 2291 (1974).
3. M. Shamma and A. S. Rothenberg, Lloydia, in press.
4. Z. F. Ismailov and S. Yu. Yunusov, Khim. Prir. Soedin., Akad. Nauk Uz. SSR, 2, 43 (1966); through Chem. Abstr., 65, 2320 (1966).
5. N. M. Mollov, P. P. Panov, Le Nhat Thuan, and L. Panova, Dokl. Bolg. Akad. Nauk, 23, 1243 (1970); through Chem. Abstr., 74, 61584t (1971).
5a. W.-t. Liao, J. L. Beal, W.-N. Wu, and R. W. Doskotch, Lloydia, in press.
6. M. P. Cava and K. Wakisaka, Tetrahedron Lett., 2309 (1972).
6a. J. Wu, J. L. Beal, W.-N. Wu, and R. W. Doskotch, Heterocycles, 6, 405 (1977).
7. N. M. Mollov and Le Nhat Thuan, Dokl. Bolg. Akad. Nauk, 24, 601 (1971); through Chem. Abstr., 75, 106055k (1971).
8. M. Shamma, J. L. Moniot, and P. Chinnasamy, Heterocycles, 6, 399 (1977).

9. An attempt to reisolate thalidoxine, which is structurally isomeric with pennsylvanine since it incorporates a 3″-methoxy-4″-hydroxy moiety, from *T. dioicum* yielded instead, a large quantity of pennsylvanine, so that thalidoxine may not be a commonly occurring base; M. Shamma, A. S. Rothenberg, and J. L. Moniot, unpublished results.

10. For the structure of (+)-thalicarpine, see M. Tomita, H. Furukawa, S.-T. Lu, and S. M. Kupchan, *Chem. Pharm. Bull.*, *Tokyo*, **15**, 959 (1967).

11. S. M. Kupchan, A. J. Liepa, V. Kameswaran, and K. Sempuku, *J. Am. Chem. Soc.*, **95**, 2995 (1975).

12. R. W. Doskotch, J. D. Phillipson, A. B. Ray, and J. L. Beal, *J. Org. Chem.*, **36**, 2409 (1971).

13. M. P. Cava and A. Afzali, *J. Org. Chem.*, **40**, 1553 (1975).

14. S. R. Johns and J. A. Lamberton, *Aust. J. Chem.*, **19**, 297 (1966).

14a. R. Ahmad, J. M. Saá, and M. P. Cava, *J. Org. Chem.*, **42**, 1228 (1977).

14b. T. Nabih, P. J. Davis, J. F. Caputo, and J. P. Rosazza, *J. Med. Chem.*, **20**, 914 (1977).

15. M. Shamma and J. L. Moniot, unpublished observations.

16. M. Shamma, J. L. Moniot, S. Y. Yao, G. A. Miana, and M. Ikram, *J. Am. Chem. Soc.*, **95**, 5742 (1973).

17. M. Shamma and J. L. Moniot, *Heterocycles*, **4**, 1817 (1976).

18. S. M. Kupchan, *Trans. N. Y. Acad. Sci.*, **32**, 85 (1970).

19. E. H. Herman and D. P. Chadwick, *Toxicol. Appl. Pharmacol.*, **26**, 137 (1973).

20. L. M. Allen and P. J. Creaven, *Cancer Res.*, **33**, 3112 (1973). See also W. A. Creasey, *Biochem. Pharmacol.*, **25**, 1887 (1976).

21. L. M. Allen and P. J. Creaven, *J. Pharm. Sci.*, **63**, 474 (1974).

22. P. J. Creaven, L. M. Allen, and C. P. Williams, *Xenobiotica*, **4**, 255 (1974).

23. P. J. Creaven, M. H. Cohen, O. S. Selawry, F. Tejada, and L. E. Broder, Abstracts of Proceedings of the 9th International Congress of Chemotherapy, London (1975); and *Cancer Chemother. Rep.*, **59**, 1001 (1975).

24. P. J. Creaven and L. M. Allen, *Cancer Treat. Rep.*, **60**, 69 (1976).

25. For a comprehensive discussion of antitumor agents in plants, including (+)-thalicarpine, see G. A. Cordell and N. R. Farnsworth, *Lloydia*, **40**, 1 (1977).

THE APORPHINE–PAVINE DIMERS

Occurrence: Ranunculaceae
Structures:

(--)-Pennsylpavine,[1] R = CH$_3$
(−)-Pennsylpavoline,[1] R = H
Thalictrum polygamum Muhl.

13.1. Introduction

Among the isoquinoline alkaloids, the aporphine–benzylisoquinolines had been considered a terminal stage of anabolic development in the biogenetic tree. The discovery of pennsylpavine and pennsylpavoline, two examples of a new group of dimeric alkaloids, the aporphine–pavines, has extended the biogenetic locus, since they are probably derived from aporphine–benzylisoquinoline dimers.

13.2. Structural Elucidation

The structural assignments of these novel dimers rest on detailed physical data including UV, mass, and PMR spectra, as well as on circular dichroism. The UV spectrum of pennsylpavine (Sec. 13.3) is essentially that of a pavine

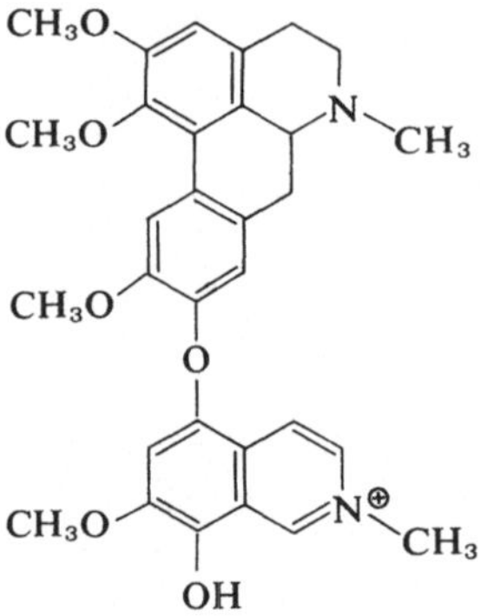

1, m/e 529 (27%)
$C_{31}H_{33}N_2O_6$

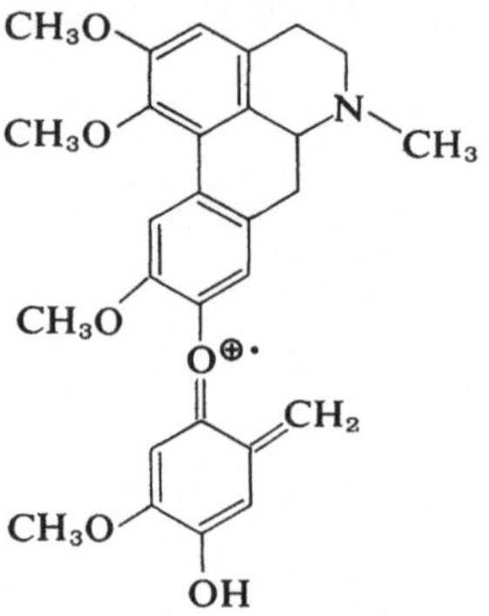

2, m/e 475 (5%)
$C_{28}H_{29}NO_6$

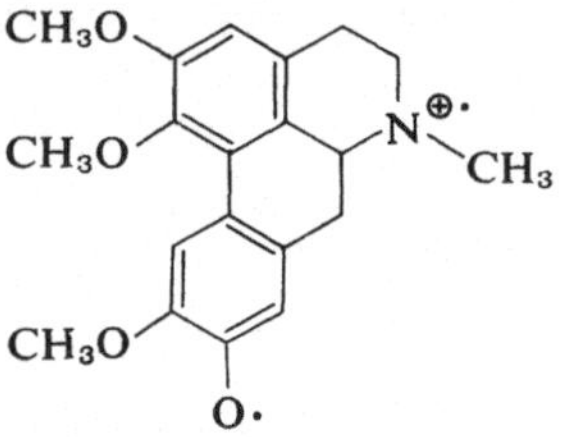

3, m/e 340 (20%)
$C_{20}H_{22}NO_4$

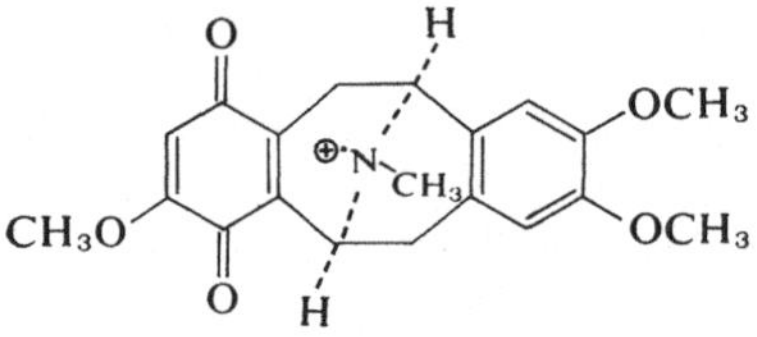

4, m/e 355 (28%)
$C_{20}H_{21}NO_5$

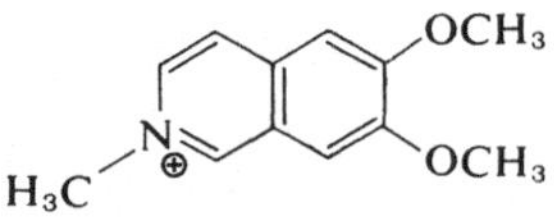

5, $m\ e$ 204 (100%)
$C_{12}H_{14}NO_2$

SCHEME 13.1

or isopavine (288 nm) superimposed on that of a 1,2,9,10-substituted aporphine (280, 308, and 320 nm). The mass spectrum contains a parent peak at
m/e 680 ($C_{40}H_{44}N_2O_8$) and is indicative of a trimethoxylated aporphine bonded
through a diaryl ether bridge to a trimethoxylated monophenolic pavine. The
fragments **3** and **4** correspond to cleavages on either side of the diaryl ether
bridge. The pavine moiety can undergo fission by two analogous pathways to
yield either fragment **2** plus the base peak **5**, or fragment **1** which includes both
nitrogen atoms (see Scheme 13.1).[1]

The most distinctive feature of the PMR spectrum of pennsylpavine (see
Table 13.1) is the presence of a pair of low-field one-proton doublets for the
C-6′ and C-12′ bridgehead protons of the pavine unit. As in the case of phenolic
aporphine–benzylisoquinolines, comparison of the PMR spectrum of the
natural product with that of the O-acetate derivative yielded critical structural
information. The magnitude of the downfield shift (0.2 ppm) of the C-9′ aromatic proton resonance upon O-acetylation is indicative of a meta relation
with the acetate group. The accompanying upfield shift of only one of the bridgehead proton doublets ($\delta 4.50 \rightarrow 4.35$) denotes a close spatial relationship between
the phenolic hydroxyl and the C-6′ bridgehead proton in the natural product.
This observation is also valid for phenolic monomeric pavines, where only the

Table 13.1. PMR Spectra of Alkaloids and Their Acetates at 60 MHz (δ Values)

Alkaloid	O-Methyl groups							Aromatic hydrogens						Methine hydrogens[a]	
	N-Me	N'-Me	C-1	C-10	Other			C-9'	Other				C-11	C-6'	C-12'
Pennsylpavine	2.50	2.57	3.71	3.91	3.76	3.76	3.88	6.23	6.48	6.48	6.52	6.60	8.15	4.50	4.06
Pennsylpavine acetate	2.49	2.55	3.70	3.90	3.68	3.77	3.84	6.43	6.48	6.46	6.51	6.60	8.15	4.35	4.15
Pennsylpavoline	2.48	2.55		3.91	3.75	3.78	3.91	6.26	6.45	6.49	6.55	6.55	8.14	4.43	4.01
Pennsylpavoline diacetate	2.50	2.55		3.86	3.67	3.78	3.79	6.42	4.46	6.52	6.56	6.66	7.60	4.21	4.11
Platycerine		2.53			3.75	3.76	3.83	6.66[b]	6.45	6.46[b]	6.61			4.40	4.00
Platycerine acetate		2.53			3.76	3.76	3.83	6.81	6.46	6.81	6.61			4.23	4.11

[a] dd J = 6 Hz. [b] dd J = 9 Hz.

unsymmetrically substituted phenolic pavines platycerine and munitagine exhibit an upfield shift of one of the bridgehead proton doublets upon acetylation.

(+)-*N*-Methyllaurotetanine   (−)-Munitagine, R = H
                                                (−)-Platycerine, R = CH$_3$

The diphenolic alkaloid, pennsylpavoline, $C_{39}H_{42}N_2O_8$, on the basis of similar spectral analyses, was shown to correspond to C-1-demethylpennsylpavine. The absolute configuration of pennsylpavine was determined by a CD study. The CD curve of (−)-pennsylpavine is nearly superimposable on that of an equimolar mixture of (+)-*N*-methyllaurotetanine and (−)-platycerine.[1]

Comparison of the structures of pennsylvanine and pennsylvanamine, two dimers of the (+)-thalicarpine series which co-occur with pennsylpavine and pennsylpavoline,[2] furnishes *prima facia* evidence that aporphine–benzylisoquinolines could act as biogenetic precursors to the aporphine–pavine dimers.

(+)-Pennsylvanine, R = CH$_3$
(+)-Pennsylvanamine, R = H

13.3. *UV and IR Spectroscopy*

Pennsylpavine λ_{max}^{EtOH} 230, 280 sh, 288, 308 sh, and 320 sh nm
                                                              (4.62, 4.38, 4.40, 4.23, and 4.15).

Pennsylpavine acetate $\nu_{max}^{CHCl_3}$ 1760 cm^{-1} (5.68 μ).

Pennsylpavoline λ_{max}^{EtOH} 230, 280 sh, 288, 306 sh, and 320 sh nm (4.47, 4.06, 4.13, 4.01, and 3.96).

Pennsylpavoline diacetate $\nu_{max}^{CHCl_3}$ 1770 and 1765 cm^{-1} (5.75 and 5.67 μ).

References

1. M. Shamma and J. L. Moniot, *J. Am. Chem. Soc.*, **96**, 3338 (1974).
2. M. Shamma and J. L. Moniot, *Tetrahedron Lett.*, 2291 (1974).

THE OXOAPORPHINES

14.1. Introduction[1,2]

A new botanical source of oxoaporphines is the family Eupomatiaceae.[3] Some recently identified oxoaporphines are shown below.

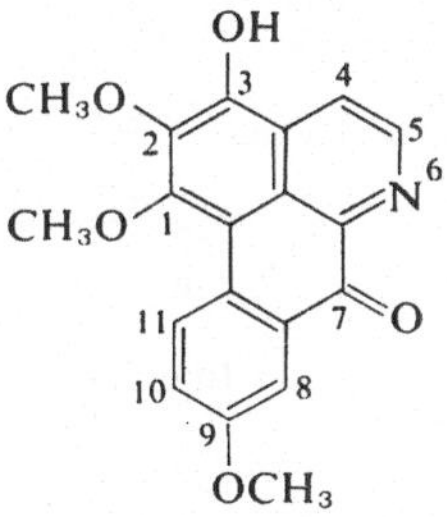

Subsessiline[4]	Corunnine, R = CH₃	Oxostephanine[6]
Guatteria subsessilis	Nandazurine,[5] R + R = CH₂	*Stephania japonica* Miers.
	Nandina domestica Thunb.	(Menispermaceae)

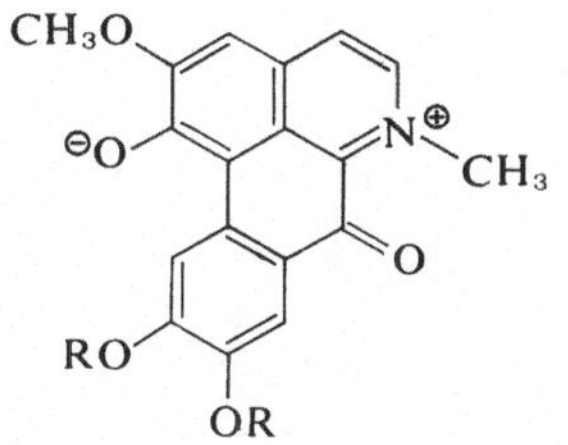

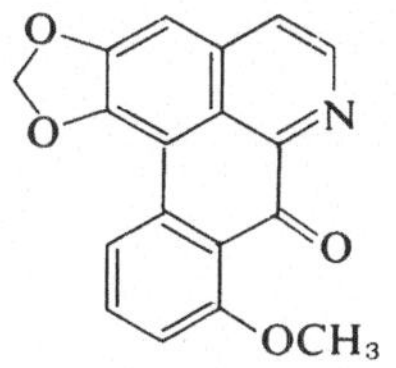

Pontevedrine, which was originally believed to be a C-7 oxoaporphine derivative, has now been shown to possess a C-4,5-dioxoaporphine structure[7] (see Sec. 16.2). The green alkaloid glauvine, found in *Glaucium* spp. (Papaveraceae),[8] is identical with corunnine.[9]

14.2. Synthesis

For a description of methods of oxidation of aporphines and noraporphines to oxoaporphines, see Sec. 10.3.1. Total syntheses of oxoaporphines usually involve the elaboration of a C-2' or a C-6' appropriately functionalized aromatic benzylisoquinoline, which is then cyclized by a Pschorr or photochemical sequence.

Imenine, an oxoaporphine oxygenated at C-4, has been synthesized by a Reissert–Pschorr sequence shown in Scheme 14.1.[10]

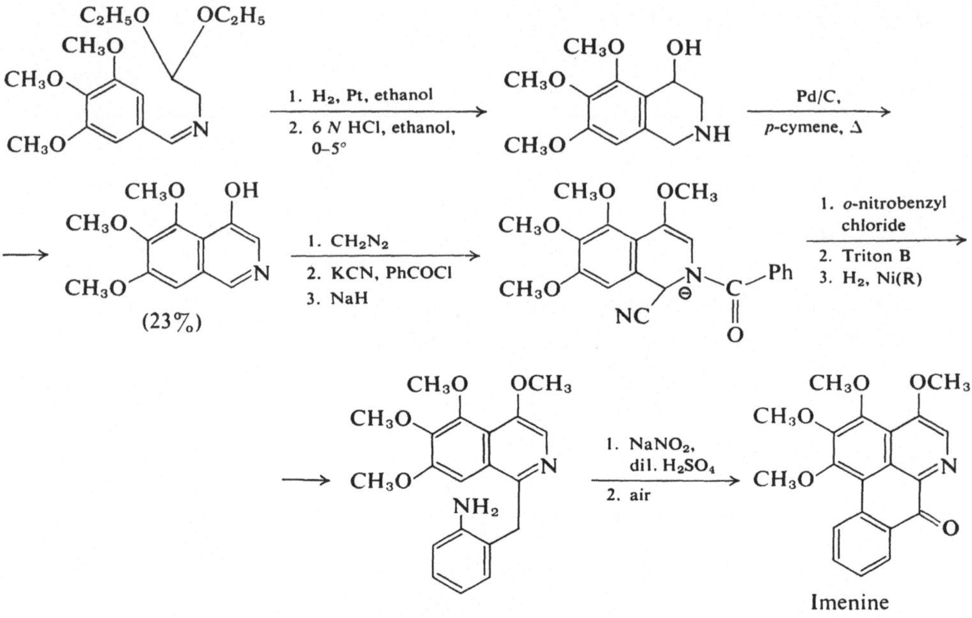

SCHEME 14.1

Corunnine and nandazurine have been prepared by oxidative photocylization of appropriately substituted phenolic benzylisoquinolines.[11]

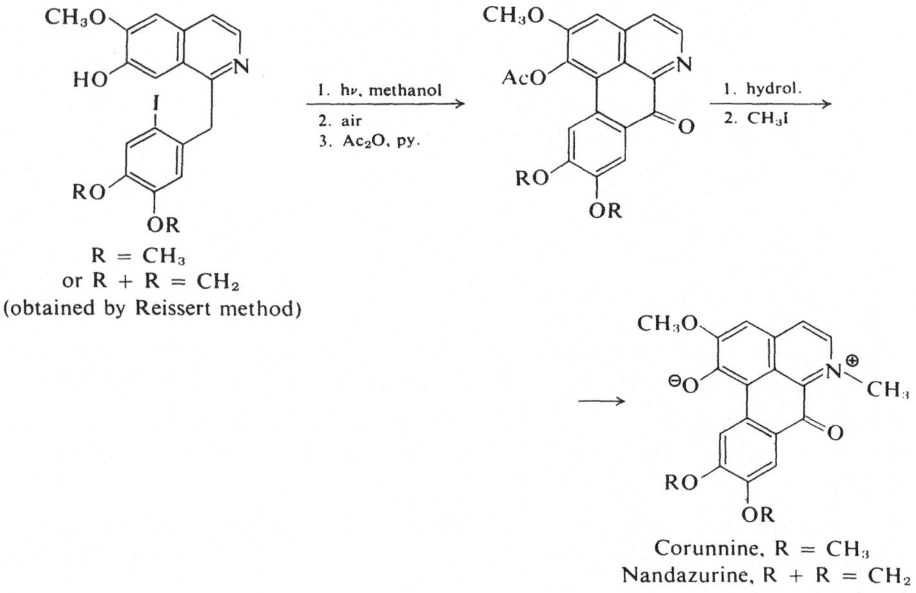

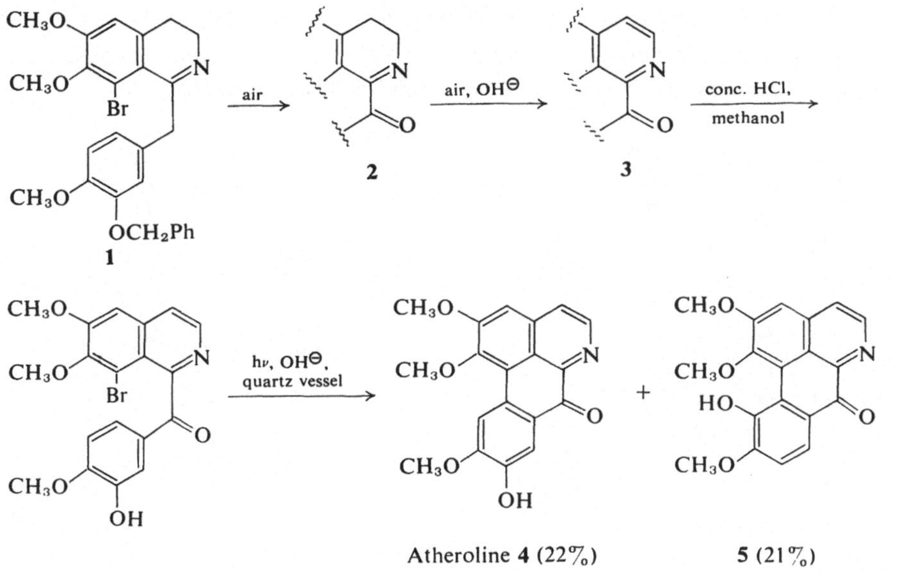

SCHEME 14.2

A synthesis of atheroline has been achieved starting from the Bischler–Napieralski product **1** which underwent air oxidation to the imino ketone **2**. In the presence of alkali and air, **2** was further oxidized to the 1-benzoylisoquinoline **3**. Debenzylation followed by photolysis in the presence of sodium hydroxide in a quartz vessel furnished atheroline (**4**), and its structural isomer **5** (see Scheme 14.2).[11a]

When the photolysis was carried out through a Vicor filter, the yield of **5** decreased substantially.[11a]

14.3. Pharmacology

Liriodenine (1,2-methylenedioxyoxoaporphine) is a broad-spectrum antimicrobial agent, active against gram-positive bacteria and yeast-like and filamentous fungi.[12,13] Liriodenine methiodide is also active,[13] and both compounds may be potentially useful as antibiotics.

14.4. Spectral Characteristics

Careful analysis of the PMR spectrum of the new oxoaporphine oxolaureline (*Laurelia novae-zelandiae* A. Cunn.) has shown that in oxoaporphines

H-11 does not necessarily appear downfield from H-8 as had been generally assumed.[14] For this alkaloid, both H-5 and H-8 absorb further downfield than H-11, regardless of whether the solvent used is TFA or $CDCl_3$.

PMR values for oxolaureline in TFA[14]

PMR values for oxolaureline in $CDCl_3$[14]

With oxonantenine (1,2-dimethoxy-9,10-methylenedioxyoxoaporphine), H-11 is furthest downfield when using TFA, but is upfield from H-5 when $CDCl_3$ is the solvent.[14]

The UV spectra of oxoaporphines are usually measured in ethanol. However, they can also be obtained in cyclohexane or in concentrated sulfuric acid. In the latter solvent, the solutions are purple-blue probably due to diprotonation, on nitrogen and on the carbonyl oxygen.[15] Not enough spectra have yet been run in these two solvents for generalizations to be drawn.

The mass spectral fragmentation of oxoaporphines has been discussed in terms of the location of ring D substituents and the resulting stabilization of the ions formed. Oxolaureline gives a molecular ion, m/e 305, which is also the base peak, the only other strong peak in the high mass region appearing at m/e 277 $(M - CO)^{\oplus}$. The isomeric lanuginosine (1,2-methylenedioxy-9-methoxy-oxoaporphine) on the other hand shows an important peak at m/e 275 $(M - CH_2O)^{\oplus}$. In one case, therefore, loss of CO is favored, whereas in the other loss of CH_2O is preferred.[14]

References and Notes

1. For a review on the oxoaporphines, see M. Shamma and R. L. Castenson, *Alkaloids*, **14**, R. H. F. Manske, ed., Academic Press, New York (1973), p. 225.
2. For a listing of oxoaporphines and a description of their spectral characteristics, see H. Guinaudeau, M. Leboeuf, and A. Cavé, *Lloydia*, **38**, 275 (1975).

3. B. F. Bowden, K. Picker, E. Ritchie, and W. C. Taylor, *Aust. J. Chem.*, **28**, 2681 (1975).
4. M. Hasegawa, M. Sojo, A. Lira, and C. Marquez, *Acta Cient. Venezuelana*, **23**, 165 (1972); through *Chem. Abstr.*, **79**, 42716z (1973).
5. J. Kunitomo, M. Ju-ichi, Y. Yoshikawa, and H. Chikamatsu, *J. Pharm. Soc. Japan*, **94**, 97 (1974).
6. Y. Watanabe, M. Matsui, M. Iibuchi, and S. Hiroe, *Phytochemistry*, **14**, 2522 (1975).
7. L. Castedo, R. Suau, and A. Mouriño, *Tetrahedron Lett.*, 501 (1976).
8. L. D. Yakhontova, V. I. Sheichenko, and O. N. Tolkachev, *Khim. Prir. Soedin.*, 214 (1972); through *Chem. Nat. Compds.*, 212 (1972).
9. L. Castedo, R. Suau, and A. Mouriño, *Heterocycles*, **3**, 449 (1975).
10. M. P. Cava and I. Noguchi, *J. Org. Chem.*, **38**, 60 (1973). For other recent syntheses of oxoaporphines, see J. G. Cannon, J. C. Kim, and M. A. Aleem, *J. Heterocycl. Chem.*, **9**, 731 (1972); M. P. Cava and I. Noguchi, *J. Org. Chem.*, **37**, 2936 (1972); M. P. Cava and S. S. Libsch, *J. Org. Chem.*, **39**, 577 (1974); M. P. Cava, K. T. Buck, I. Noguchi, M. Srinivasan, M. G. Rao, and A. I. DaRocha, *Tetrahedron*, **31**, 1667 (1975).
11. S. M. Kupchan and P. F. O'Brien, *Chem. Commun.*, 915 (1973). For other syntheses of corunnine, see I. Ribas, J. Saá, and L. Castedo, *Tetrahedron Lett.*, 3617 (1973); and Ref. 9 above.
11a. T. Kametani, R. Nitadori, H. Terasawa, K. Takahashi, M. Ihara, and K. Fukumoto, *Tetrahedron*, **33**, 1069 (1977).
12. C. R. Chen, J. L. Beal, R. W. Doskotch, L. A. Mitscher, and G. H. Svoboda, *Lloydia*, **37**, 493 (1974).
13. C. D. Hufford, M. J. Funderburk, J. M. Morgan, and L. W. Robertson, *J. Pharm. Sci.*, **64**, 789 (1975).
14. A. Urzúa and B. K. Cassels, *Heterocycles*, **4**, 1881 (1976); C. C. Hsu, R. H. Dobberstein, G. A. Cordell, and N. R. Farnsworth, *Lloydia*, **40**, 152 (1977).
15. C.-L. Chen, H. Chang, and E. B. Cowling, *Phytochemistry*, **15**, 547 (1976).

THE PHENANTHRENES

The structures of some recently isolated phenanthrene alkaloids are shown in Scheme 15.1.

15.1. Structural Elucidation and Synthesis

The phenanthrene alkaloids are a small group of optically inactive bases derived *in vitro* and probably *in vivo* by Hofmann degradation of quaternary aporphine salts. Of the more than one hundred aporphine alkaloids presently known, only three possess a methylenoxy bridge.[4] However, five of the fourteen known phenanthrene alkaloids incorporate this bridge.

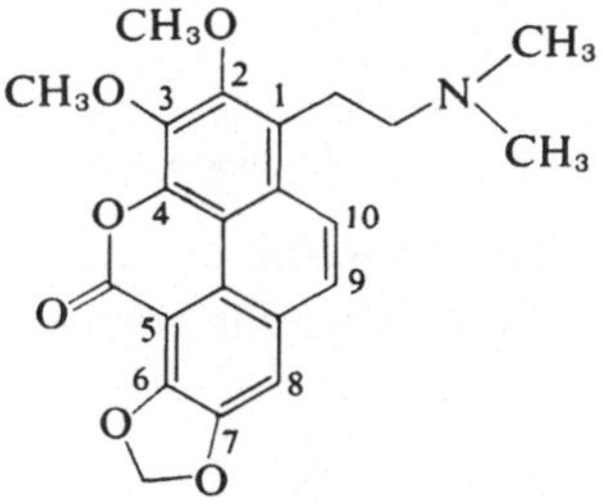

Thalflavidine[1]
Thalictrum flavum L.

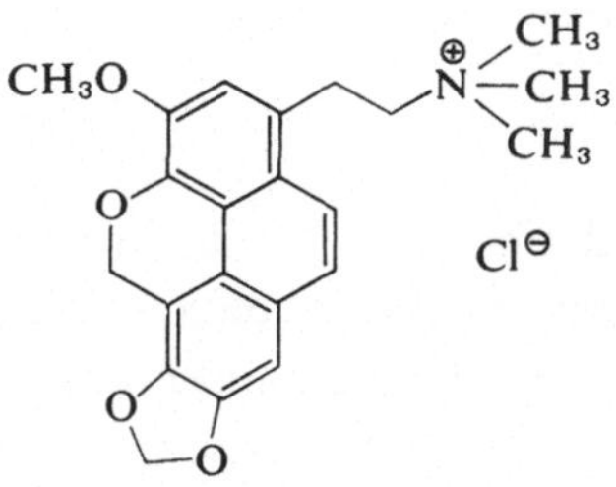

Thaliglucine methochloride[2]
T. polygamum Muhl.

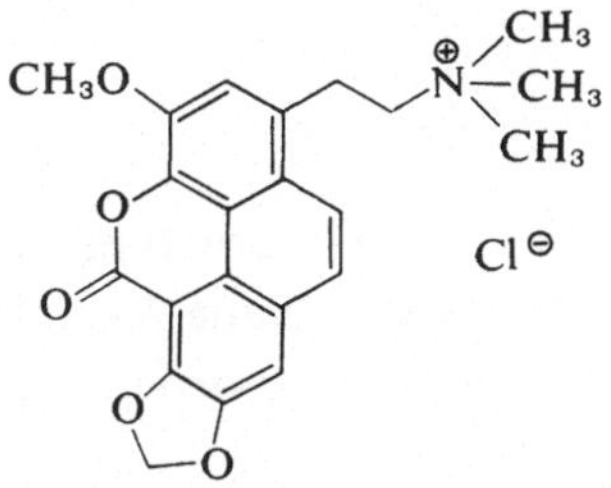

Thaliglucinone methochloride[2]
T. polygamum

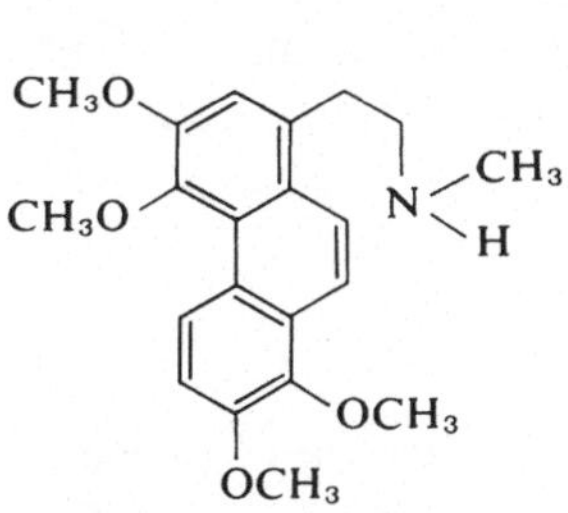

Noruvariopsamine[3]
Uvariopsis guineensis
(Annonaceae)

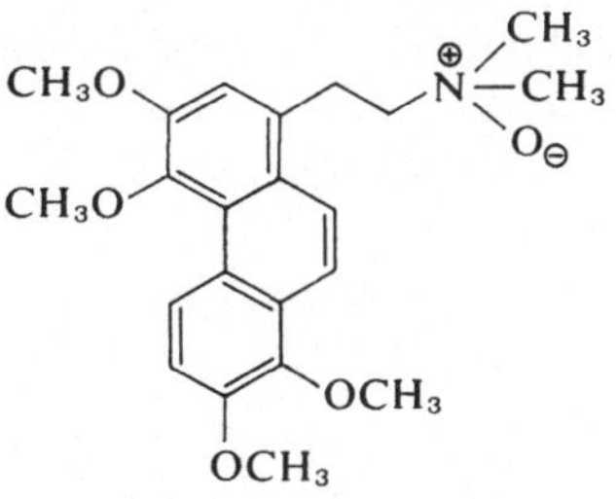

Uvariopsamine *N*-oxide[3]
U. guineensis

SCHEME 15.1

The structures of thaliglucine methochloride and thaliglucinone metho-chloride were confirmed by a standard chemical sequence. Northalphenine prepared by a photochemical route (see Sec. 10.2.1) was quaternized with methyl iodide to furnish thalphenine and subsequent Hofmann degradation produced the phenanthrene base thaliglucine[5] (see below). Dichromate oxida-tion of thaliglucine leads to the phenanthrene lactone thaliglucinone.[6]

Northalphenine   Thalphenine

Thaliglucine

The structural elucidations of thalflavidine,[1] noruvariopsamine[3] and uvariopsamine *N*-oxide[3] rest on spectral data, especially PMR, since the corre-sponding aporphine quaternary salts are unknown.[7]

15.2. Pharmacology

Thaliglucinone has been shown to possess *in vitro* potency against *Myco-bacterium smegmatis* strains by the agar dilution streak method.[8]

15.3. Mass Spectroscopy

The most characteristic feature in the mass spectra of phenanthrene alka-loids is the mode of cleavage of the amino-containing side chain. Free bases and quaternary salts in this series generally undergo fragmentation *beta* to the nitro-gen to generate a tropylium-type ion **1** and a mono-, di-, or trimethylamino-methylene ion as the base peak **2**. In the case of the *N*-oxide of uvariopsamine,

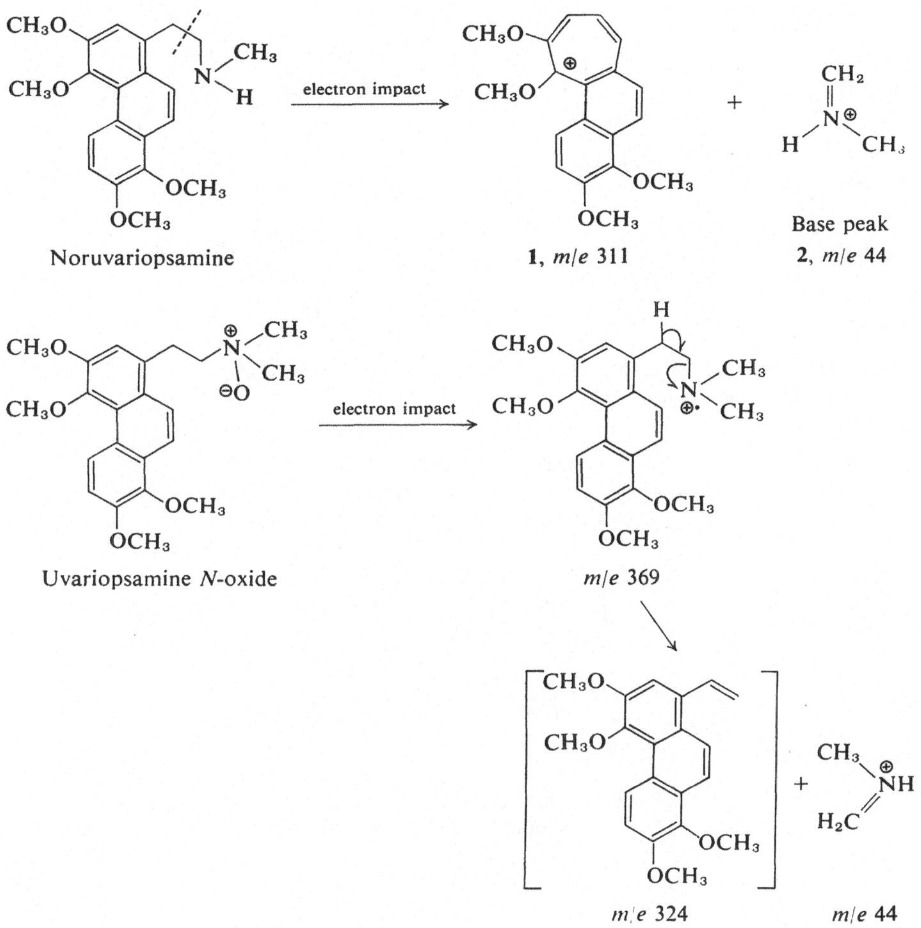

Noruvariopsamine   1, m/e 311   Base peak 2, m/e 44

Uvariopsamine N-oxide   m/e 369

m/e 324   m/e 44

SCHEME 15.2

however, the oxygen atom is lost first, followed by β-elimination of the mono-methylaminomethylene cation (see Scheme 15.2).[3]

15.4. PMR Spectroscopy

The C-5 proton of the phenanthrene alkaloids gives rise to a signal appreciably downfield and apart from the other aromatic proton resonances. The presence of a methylenedioxy group at C-3,4 in this series results in an upfield shift of the C-5 proton signal to ≈δ9.00. Other phenanthrene bases with methoxyl and/or hydroxyl substituents at C-3,4 exhibit PMR spectra with H-5 signals downfield between δ9.3 and 9.90.[9]

PMR chemical shifts for
thaliglucine methochloride[2]

PMR chemical shifts for thalflavidine[1]

PMR chemical shifts for thaliglucinone
methochloride[2]

15.5. UV Spectroscopy

Thalflavidine[1]	λ_{max}^{EtOH}	254, 263, 296, and 400 nm
		(4.38, 4.49, 3.94, and 3.60).
Thaliglucine methochloride[2]	λ_{max}^{EtOH}	233, 260, 272, 282, 295 sh, 326, 340, 358, and 370 nm
		(4.22, 4.42, 4.46, 4.45, 4.23, 3.82, 3.71, 3.32, and 3.32).
Thaliglucinone methochloride[2]	λ_{max}^{EtOH}	225, 237, 257 sh, 267, 288, 313, 333 sh, and 400 nm
		(4.15, 4.26, 4.36, 4.50, 3.85, 3.99, 3.71, and 3.61).

References and Notes

1. Kh. S. Umarov, Z. F. Ismailov, and S. Yu. Yunusov, *Khim. Prir. Soedin.*, 683 (1973).
2. M. Shamma and J. L. Moniot, *Heterocycles*, **2**, 427 (1974).
3. M. Leboeuf and A. Cavé, *Phytochemistry*, **11**, 2833 (1972).
4. M. Shamma, J. L. Moniot, S. Y. Yao, and J. A. Stanko, *Chem. Commun.*, 408 (1972).
5. M. Shamma and D.-Y. Hwang, *Tetrahedron*, **30**, 2279 (1974).
6. N. M. Mollov, L. N. Thuan, and P. P. Panov, *C. R. Acad. Bulg. Sci.*, **24**, 1047 (1971).
7. For a recent listing of the phenanthrene alkaloids, their physical constants and their spectral data, see H. Guinaudeau, M. Leboeuf, and A. Cavé, *Lloydia*, **38**, 275 (1975).
8. W.-N. Wu, J. L. Beal, G. W. Clark, and L. A. Mitscher, *Lloydia*, **39**, 65 (1976).
9. M. Shamma and J. L. Moniot, *Experientia*, **32**, 282 (1976).

THE 4,5-DIOXOAPORPHINES

Occurrence: Menispermaceae, Papaveraceae, and Piperaceae (?)

Structures:

Cepharadione-A[1]

Cepharadione-B, R = CH₃[1]
Norcepharadione-B, R = H[2]

Pontevedrine[3]

16.1. Introduction

It is only recently that the reddish-orange 4,5-dioxoaporphines have been recognized as a distinct group of isoquinoline alkaloids. This group includes cepharadione-A and -B, norcepharadione-B, and pontevedrine which was previously believed to be a 5,7-dioxoaporphine. There is a report of the occurrence of cepharadione-A and -B in *Piper auritum* (Piperaceae).[4] This finding needs to be reconfirmed, however, since 4,5-dioxoaporphines clearly originate biogenetically from oxidation of aporphines, and aporphines do not as a rule occur in the family Piperaceae.

16.2. A Revised Structure for Pontevedrine

Pontevedrine is a red, crystalline, alkaloid accompanying corunnine in *Glaucium flavum* Cr. (Papaveraceae). Its structure had been originally formulated as the 5,7-dioxoaporphine **1**. However, the isolation of (+)-cataline from the same source, and its easy oxidation to pontevedrine, using iodine or DDQ (dichlorodicyanoquinone), suggested an alternate formula for pontevedrine, namely the 4,5-dioxoaporphine structure **2**. This assumption was borne out when it was found that pontevedrine undergoes a base-catalyzed benzylic acid rearrangement. Treatment of pontevedrine (**2**) with sodium hydroxide in methanol supplied the yellow aristolactam **4** through the intermediacy of the α-hydroxyacid anion **3** which was not isolated.[3]

1   (+)-Cataline

Pontevedrine **2**   3   4

Interestingly, when the benzylic acid rearrangement was run in ethanolic sodium hydroxide, the product was the analog of **4** with an ethoxyl rather than a methoxyl group at C-4. Benzylic rearrangements within this context may have some analogy in nature, since aristolactams are found as natural products.[3] This topic will be discussed in greater detail in Sec. 17.4.

16.3. Cepharadione-A and -B

The orange colored cepharadiones-A and -B, found in *Stephania cepharantha* Y. Hayata (Menispermaceae), were the first 4,5-dioxoaporphines to be fully characterized. The structural assignments were based on the spectral data indicated.[1]

The mass spectra of cepharadione-A and -B show molecular ions at m/e 305 (88) and 321 (100), respectively; with intense peaks for the loss of carbon monoxide at m/e 277 (100) and 293 (84), respectively.[1]

Cepharadione-A
ν_{max}^{KBr} 1650 and 1675 cm^{-1}
(6.06 and 5.97 μ)

λ_{max}^{EtOH} 219, 238, 265, 279 sh, 290, 303, 315,
and 439 nm
(4.59, 4.61, 4.21, 4.20, 4.21, 4.24,
4.28, and 4.26)

Cepharadione-B
ν_{max}^{KBr} 1650 and 1667 cm^{-1}
(6.06 and 6.00 μ)

λ_{max}^{EtOH} 213, 244, 273 sh, 303, 315, and 440 nm
(4.57, 4.61, 4.20, 4.26, 4.29, and 4.23)

16.4. Synthesis

Photooxidation of dehydronuciferine in hexane was found to give cepharadione-B in 7–9% yield, together with smaller amounts of the oxoaporphine lysicamine.[5]

Dehydronuciferine

Lysicamine

References

1. M. Akasu, H. Itokawa, and M. Fujita, *Tetrahedron Lett.*, 3609 (1974).
2. M. Akasu, H. Itokawa, and M. Fujita, *Phytochemistry*, **14**, 1673 (1975).
3. L. Castedo, R. Suau, and A. Mouriño, *Tetrahedron Lett.*, 501 (1976).
4. R. Hänsel, A. Leuschke, and A. Gomez-Pompa, *Lloydia*, **38**, 529 (1976). See also R. Hänsel and A. Leuschke, *Phytochemistry*, **15**, 1323 (1976).
5. J. M. Saá, M. J. Mitchell, and M. P. Cava, *Tetrahedron Lett.*, 601 (1976).

THE ARISTOLOCHIC ACIDS AND ARISTOLACTAMS

Occurrence: Aristolochiaceae, Menispermaceae, Monimiaceae
The 5 aristolochic acids and 11 aristolactams are shown in Scheme 17.1.[1]

Aristolochic Acids

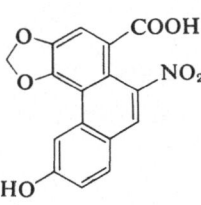

Aristolochic acid-I[2,3]

Aristolochic acid-B[4]

Aristolochic acid-C[4]

Aristolochic acid-D[5]
(aristolochic acid-IVa)[6]

Aristolochic acid-II[7]

Aristolactams

Aristolactam[4,5]

Aristored[3]

Taliscanine[8]

SCHEME 17.1

Aristolochic acid-D
methyl ether lactam[5]

Aristolactam
β-D-glucoside[5]

Aristolactam-AII[9]

Aristolactam-AIII[9]

Aristolactam-BII[9]
(cepharanone-B)[10]

Aristolactam-BIII[9]

Cepharanone-A[10]

Doryflavine[11]

SCHEME 17.1 (Continued)

17.1. Introduction

Aristolochic acids and aristolactams occur mostly within the family of Aristolochiaceae, although cepharanone-A and aristolactam-BII were found in *Stephania cepharantha* Y. Hayata (Menispermaceae),[10] and doryflavine was obtained from *Doryphora sassafras* Endlicher (Monimiaceae).[11] There is a possibility that aristolochic acid-B may correspond to aristolochic acid-D.

17.2. Structural Elucidation

The structural elucidation of the yellow colored aristolochic acid-I was carried out in the 1950s mainly by Pailer and his students.[3] The main degrada-

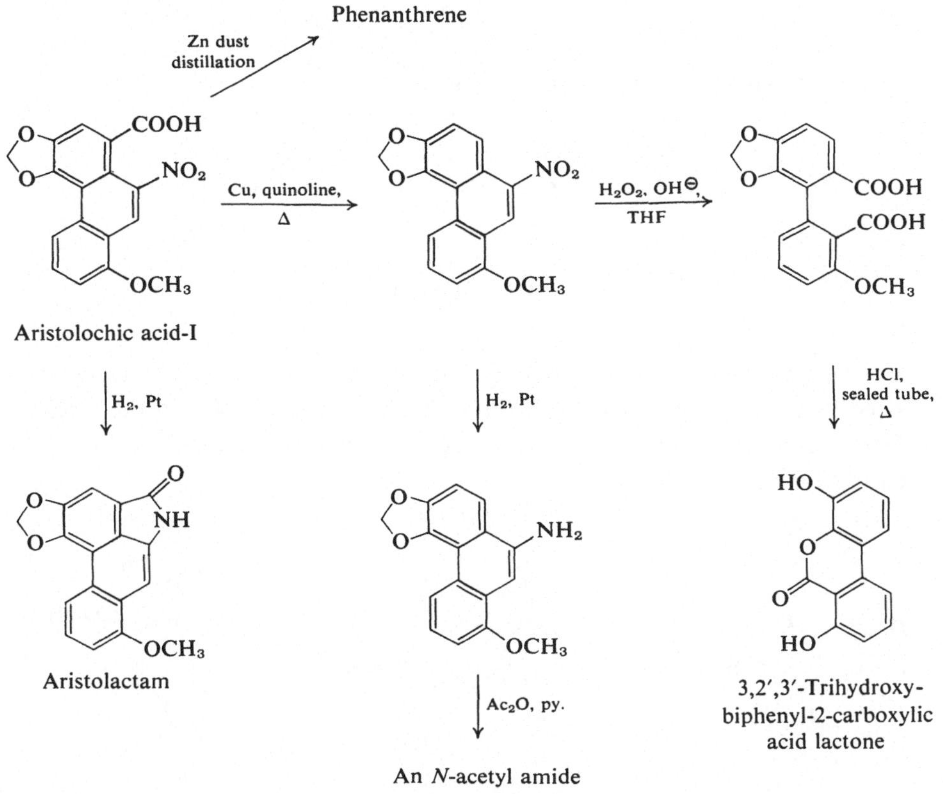

SCHEME 17.2

tion sequences are outlined in Scheme 17.2. The characterization of such products as phenanthrene and 3,2',3'-trihydroxybiphenyl-2-carboxylic acid was critical in the structural assignment. It should be noted that catalytic reduction of aristolochic acid-I gives aristolactam which was later found to be a natural product.

17.3. Synthesis

Acid-catalyzed condensation of the imine **2** with the substituted nitromethane **1** gave yellow crystals of the *cis*-stilbene **3** which upon irradiation provided the methyl ester of aristolochic acid-I. Hydrolysis of the ester produced the natural product.[2]

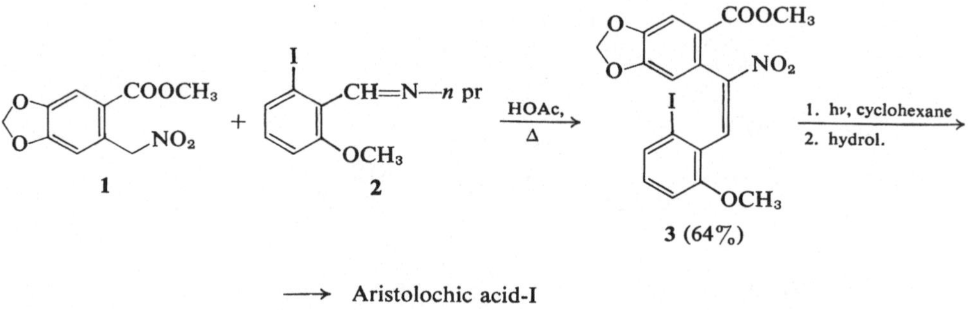

1   **2**

3 (64%)

⟶   Aristolochic acid-I

Degradation product **7**, obtained from aristolochic acid-D, was synthesized by condensation of the substituted phenylacetic acid **4** with the aldehyde **5** to supply the stilbene **6**, which was converted to the desired product **7** by classical means[6]:

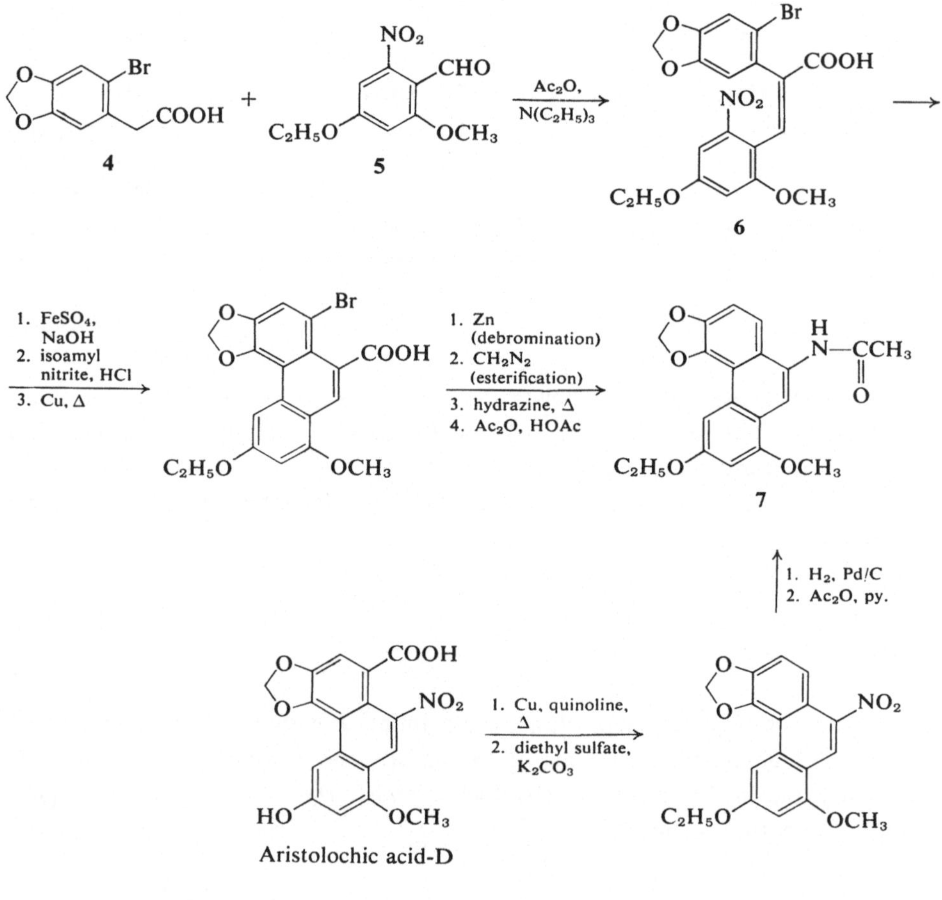

4   **5**   **6**

1. FeSO$_4$, NaOH
2. isoamyl nitrite, HCl
3. Cu, Δ

1. Zn (debromination)
2. CH$_2$N$_2$ (esterification)
3. hydrazine, Δ
4. Ac$_2$O, HOAc

7

1. H$_2$, Pd/C
2. Ac$_2$O, py.

1. Cu, quinoline, Δ
2. diethyl sulfate, K$_2$CO$_3$

Aristolochic acid-D

A partial synthesis of a tetrasubstituted aristolactam has been achieved from the tetramethoxyphenanthrene carboxylic acid **8**, which had been derived from natural (+)-glaucine via Hofmann degradation followed by oxidation.[12]

(+)-Glaucine

8

17.4. Biogenesis

The biosynthesis of aristolochic acid-I in *Aristolochia sipho* has been studied. Tyrosine, dopa, dopamine, noradrenaline,[13] and norlaudanosoline[14] can serve as precursors; the nitro group of aristolochic acid-I is derived from the amino group of tyrosine.

There is a possibility that a 1,2,4-trioxygenated aporphine is oxidized to a 4,5-dioxoaporphine such as cepharadione-A which can undergo net overall decarbonylation, presumably through a benzylic acid rearrangement, to yield the corresponding aristolactam, in the present case cepharanone-A. Further oxidation would then yield aristolochic acid-II.[15,16] Significantly, cepharadione-A and cepharanone-A occur in the same plant, *Stephania cepharantha* Y. Hayata (Menispermaceae).[10] Noteworthy also is the observation that aristolochic acids and aristolactams are never found in nature oxygenated at C-7. Since aristolochic acid-D is oxygenated at C-6 and C-8, it could be that a dienol–benzene rearrangement is involved in the formation of this and other aristolochic acids.

Five phenanthrene derivatives, **9–13**, have recently been found in *Aristolochia indica*. Compound **11** was labeled aristolamide, **12** is aristolinic acid methyl ester, and **13** is methyl aristolochate. All five must be of aporphinoidal origin.[17]

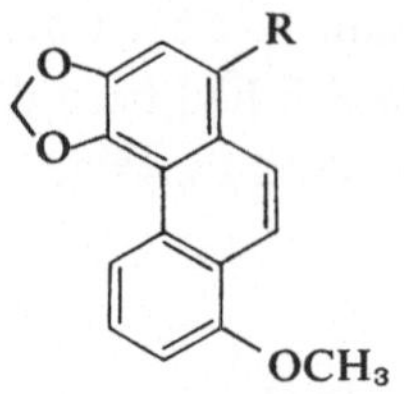

9, R = COOCH$_3$
10, R = COOH
11, R = CONH$_2$

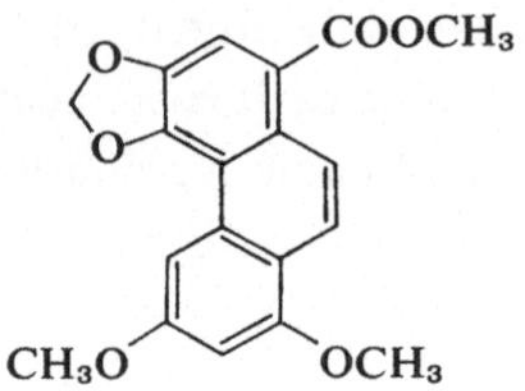

12

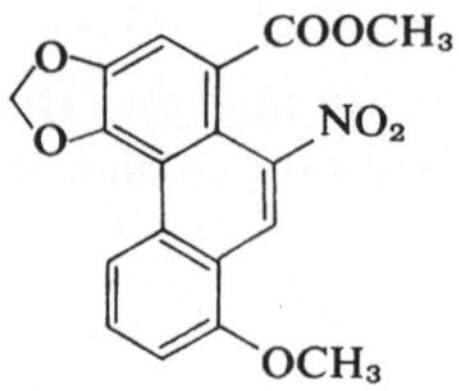

13

17.5. Pharmacology

Aristolochic acid-I exhibits tumor inhibitory activity against the adeno-carcinoma 755 test system. Subsequently, however, when tested on mice, it was found to produce papillomas.[2] Aristolochic acid-I is the probable causative agent in Balkan endemic nephropathy. A convenient fluorometric assay for its determination is based on hydrosulfite reduction to the lactam, and measurement of the intensity of fluorescence.[18]

17.6. PMR Spectroscopy

The PMR spectra of the aristolochic acids and aristolactams are usually run in DMSO-d$_6$. Pyridine-d$_5$ has also been used in connection with the aristolactams.

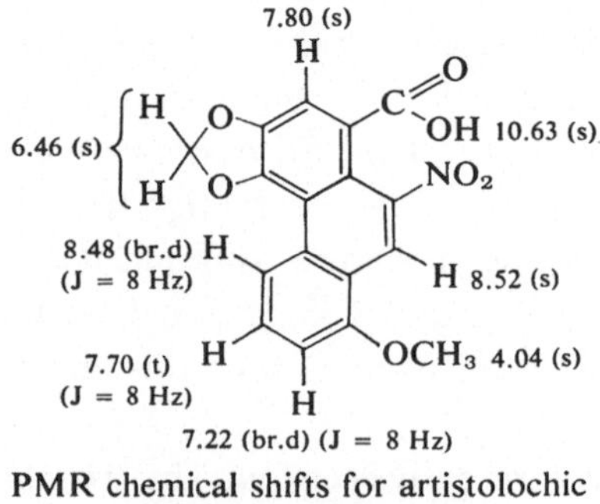

PMR chemical shifts for artistolochic acid-I in DMSO-d$_6$.[5]

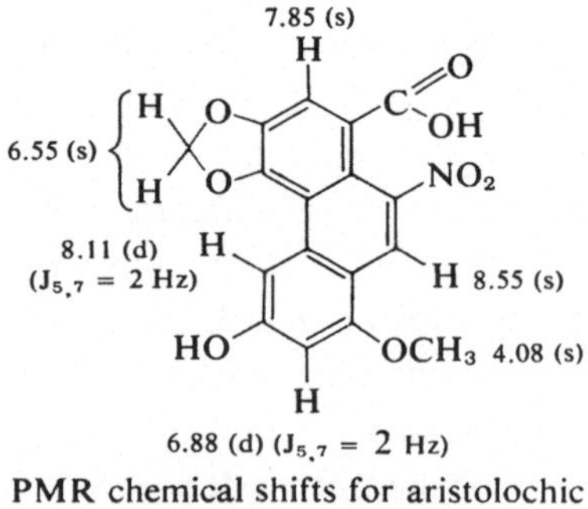

PMR chemical shifts for aristolochic acid-D in DMSO-d$_6$.[5]

7.52 (s)
H

6.40 (s) { H, O
           H, O

NH 10.32 (br. s)

O

7.60 (d)
($J_{5,7}$ = 2 Hz) H   H 7.27 (s)

3.92 (s) CH$_3$O   OCH$_3$ 3.92 (s)
or 3.98 (s)   or 3.98 (s)
H
6.77 (d)
($J_{5,7}$ = 2 Hz)

PMR chemical shifts for aristolochic
acid-D methyl ether lactam
in DMSO-d$_6$.[5]

7.22 (s)
H

3.96 (s) CH$_3$O   O

NH

HO

9.22 (d) H
($J_{5,7}$ = 2 Hz)   H 8.11 (s)

HO   H
          7.86 (d)
H   ($J_{7,8}$ = 8 Hz)
7.48 (d)
($J_{7,8}$ = 8 Hz)
($J_{5,7}$ = 2 Hz)

PMR chemical shifts for doryflavine
in py.-d$_5$.[11]

17.7. UV and IR Spectroscopy

Aristolochic acid-I[4]	λ_{max}^{EtOH}	250, 318, and 390 nm (4.43, 4.08, and 3.81)
Aristolochic acid-D[5]	λ_{max}^{EtOH}	220, 242, 252, 292, and 325 nm (4.47, 4.58, 4.58, 4.14, and 4.05)
Aristolactam[4]	λ_{max}^{EtOH}	241, 250, 259, 291, and 300 nm (4.50, 4.47, 4.56, 4.16, and 4.15)
Aristored[3,4]	λ_{max}^{EtOH}	253, 265, and 294 nm (4.63, 4.50, and 4.29)
Cepharanone-A[10]	λ_{max}^{EtOH}	225, 232, 265, 277, 288, 328, 341, 376, and 393 nm (4.45, 4.49, 4.46, 4.53, 4.51, 4.06, 4.04, 3.93, and 3.93)
Doryflavine[11]	λ_{max}^{MeOH}	207, 230, 251, 260, 277, 290, 320, and 400 nm (4.42, 4.59, 4.45, 4.41, 4.32, 4.34, 4.11, and 3.91)

The IR spectra of aristolochic acids obtained as Nujol mulls usually show bands near 1689 cm^{-1} (5.92 μ) (COOH), and 1348 cm^{-1} (7.42 μ) (NO$_2$). Aristolactams exhibit a band near 1692 cm^{-1} (5.91 μ) (lactam C=O).

References and Notes

1. For a review on this dual topic, see M. Shamma in *Specialist Periodical Reports, The Alkaloids*, Vol. 6, M. F. Grundon, Sr. Reporter, The Chemical Society, London (1976), p. 183.

2. S. M. Kupchan and H. C. Wormser, *J. Org. Chem.*, **30**, 3792 (1965).
3. M. Pailer, L. Belohlav, and E. Simonitsch, *Monatsh. Chem.*, **87**, 249 (1956). R. T. Coutts, J. B. Stenlake, and W. D. Williams, *J. Chem. Soc.*, 4120 (1957).
4. S. Sasagawa, *J. Pharm. Soc. Japan*, **82**, 921 (1959); and M. Tomita and S. Sasagawa, *J. Pharm. Soc. Japan*, **79**, 973, and 1470 (1959).
5. S. M. Kupchan and J. J. Merianos, *J. Org. Chem.*, **33**, 3735 (1968).
6. E. A. Rúveda, S. M. Albonico, H. A. Priestap, V. Deulofeu, M. Pailer, E. Gössinger, and P. Bergthaller, *Monatsh. Chem.*, **99**, 2349 (1968).
7. M. Pailer and A. Schleppnik, *Monatsh. Chem.*, **89**, 175 (1958).
8. L. A. Maldonado, J. Herran, and J. Romo, *Ciencia (Mexico)* **24**, 237 (1966); through *Chem. Abstr.*, **65**, 15438e (1966).
9. R. Crohare, H. A. Priestap, M. Farina, M. Cedola, and E. A. Rúveda, *Phytochemistry* **13**, 1957 (1974).
10. M. Akasu, H. Itokawa, and M. Fujita, *Tetrahedron Lett.*, 3609 (1974).
11. C. R. Chen, J. L. Beal, R. W. Doskotch, L. A. Mitscher, and G. H. Svoboda, *Lloydia*, **37**, 493 (1974).
12. P. Gorecki and H. Otta, *Pharmazie*, **30**, 337 (1975).
13. F. Comer, H. P. Tiwari, and I. D. Spenser, *Can. J. Chem.*, **47**, 481 (1969).
14. H. R. Schuette, U. Orban, and K. Mothes, *Eur. J. Biochem.*, **1**, 70 (1967).
15. M. Shamma and J. L. Moniot in *Specialist Periodical Reports, The Alkaloids*, Vol. 6, M. F. Grundon, Sr. Reporter, The Chemical Society, London (1976), pp. 179 and 185.
16. L. Castedo, R. Suau, and A. Mouriño, *Tetrahedron Lett.*, 501 (1976).
17. S. C. Pakrashi, P. Ghosh-Dastidar, S. Basu, and B. Achari, *Phytochemistry*, **16**, 1103 (1977).
18. K. V. Rao, Y. Tanrikut, and K. Killion, *J. Pharm. Sci.*, **64**, 345 (1975).

THE DIBENZAZONINES

Occurrence: Menispermaceae and Leguminosae

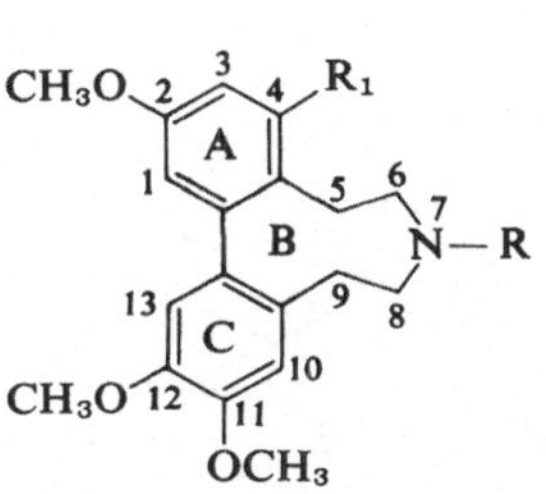

Laurifine,[1] R = R$_1$ = H
Laurifonine,[1] R = CH$_3$
R$_1$ = H
Protostephanine,[3] R = CH$_3$,
R$_1$ = OCH$_3$

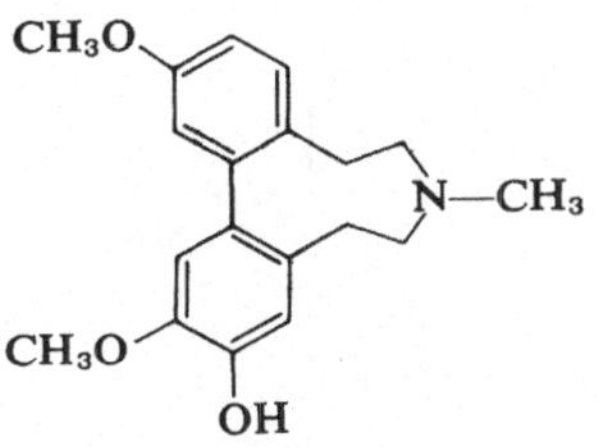

Laurifinine[7]
Cocculus laurifolius DC.

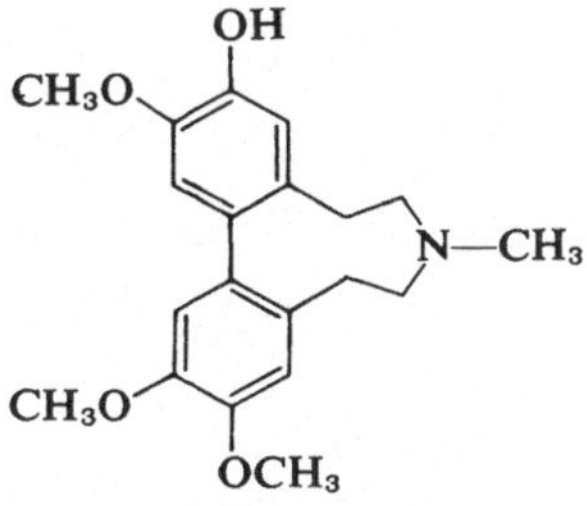

Erybidine[2]
Erythrina X *bidwillii* Lindl.

18.1. Introduction

The dibenzazonines can be divided into two types, (a) the fully oxygenated bases such as erybidine, and (b) those which have undergone a net deoxygenation with respect to their precursors, e.g., protostephanine, laurifinine, laurifine, and laurifonine.

18.2. Structural Elucidation

Zinc dust distillation of laurifonine, $C_{20}H_{25}NO_3$, produced 2,3,6-tri-methoxyphenanthrene (**1**); and two successive Hofmann degradations followed by oxidative cleavage generated the biphenyl dialdehyde **2** (see Scheme 18.1). The location of the substituents in **1** and **2** was confirmed by short syntheses.[1]

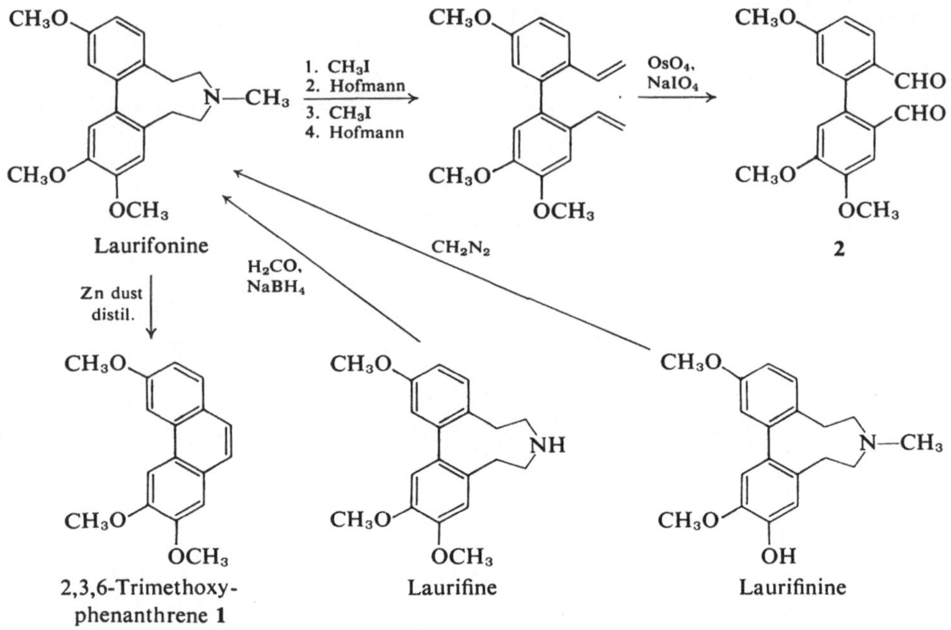

SCHEME 18.1

The related bases laurifine and laurifinine were converted by reductive *N*-methylation in one case, and *O*-methylation in the other, into laurifonine. The phenolic function of laurifinine was suggested to be adjacent to a methoxyl group on the basis of alkali-catalyzed deuterium exchange experiments.[1]

18.3. Synthesis

18.3.1. Through a Benzyne Intermediate

The biphenyl bond of dibenzazonine and dibenzazecine systems has been formed by the benzyne reaction—induced with dimsyl sodium—of 2′-halobenzyl- and 2′-halophenethyltetrahydroisoquinolines, respectively. (See Sec. 8.1.1 for related conversions.) Reductive deoxygenation of product **3** to the sulfide stage followed by Raney nickel desulfurization afforded dibenzazonine **4** and dibenzazecine **5** in unspecified yields (see Scheme 18.2).[4]

18.3.2. By Photolysis

Erybidine has been prepared by a route which includes as a key step the photochemical formation of the biphenyl bond. The amide **6** was photolyzed

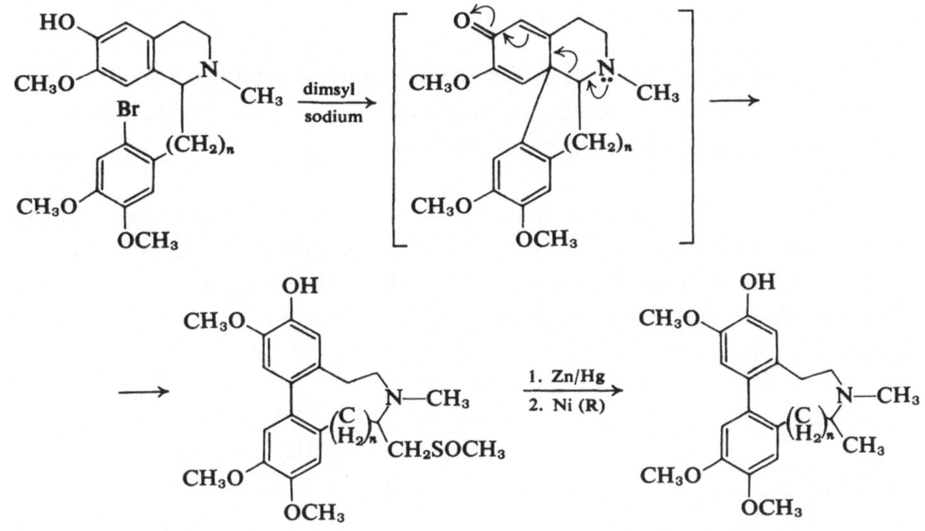

3

4, $n = 1$
5, $n = 2$

SCHEME 18.2

in base to supply dibenzazonines **7** and **8**. Borohydride reduction and *N*-methylation of **8** afforded erybidine in moderate yield[5]:

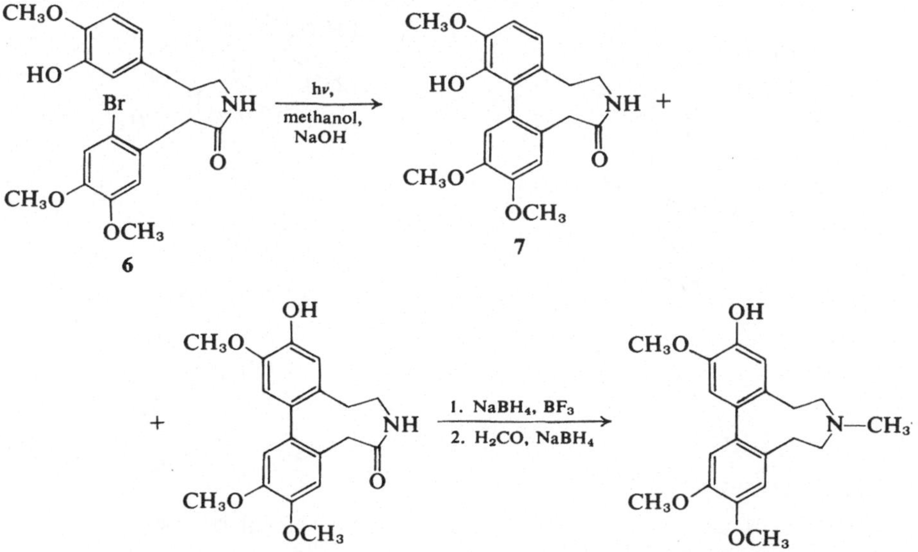

6, 7, 8

Erybidine
(≈18% overall from **6**)

18.3.3. Through Rearrangements

For a detailed discussion of the acid-catalyzed rearrangement of morphin-andienones and the base-catalyzed rearrangement of neoproaporphine–borane complexes to the dibenzazonine skeleton, see Sec. 10.2.3.

18.3.3.1. Of Neospirenes. The nonphenolic benzylisoquinoline *N*-formyl-norlaudanosine was oxidized with vanadium oxytrifluoride in trifluoroacetic acid to the neospirene **9** whose structure was established by X-ray analysis. Lithium aluminum hydride reduction of the corresponding ketal **10** supplied *O*-methylerybidine[6]:

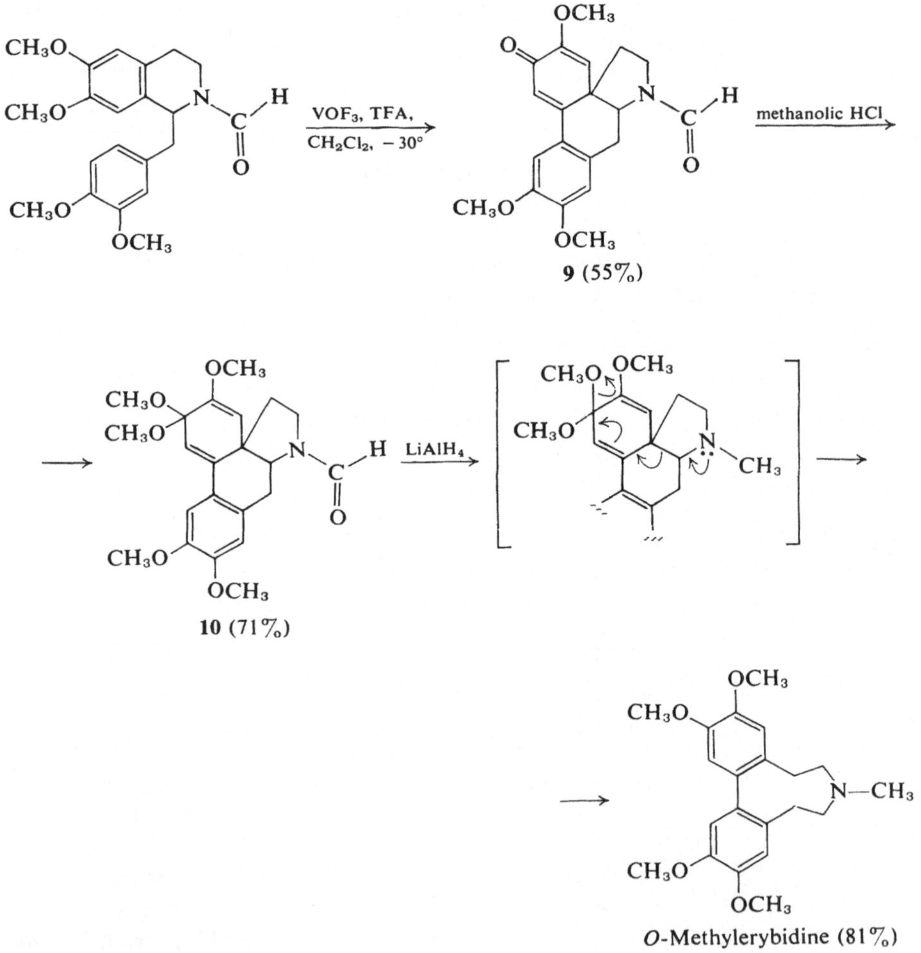

9 (55%)

10 (71%)

O-Methylerybidine (81%)

18.3.3.2. Of Neoproaporphines. Oxidation of norprotosinomenine trifluoracetamide with vanadium oxytrifluoride gave the phenolic neoproaporphine **11**. Alkaline hydrolysis and bond cleavage, followed by borohydride reduction of the imine bond, supplied the dibenzazonine base **12** in good yield.[7]

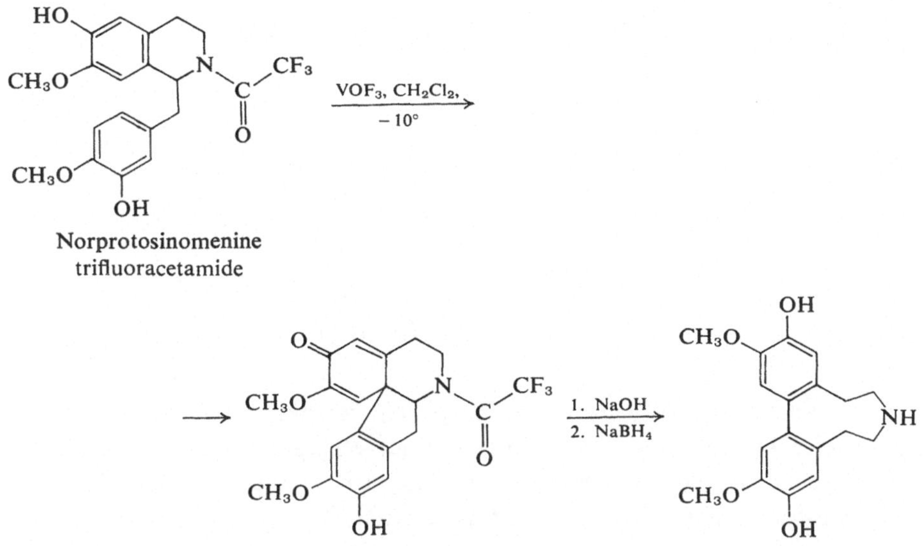

Norprotosinomenine
trifluoracetamide

11 (40%)

12 (80%)

18.3.3.3. Of Morphinandienols. The natural base laurifonine has been synthesized by a route that may parallel the biogenetic pathway for the formation of this and related alkaloids. *O*-Methylflavinantine was reduced with sodium borohydride to supply a separable mixture of epimeric dienols. Skeletal rearrangement of the mixture to the neospirene **13** was effected by boron trifluoride–etherate, and subsequent catalytic reduction produced laurifonine in excellent yield[8]:

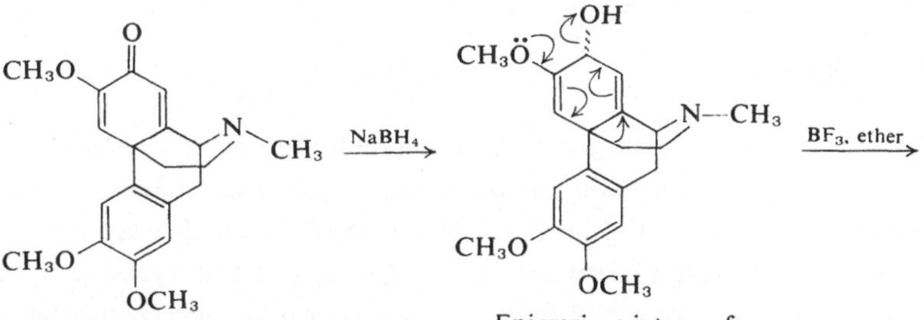

O-Methylflavinantine

Epimeric mixture of
dienols (83%)

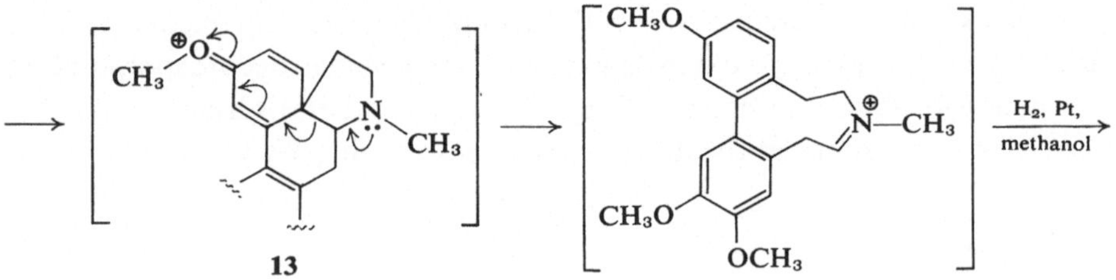

13

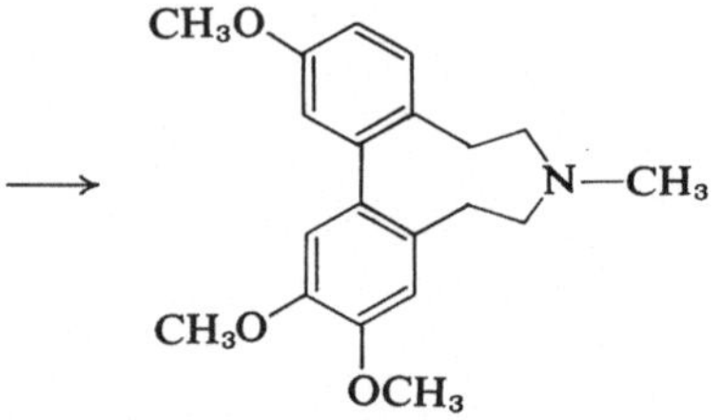

Laurifonine (81%)

A related dienol–benzene rearrangement is that of the dihydroneopro-aporphine **14**. Reaction with boron trifluoride–etherate or concentrated hydrochloric acid at room temperature gave exclusively the deoxyaporphine **15**.[8] No laurifonine was obtained from this reaction.[1]

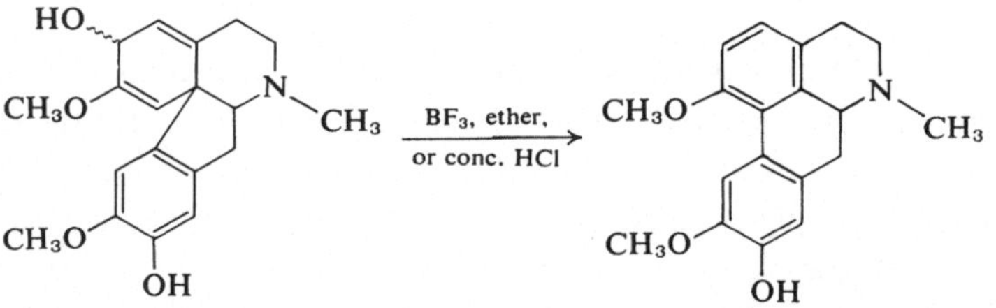

14   15 (85%)

18.4. Biogenesis

Some progress has been made in understanding the biosynthetic pathway to protostephanine in *Stephania japonica* Miers (Menispermaceae). This plant was treated with isotopically labeled amines **16–22**, and all were found to be incorporated to an insignificant extent. On the other hand, tyrosine, tyramine, dopa, and dopamine were specifically incorporated into protostephanine. Analysis of these results indicate that (a) protostephanine is constructed from two C_6–C_2 units derived from tyrosine, and (b) only the ring-A unit, including the ethylamine residue, is labeled by [2-^{14}C]-dopa, [2-^{14}C]-dopamine, and [2-^{14}C]-tyramine. Furthermore, feeding studies with phenethylamines **23–26** revealed that catechols **23** and **24**, but not monophenols **25** and **26**, acted as precursors for protostephanine, so that the phenethylamine which forms ring A is probably a trioxygenated monomethyl derivative such as **24**[9]:

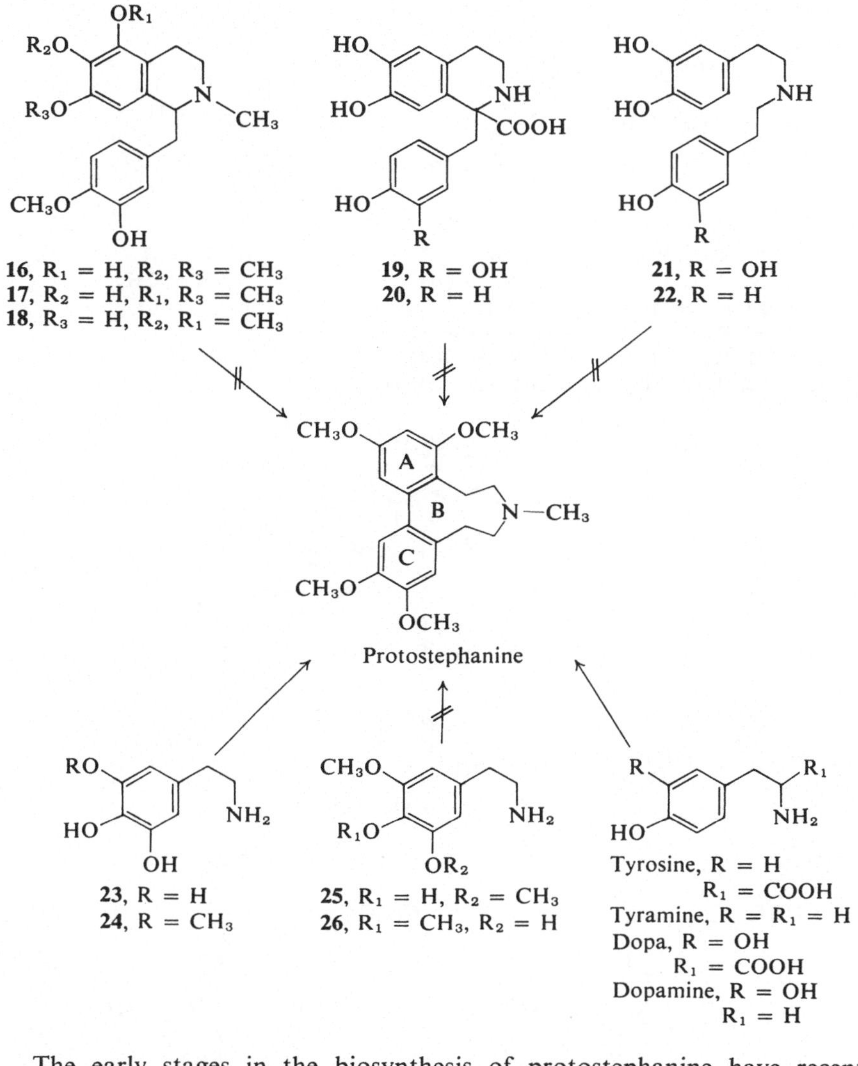
OR₁
R₂O
R₃O
N
CH₃
CH₃O
OH
16, R₁ = H, R₂, R₃ = CH₃
17, R₂ = H, R₁, R₃ = CH₃
18, R₃ = H, R₂, R₁ = CH₃
HO
HO
NH
COOH
HO
R
19, R = OH
20, R = H
HO
HO
NH
HO
R
21, R = OH
22, R = H
CH₃O
OCH₃
A
B
N—CH₃
C
CH₃O
OCH₃
Protostephanine
RO
HO
NH₂
OH
23, R = H
24, R = CH₃
CH₃O
R₁O
NH₂
OR₂
25, R₁ = H, R₂ = CH₃
26, R₁ = CH₃, R₂ = H
R
R₁
NH₂
HO
Tyrosine, R = H
R₁ = COOH
Tyramine, R = R₁ = H
Dopa, R = OH
R₁ = COOH
Dopamine, R = OH
R₁ = H

The early stages in the biosynthesis of protostephanine have recently been delineated by Battersby and his school. The labeled amines **27–32** were all incorporated specifically and efficiently by *S. japonica* Miers into protostephanine, whereas amines **33–36** were not. The results indicate that ring A of the tetrahydrobenzylisoquinoline precursor must be trioxygenated at C-6, C-7, and C-8, and with phenolic functions at C-7 and C-8. The benzylic ring, when monooxygenated, must be phenolic at C-4′, but when dioxygenated, must be phenolic at C-3′. The exact timing of *N*-methylation is not critical. Thus, the pathway shown in Scheme 18.3 and the classification of protostephanine as an isoquinoline-derived alkaloid are firmly established.[10a]

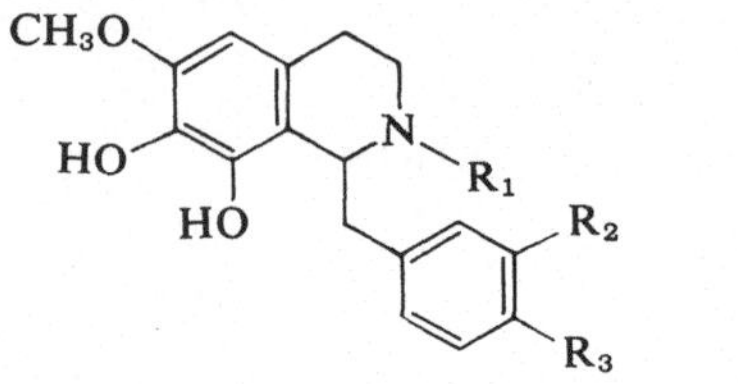

27, R₁ = R₂ = H, R₃ = OH

27, $R_1 = R_2 = H, R_3 = OH$
28, $R_1 = CH_3, R_2 = H, R_3 = OH$
29, $R_1 = H, R_2 = R_3 = OH$
30, $R_1 = CH_3, R_2 = R_3 = OH$
31, $R_1 = H, R_2 = OH, R_3 = OCH_3$
32, $R_1 = CH_3, R_2 = OH, R_3 = OCH_3$

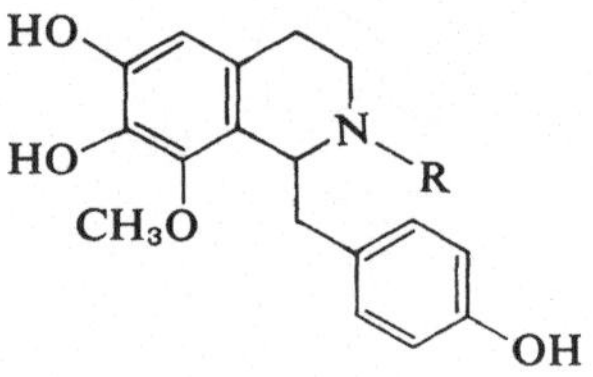

33, R = H,
34, R = CH_3

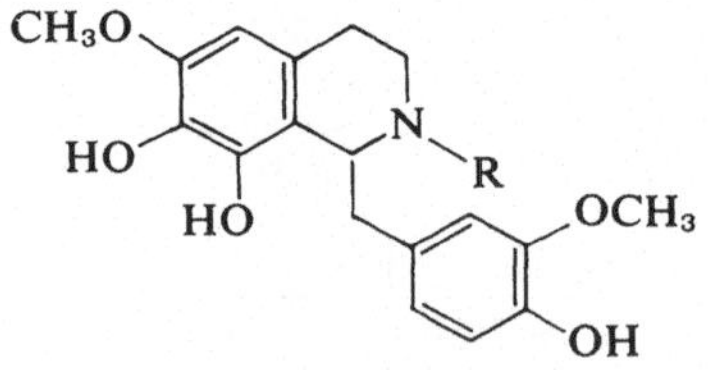

35, R = H
36, R = CH_3

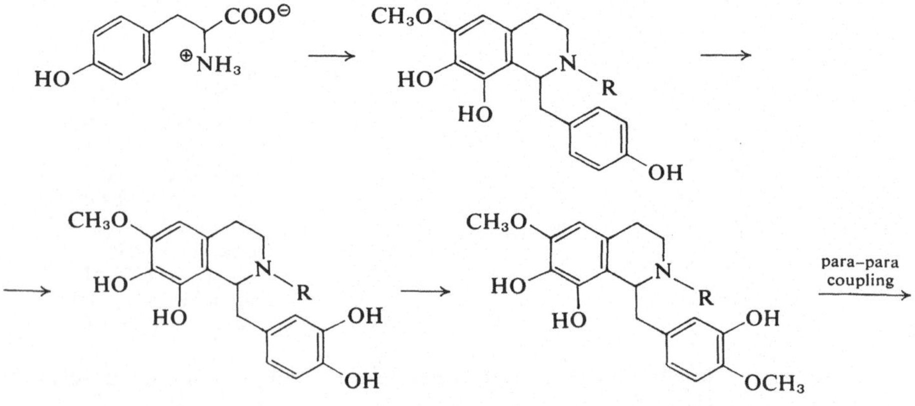

SCHEME 18.3

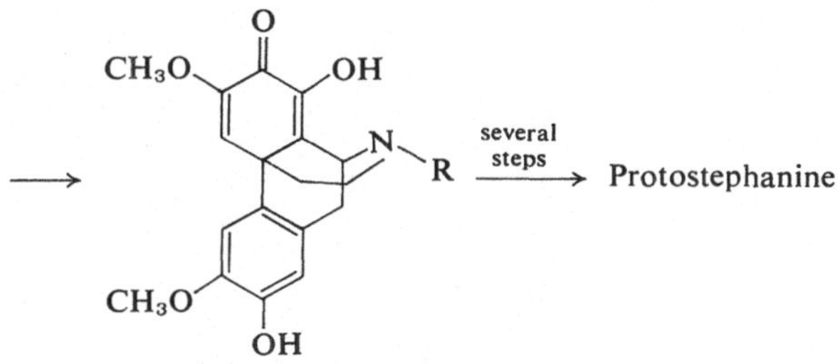

SCHEME 18.3 *(Continued)*

Turning now to the later stages of biosynthesis, an obvious pathway to the dibenzazonine system proceeds through the rearrangement of a neoproaporphine, a process which is also a facile *in vitro* transformation. This route is probably operative in the biosynthesis of the fully oxygenated base erybidine.

It is tempting to speculate that some of the dibenzazonines of the Menispermaceae could be formed by dehydrative rearrangement of neoproaporphine-dienol precursors. In the laboratory, however, such an attempt led exclusively to aporphinoids (see Sec. 18.3.3); and no data derived from the feeding of labeled precursors is available to support such a thesis.

Barton's valid proposal for the biogenesis of protostephanine,[10] by rearrangement of a morphinandienol through a neospirene to the dibenzazonine system, has been extended in an *in vitro* series of experiments to the synthesis of the Menispermacious alkaloid laurifonine (see Sec. 18.3.3.3). The oxygenation pattern differences between erybidine found in the Leguminosae and the other dibenzazonines, present in the Menispermaceae, may thus reflect two quite different biogenetic pathways (see Scheme 18.4).

It is of chemotaxonomic interest that the bases cocculine and cocculidine, which closely resemble some of the *Erythrina* alkaloids, have been isolated from *Cocculus* species (Menispermaceae).[11-13] The recent isolation from *C. trilobus* of coccutrine,[13] an apparent relative of protostephanine, as reflected by the ring-A oxygenation pattern, suggests that the dibenzazonines are not an end point in alkaloid biosynthesis.

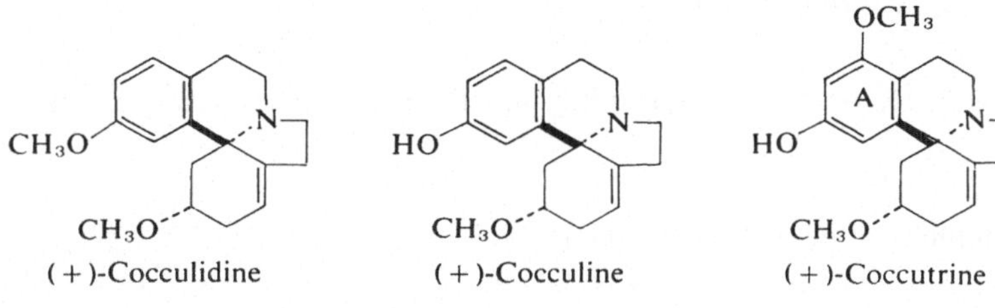

(+)-Cocculidine   (+)-Cocculine   (+)-Coccutrine

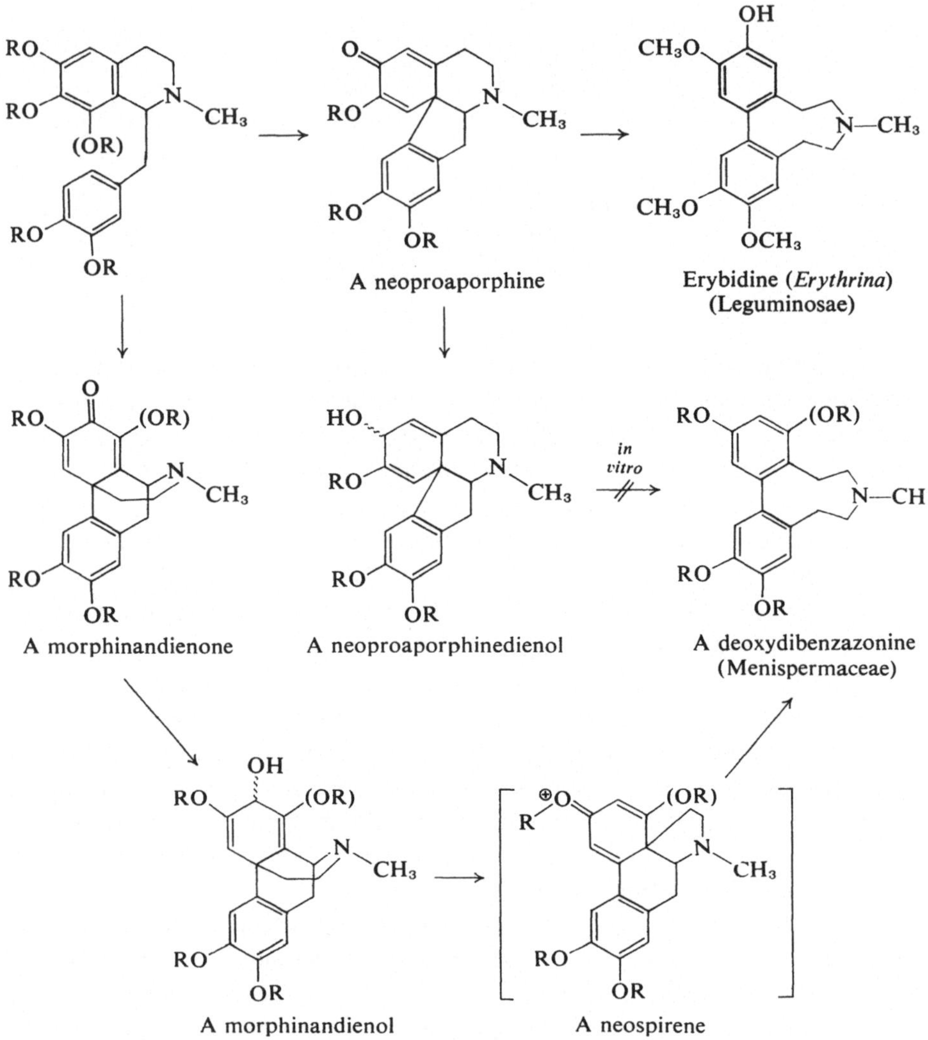

SCHEME 18.4

18.5. PMR Spectroscopy

The PMR data for laurifonine, laurifine, and laurifinine are summarized below.[1]

Laurifonine: $\delta 2.32$ (s) (N–CH$_3$)
                     2.5–2.7 (m) (4 × CH$_2$)

6.72 (s) and 6.78 (s) (ring C ArH)
6.80–7.27 (m) (ring A ArH)
3.76, 3.80 and 3.90 (s) (3 × OCH₃)

Laurifine: 2.3–3.2 (m) (4 × CH_2)
6.70 (s) and 6.72 (s) (ring C ArH)
6.72–7.26 (m) (ring A ArH)
3.82, 3.90 and 3.94 (s) (3 × OCH₃)

Laurifinine: 2.32 (s) (N–CH₃)
2.58–3.25 (m) (4 × CH_2)
6.66 (s) and 6.76 (s) (ring C ArH)
6.74–7.24 (m) (ring A ArH)
3.80 (s) and 3.82 (s) (2 × OCH₃)

In contrast to other isoquinoline alkaloids incorporating a biphenyl system, dibenzazonines do not exhibit downfield signals for aromatic protons ortho to the biphenyl bond. This spectral characteristic may indicate a strongly skewed conformation of the biphenyl system in the dibenzazonines.

18.6. UV Spectroscopy

The UV data for laurifonine, laurifine, and laurifinine[1] are given below.

Laurifonine λ_{max}^{MeOH} 221 and 283 nm
(4.30 and 3.90)

Laurifine λ_{max}^{MeOH} 221 and 284 nm
(4.34 and 3.96)

Laurifinine λ_{max}^{MeOH} 223 and 284 nm
(4.24 and 3.84)

References

1. H. Uprety and D. S. Bhakuni, *Tetrahedron Lett.*, 1201 (1975); and H. Pande and D. S. Bhakuni, *J. Chem. Soc. Perkin I*, 2197 (1976).
2. K. Ito, H. Furukawa, H. Tanaka, and T. Rai, *J. Pharm. Soc. Japan*, **93**, 1218 (1973).
3. A. R. Battersby, A. K. Bhatnagar, P. Hackett, C. W. Thornber, and J. Staunton, *Chem. Commun.*, 1214 (1968).
4. S. Kano, T. Ogawa, T. Yokomatsu, E. Komiyama, and S. Shibuya, *Tetrahedron Lett.*, 1063 (1974); see also S. Kano, E. Komiyama, Y. Takahagi, and S. Shibuya, *Chem. Pharm. Bull.*, Tokyo, **24**, 648 (1976).
5. K. Ito and H. Tanaka, *Chem. Pharm. Bull.*, Tokyo, **22**, 2108 (1974).

6. S. M. Kupchan, A. J. Liepa, V. Kameswaran, and R. F. Bryan, *J. Am. Chem. Soc.*, **95**, 6861 (1973).
7. S. M. Kupchan, C. K. Kim, and J. T. Lynn, *Chem. Commun.*, 86 (1976).
8. S. M. Kupchan, C. K. Kim, and K. Miyano, *Heterocycles*, **4**, 235 (1976).
9. A. R. Battersby, R. C. F. Jones, R. Kazlauskas, C. Poupat, C. W. Thornber, S. Ruchirawat, and J. Staunton, *Chem. Commun.*, 773 (1974).
10. D. H. R. Barton, *Pure Appl. Chem.*, **9**, 35 (1964).
10a. A. R. Battersby, A. Minta, A. P. Ottridge, and J. Staunton, *Tetrahedron Lett.*, 1321 (1977).
11. R. Razakov, S. Y. Yunusov, S. M. Nasyrov, A. N. Chekhlov, V. G. Adrianov, and Y. T. Struckhov, *Chem. Commun.*, 150 (1974).
12. D. S. Bhakuni, H. Uprety, and D. A. Widdowson, *Phytochemistry*, **15**, 739 (1976); Norprotosinomenine is a true precursor of isococculidine, in *Cocculus laurifolius* (Menispermaceae); D. S. Bhakuni, A. N. Singh, and R. S. Kapil, *Chem. Commun.*, 211 (1977).
13. A. T. McPhail, K. D. Onan, H. Furukawa, and M. Ju-ichi, *Tetrahedron Lett.*, 485 (1976).

THE PROTOBERBERINES AND RETROPROTOBERBERINES

Some protoberberines and retroprotoberberines of interest are shown in Scheme 19.1.

19.1. Introduction

Most protoberberine alkaloids occur in nature as tetrahydroprotoberines or as quaternary protoberberinium salts. However, in the last few years a number of quaternary *N*-methyltetrahydroprotoberberine salts as well as *N*-oxides have been reported. In addition to the usual substitution pattern,[9] the C-8 position may bear a methyl group as in the naturally occurring protoberberine alkaloid (−)-corytenchirine. Two recently reported retroprotoberberines are *N*-methyl-[7] and *N,O*-dimethylmecambridine[8]; the chemistry,[10] synthesis,[10] pharmacology,[11] and biogenesis[12] of the protoberberine alkaloids have been reviewed.

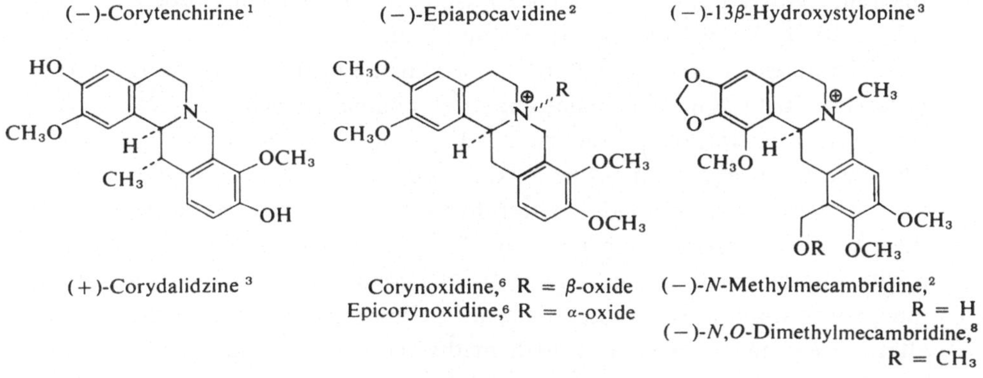

(−)-Corytenchirine[1]

(−)-Epiapocavidine[2]

(−)-13β-Hydroxystylopine[3]

(+)-Corydalidzine[3]

Corynoxidine,[6] R = β-oxide
Epicorynoxidine,[6] R = α-oxide

(−)-*N*-Methylmecambridine,[2]
R = H
(−)-*N,O*-Dimethylmecambridine,[8]
R = CH₃

SCHEME 19.1

19.2. Synthesis

19.2.1. Mannich Condensation

The classical preparation of the tetrahydroprotoberberine skeleton involves Mannich condensation of a benzylisoquinoline such as **1** with formaldehyde, giving products possessing the 2,3,9,10- and 2,3,10,11-oxygenation patterns.[13]

Nandinine   **2**

A study of the effect of pH on the product distribution from the Mannich condensation of **1** revealed that as the pH increases from 1.2 to 7.8 the ratio of nandinine to **2** changes from 1:1 to 1:3.8.[14] In the absence of an activating phenolic hydroxyl at C-3' of the benzylisoquinoline unit, only products having the 2,3,10,11-oxygenation pattern are obtained.

The above observations were applied in the synthesis of the retroprotoberberine alkaloids mecambridine and orientalidine (see Scheme 19.2). Bischler–Napieralski cyclization of the amide **3**, followed by borohydride reduction furnished the benzylisoquinoline **4**. Mannich condensation of **4** with formalin and debenzylation afforded the tetrahydroprotoberberine **5** which possesses the phenol activating group necessary for hydroxymethylation. Treatment of **5** with formaldehyde and base gave the key compound 11-demethylmecambridine, which upon methylation with diazomethane afforded mecambridine,[15] or alternatively on methylenation with methylene chloride–sodium hydride gave orientalidine.[16]

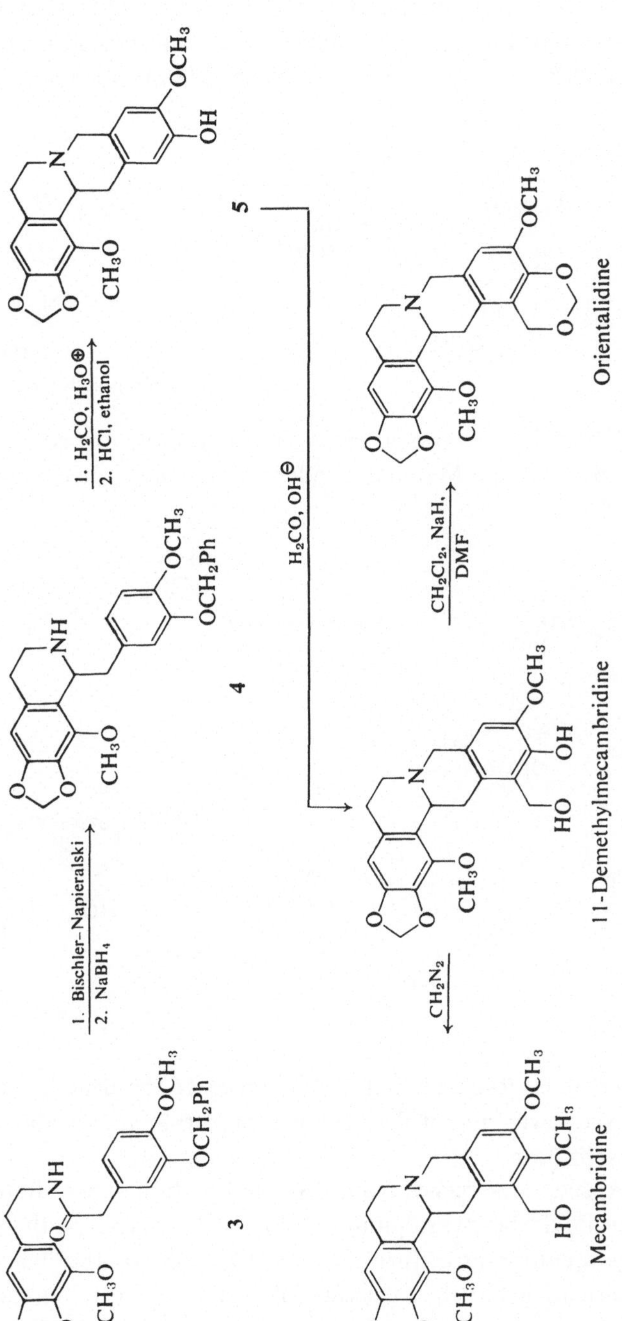
3
1. Bischler–Napieralski
2. NaBH4
4
1. H2CO, H3O⊕
2. HCl, ethanol
5
H2CO, OH⊖
CH2N2
CH2Cl2, NaH,
DMF
Mecambridine
11-Demethylmecambridine
Orientalidine

SCHEME 19.2

Homoxylopinine, a synthetic B-homotetrahydroprotoberberine, was obtained along standard lines from the amide of a γ-substituted propylamine via Bischler–Napieralski cyclization and subsequent Mannich condensation.[17]

B-Homoxylopinine

The use of bromine as a protecting group to promote formation of 2,3,9,10-substituted products in the Mannich condensation is illustrated in the synthesis of capaurine:[18]

Capaurine

Often, however, it is observed that the presence of the deactivating bromine substituent in the lower ring of the Mannich precursor appreciably reduces the yields in the cyclization.

In an unexpected divergence from the usual pattern, it was found that the brominated tetrahydrobenzylisoquinoline **5a**, when treated with 30% formaldehyde in hydrochloric acid, furnished a 75% yield of the benzoxazepino-isoquinoline derivative **5b**, the structure of which was established by X-ray crystallography. A mechanism involving a protoberbinoid intermediate was advanced to explain this transformation (see Ref. 18a).

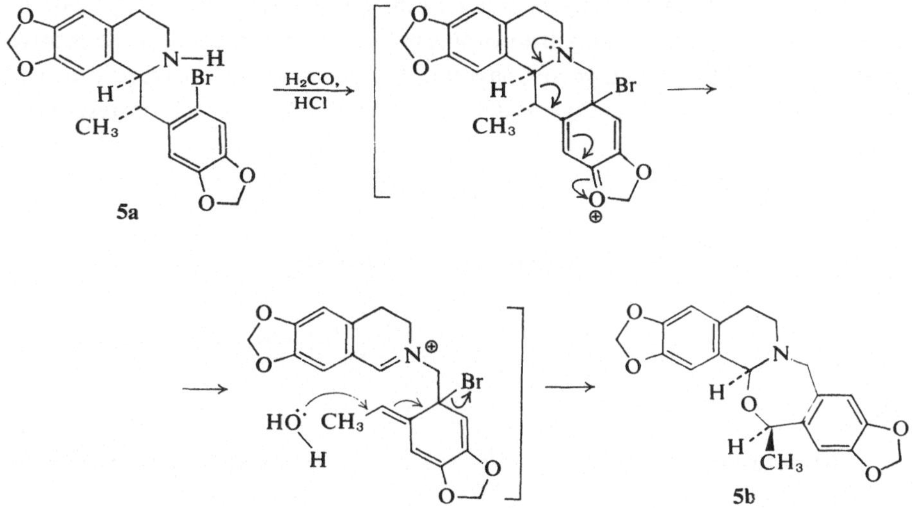

5a

5b

The Mannich cyclization has also been used in the preparation of 13-hydroxylated tetrahydroprotoberberines. Air oxidation of 3,4-dihydropapaverine gave 3,4-dihydropapaveraldine which was reduced with sodium borohydride to furnish only one α-hydroxylated tetrahydrobenzylisoquinoline. Mannich cyclization of this material with formalin gave a 60% yield of 13α-hydroxyxylopinine.[19a]

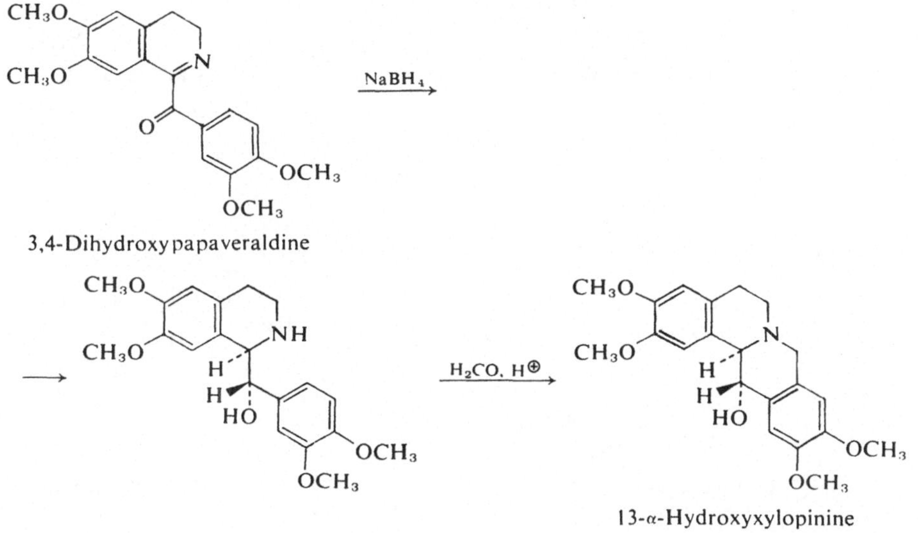

3,4-Dihydroxypapaveraldine

13-α-Hydroxyxylopinine

Utilization of acetaldehyde in the Mannich reaction with benzylisoquinolines leads to 8-methyltetrahydroprotoberberines. Condensation of the resolved enantiomers of ($\pm$)-tetrahydropapaverine with acetaldehyde gave rise to ($+$)-coralydine and ($+$)-*O*-methylcorytenchirine, as well as to the corresponding antipodes[19]:

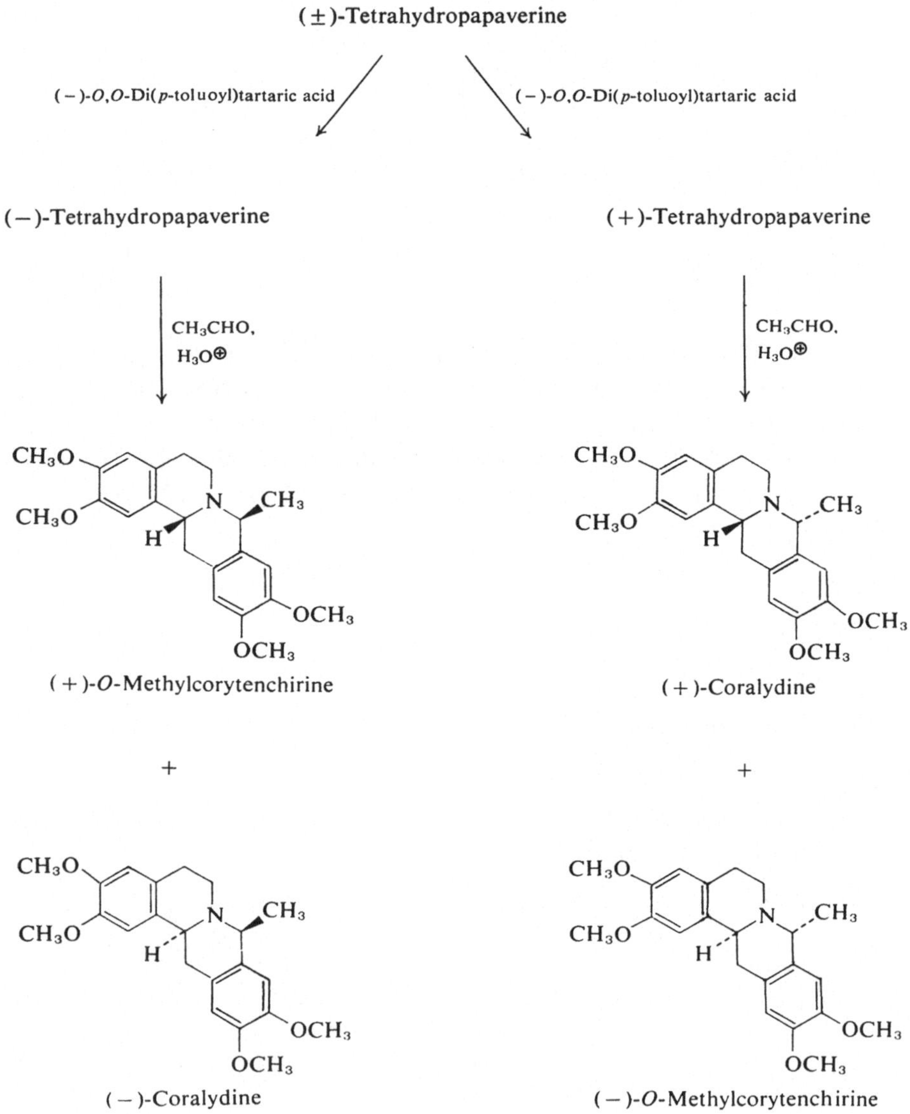

($\pm$)-Tetrahydropapaverine

($-$)-*O*,*O*-Di(*p*-toluoyl)tartaric acid   ($-$)-*O*,*O*-Di(*p*-toluoyl)tartaric acid

($-$)-Tetrahydropapaverine   ($+$)-Tetrahydropapaverine

CH$_3$CHO,   CH$_3$CHO,
H$_3$O$^{\oplus}$ H$_3$O$^{\oplus}$

(+)-*O*-Methylcorytenchirine

(+)-Coralydine

+   +

(−)-Coralydine

(−)-*O*-Methylcorytenchirine

The absolute configurations of (+)-coralydine and (+)-*O*-methylcory-
tenchirine were confirmed by X-ray crystallography.[19]

Alternatively, sodium borohydride reduction of the 1-benzoyl-3,4-dihydro-
isoquinoline **6** produced the alcohol **7** whose debenzylation with hot ethanolic
hydrochloric acid afforded the epimeric phenolic alcohol **8**. Phenolic cyclization of this compound with formalin generated two 13β-hydroxytetrahydro-
berberines in a 1:1 ratio[19a]:

These results were then extended to a synthesis of the alkaloid ophiocarpine.[19a]

Ophiocarpine

19.2.2. Variants of the Bischler–Napieralski Cyclization

The Bischler–Napieralski cyclization of a brominated tetrahydrobenzyl–isoquinoline formamide was employed to obtain the desired 2,3,9,10-substitution pattern required for a synthesis of cheilanthifoline[20]:

Cheilanthifoline, $R_1 = CH_3$, $R_2 = H$
Tetrahydrogroenlandicine, $R_1 = H$, $R_2 = CH_3$

Tetrahydrogroenlandicine was synthesized in low yield by a similar route.[21]

An interesting approach to protoberberines has been reported which utilizes the ortho hydroxymethylation of phenols with boronic acids[22] to generate the δ-lactone **9** in over 80% yield. Subsequent *O*-methylation and condensation with a phenethylamine supplied an amidoalcohol which upon Bischler–Napieralski cyclization and borohydride reduction afforded the desired tetrahydroprotoberberine **10** in 65% overall yield (see Scheme 19.3).[23] For a related conversion of an isoquinolone to a protoberberine lactam, see Sec. 3.3.

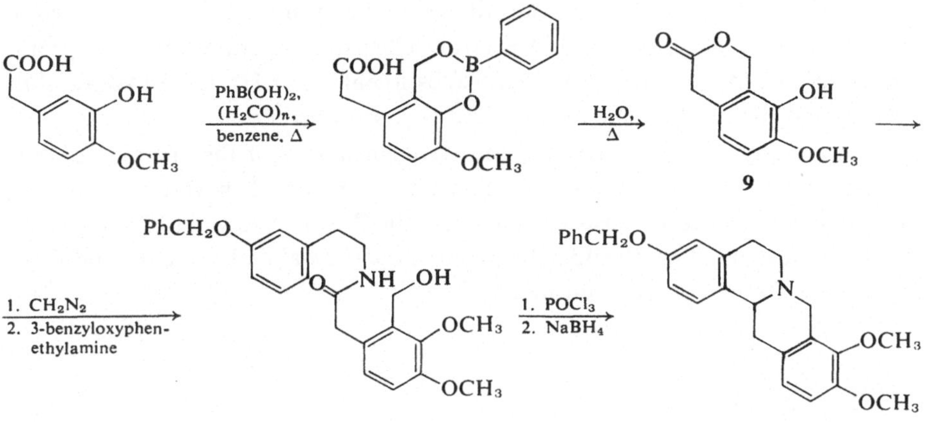

A tetrahydroprotoberberine

10

SCHEME 19.3

19.2.3. Enamide Photocyclization

The aroylation of 1-alkyl-3,4-dihydroisoquinolines produces enamides which readily undergo photocyclization to oxoberberine derivatives.[24] The reaction is analogous to the hexatriene–cyclohexadiene rearrangement and proceeds most favorably (50–80% yield) when the irradiations are performed on degassed solutions[25]:

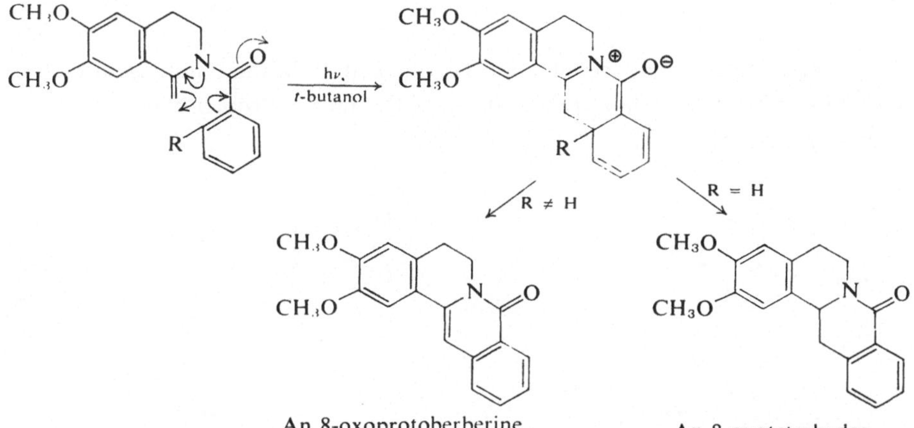

An 8-oxoprotoberberine

An 8-oxotetrahydro-
protoberberine

If the aroyl group is ortho substituted by the group R, that group is eliminated in the reaction as HR, and electron redistribution affords an 8-oxoprotoberberine. In the absence of an ortho substituent (R = H), a 1,5 hydride shift occurs and the product is an 8-oxotetrahydroprotoberberine.

In a study of the mechanism and stereochemistry of the enamide photocyclization, Lenz has demonstrated that ethylidene and benzylidene isoquinoline enamides cyclize stereospecifically from the Z-isomer to yield 13-substituted oxoberberines in which the 13-substituent is quasi-axial and the C-13 and C-14 hydrogens are in a cis relationship.[26]

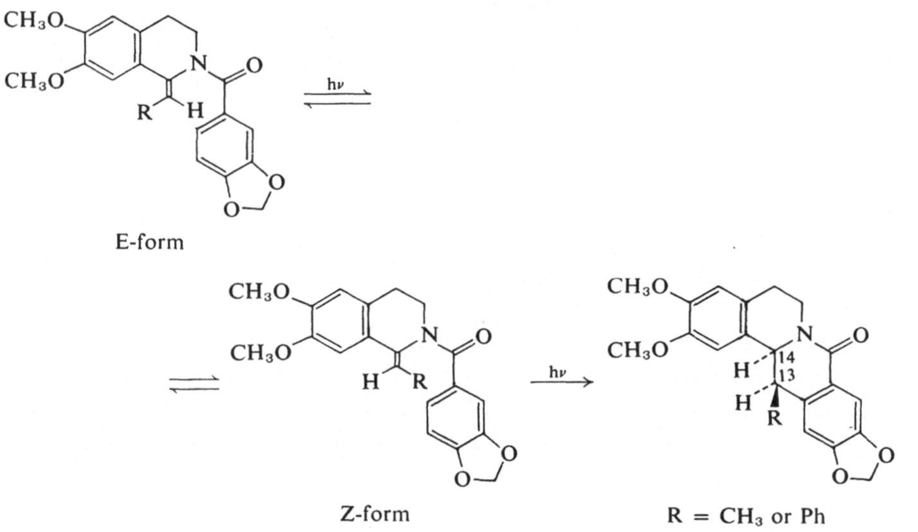

E-form

Z-form   R = CH₃ or Ph

Photocyclization of the enamide **11** followed by lithium aluminum hydride reduction provided a short preparation of the tetrahydroprotoberberine xylopinine in excellent yield.[27,28]

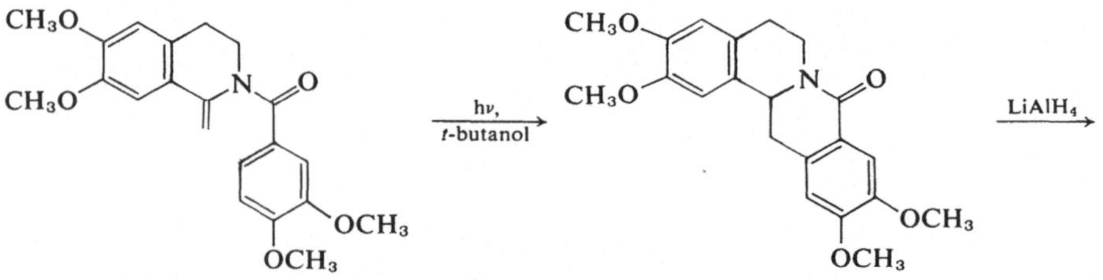

11

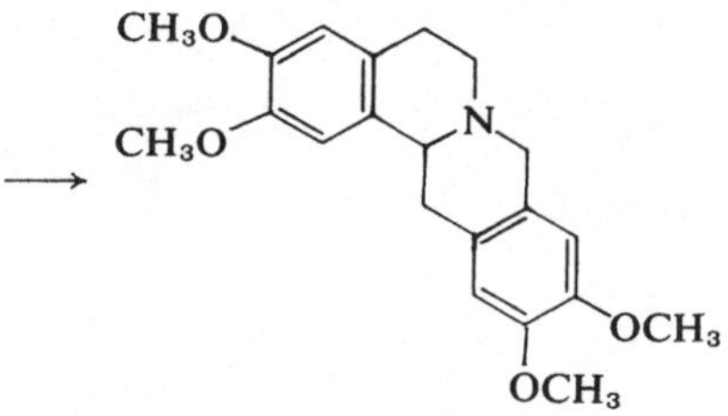

Xylopinine

The 13-methyltetrahydroprotoberberine base cavidine was synthesized by aroylation of a 1-ethyl-3,4-dihydroisoquinoline followed by enamide photocyclization which gave, in addition to the desired product **13**, an undesired side product **12**. Subsequent two-step reduction of **13** afforded cavidine[29]:

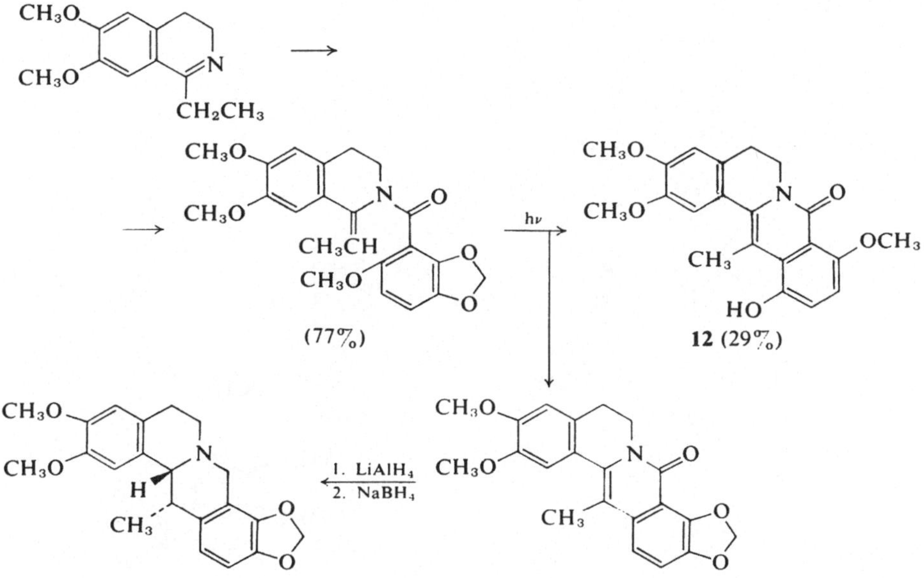

(77%)

12 (29%)

13 (41%)

Cavidine

Tetrahydropalmatine and sinactine have been similarly prepared.[30]

N-Formylation of 3,4-dihydrobenzylisoquinolines with mixed formic–acetic anhydride yields a mixture of E- and Z-enamides which can be differentiated by NMR and UV spectroscopy. The predominant isomer is the thermodynamically more stable Z compound.

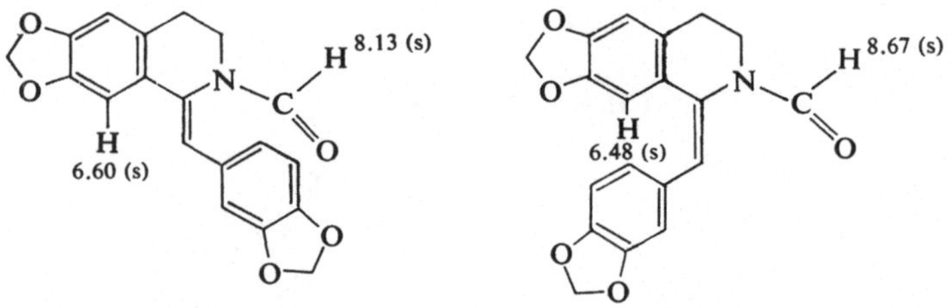

Z-form
λ_max 337 nm (4.33)

E-form
λ_max 298 nm (4.13)

Irradiation of the Z- or of a mixture of the E- and Z-isomers of the *N*-formyl enamides in the presence of hydriodic acid furnished the corresponding protoberberine salts in good yields. The photocyclization is regiospecific, forming 10,11-substituted protoberberine salts from 3′,4′-disubstituted benzylidene enamides. A suggested mechanism is[30a]

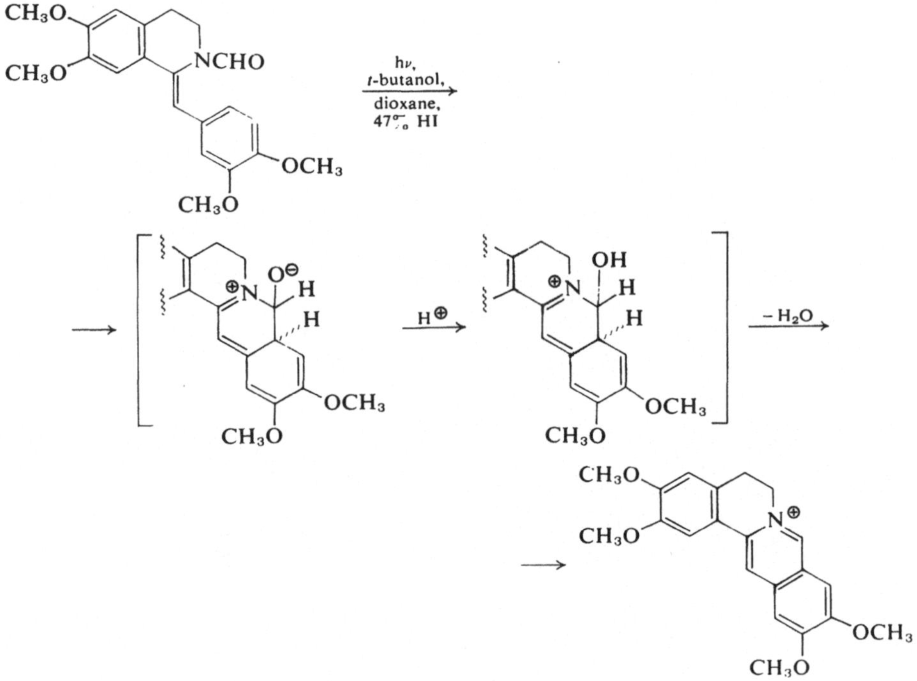

These transformations represent the first direct photochemical preparation of a quaternary protoberberine salt from an enamide.

9,10-Substituted protoberberines have been synthesized in good yields by photolysis of the corresponding bromoenamides (R = Br).[30b]

Corytenchirine, found in *Corydalis ochotensis* Turcz. (Fumariaceae) has been prepared in the racemic form by irradiation of an *N*-acetylenamide followed by reduction of the resulting protoberberinium salt and debenzylation.[30c]

An alternate synthesis of corytenchirine involves Mannich cyclization of the required tetrahydrobenzylisoquinoline precursor with acetaldehyde. A diastereomer of corytenchirine was also produced in this instance, but in minor amounts.[30c] (See also Sec. 19.2.1.)

Corytenchirine
(minor product)

(major product)

19.2.4. Thermolysis of Benzocyclobutenes

Benzocyclobutene derivatives, upon heating, open in a conrotatory fashion to give reactive *o*-quinodimethides which react with enamines or imines to produce isoquinolines.[31] This approach has been utilized in the synthesis of the protoberberines xylopinine[32] and discretine.[33] In the synthesis of coreximine, the bromobenzaldehyde **14** was treated with cyanoacetic acid. Subsequent borohydride reduction and decarboxylation afforded the phenylpropionitrile **15**. Sodamide treatment of **15** produced the cyanobenzocyclobutene **16** through a benzyne intermediate. Hydrolysis of the cyano group, condensation with the appropriately substituted phenethylamine, and Bischler–Napieralski cyclization of the resulting amide, generated the thermolysis precursor **17**. Unexpectedly, the thermolysis product was the D-ring monophenolic protoberberine salt **18**, debenzylation of which afforded coreximine. The pyrolysis reaction initially generates a dihydroprotoberberine intermediate, but rapid oxidation to the aromatic protoberberine must occur[34]:

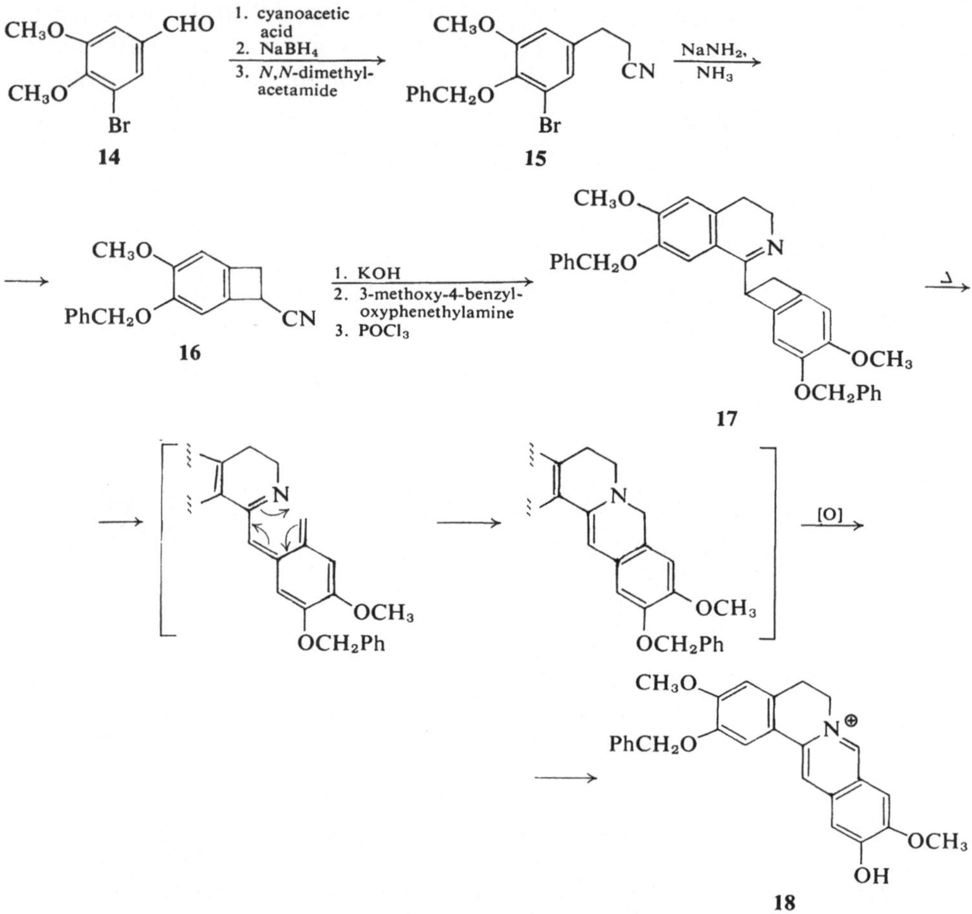

In a further development, it was found that simple admixture and heating of bromobenzocyclobutene and the 3,4-dihydroisoquinoline **19** gave directly the aromatic protoberberinium salt **20**.[35]

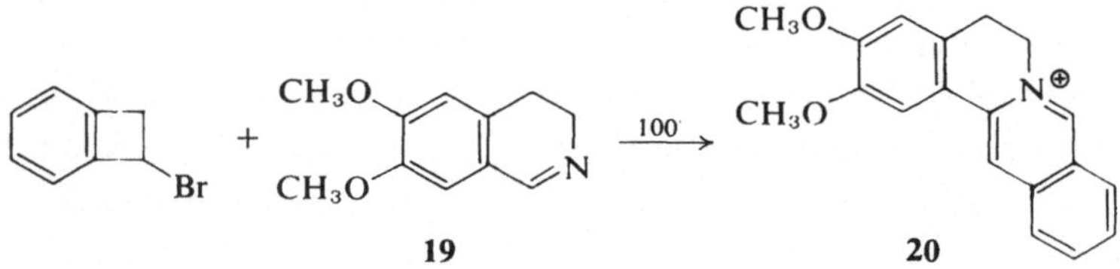

The C-8 methylated tetrahydroprotoberberine alkaloids coralydine and *O*-methylcorytenchirine have also been synthesized via benzocyclobutene thermolysis followed by reduction.[35]

19.2.5. Pomeranz–Fritsch Cyclization

Dyke and Tiley have successfully used the Bobbitt modification of the Pomeranz–Fritsch cyclization in a synthesis of berberastine. The deoxybenzoin **21** was condensed with aminoacetaldehyde dimethyl acetal and the resulting Schiff base was reduced with borohydride to give amine **22**. The remaining key steps were a Mannich condensation and a Pomeranz–Fritsch acid-catalyzed cyclization[36]:

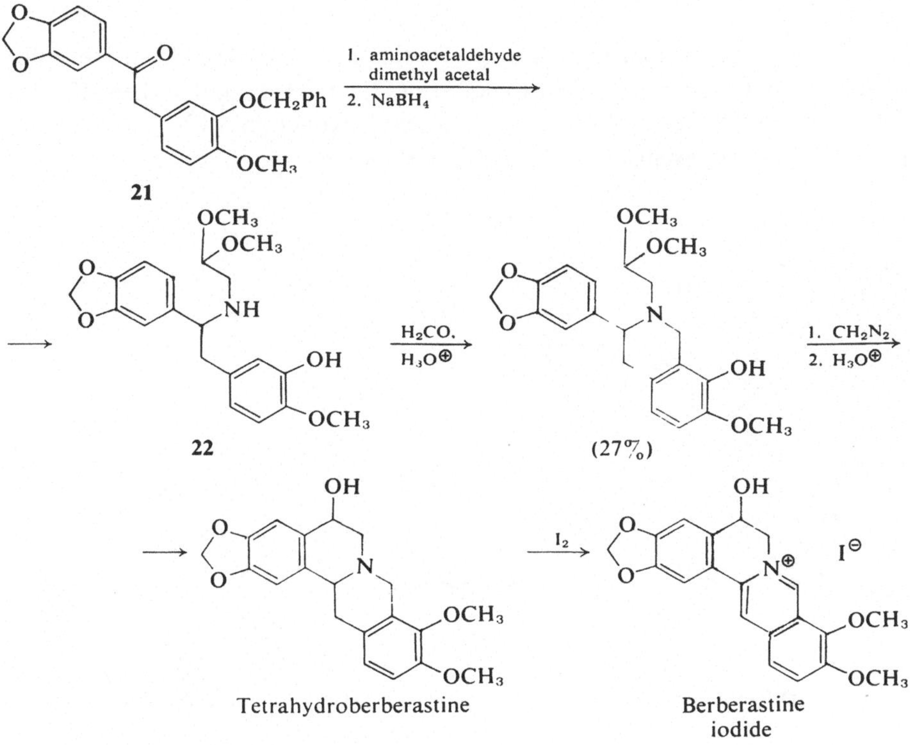

Tetrahydroberberastine

Berberastine iodide

19.2.6. From Immonium Salts

Rapoport and co-workers have established that brief heating of an α-tertiary amino acid, e.g., **24**, in phosphorus oxychloride leads regiospecifically to a high yield of the corresponding immonium salt (**25**). This salt can be cyclized intramolecularly to form a tetrahydroprotoberberine[37]:

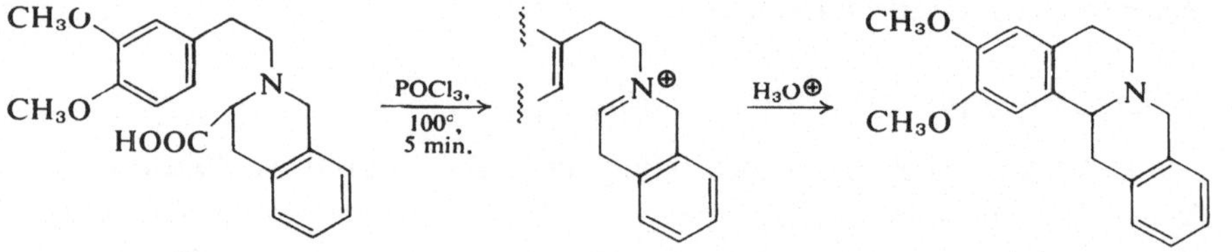

23   **24**   (79% overall)

19.2.7. From Protopines

The unusual alkaloid 13β-hydroxystylopine was synthesized from protopine which was first converted to dihydrocoptisine. Enamine oxidation then afforded the corresponding phenolbetaine which, upon reduction with sodium borohydride, provided exclusively 13β-hydroxystylopine[3]:

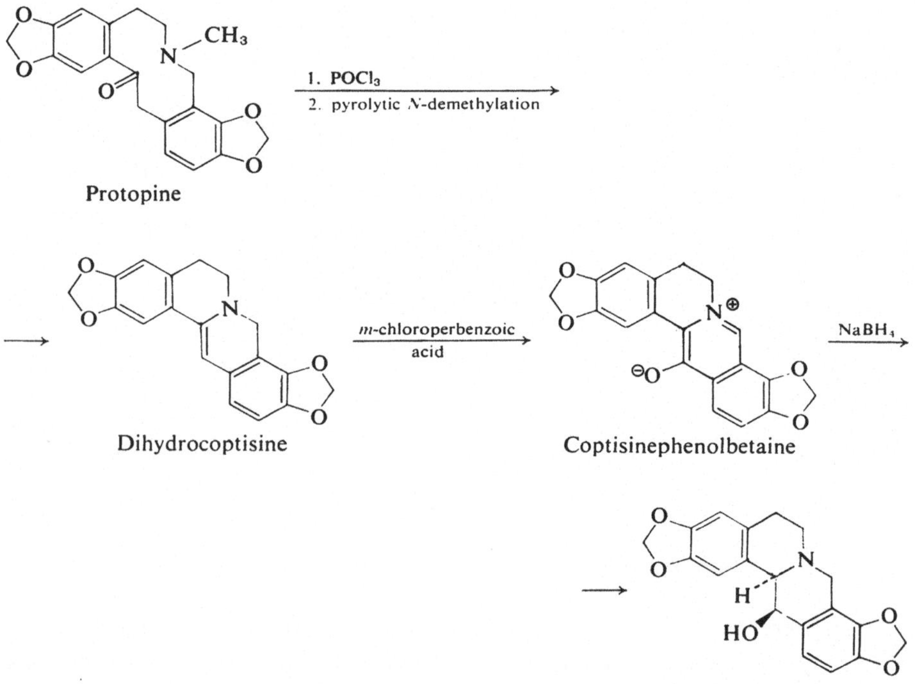

Protopine

1. POCl₃
2. pyrolytic *N*-demethylation

Dihydrocoptisine

m-chloroperbenzoic acid

Coptisinephenolbetaine

NaBH₄

13β-Hydroxystylopine

19.2.8. From Spirobenzylisoquinolines

Irie and co-workers have prepared xylopinine in excellent yield by a photochemical route.[38] 3,4-Dihydroxyphenethylamine was condensed with the indandione 25 under Pictet–Spengler conditions, and methylation of the product produced the spirobenzylisoquinoline ketone 26. Irradiation of 26 furnished the quaternary protoberberinium salt 27. Sodium borohydride reduction to xylopinine proceeded quantitatively:

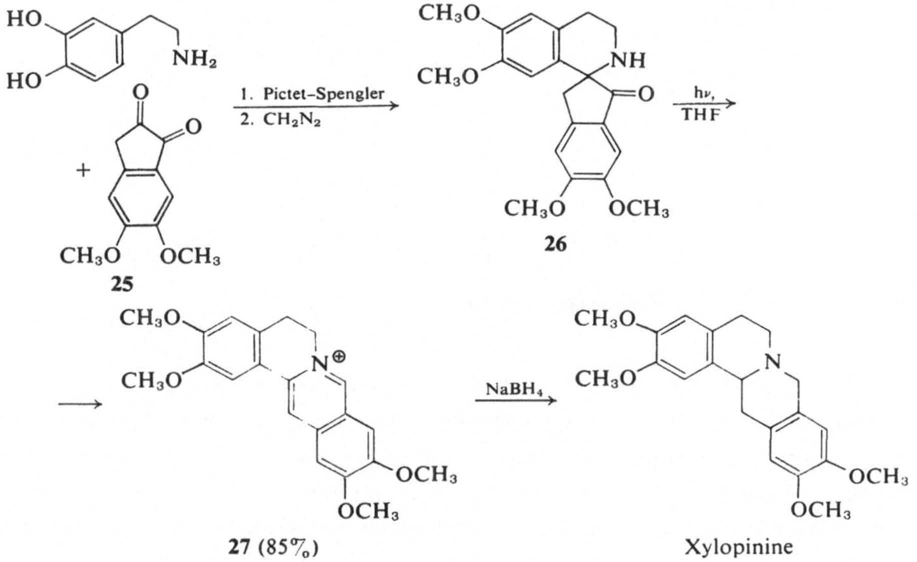

In a separate study, irradiation of 28 in ether gave rise to the pyridone 29; but when the irradiation was performed in the presence of concentrated hydrochloric acid, the major product was the protoberberinium salt 30.[38a]

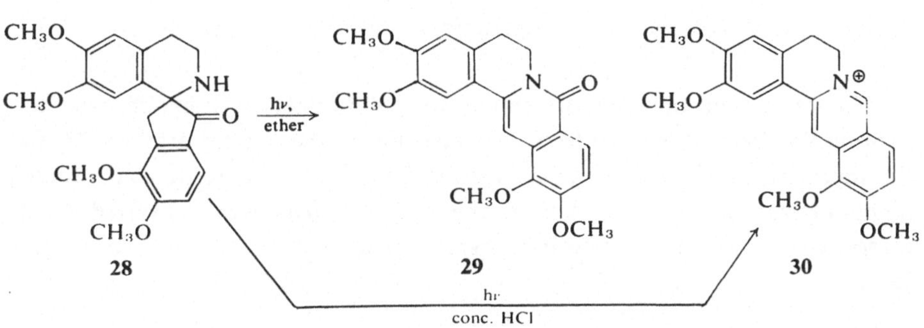

The rearrangement of a ketonic spirobenzylisoquinoline to a protoberbinoid derivative can also be induced by base. Thus the spirobenzylisoquinoline **28** led to the lactam **31** through the intermediacy of an aziridinol. No rearrangement occurred if the nitrogen atom of the starting material was tertiary.[38a]

31 (41%)

19.2.9. From Norphthalideisoquinolines

The synthesis of 13-hydroxylated tetrahydroprotoberberines from norphthalideisoquinolines has been achieved. Catalytic reduction of the phthalideisoquinoline **32**, derived from papaverine (see Sec. 24.2.3), afforded a separable mixture or *erythro-* and *threo*-norphthalideisoquinolines. Base-catalyzed lactamization followed by lithium aluminum hydride reduction of the *threo*-nor and *erythro*-norphthalides yielded the 13β- and 13α-hydroxytetrahydroprotoberberines, respectively. The hydroxylactam derived from the *threo*-norphthalide tends to dehydrate readily to the oxoprotoberberine system (see Scheme 19.4).[39]

19.2.10. From Homophthalic Anhydrides

Condensation of norhydrastinine with 4,5-dimethoxyhomophthalic anhydride has been shown to lead to a mixture of the kinetic product **33** and the thermodynamic product **33a**. Heating **33** in refluxing acetic acid resulted in epimerization to **33a**. Each of the two products was converted to its respective 13-methyltetrahydroprotoberberine derivative as shown in Scheme 19.5.[39a]

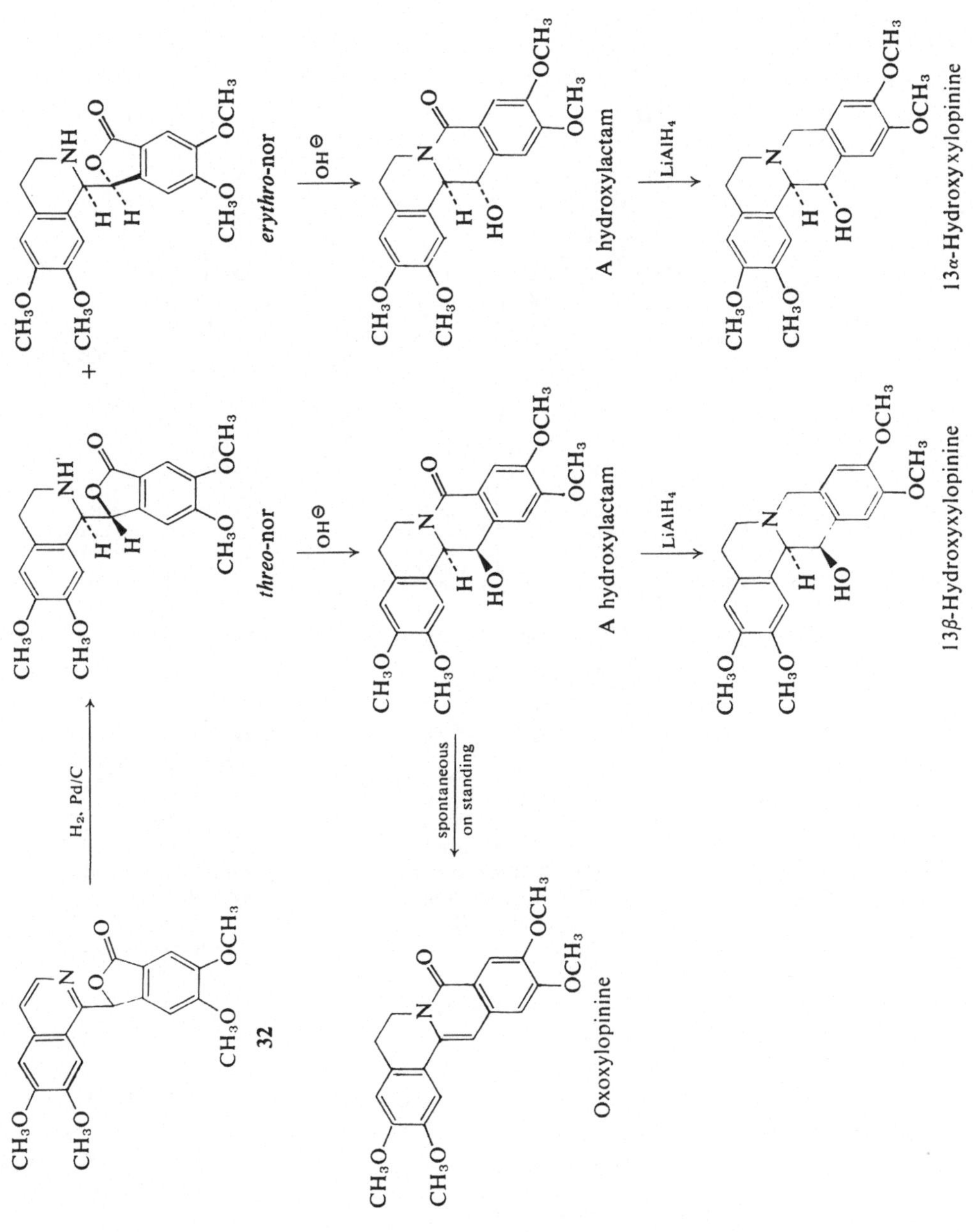
32
H2, Pd/C
threo-nor
erythro-nor
OH⊖
spontaneous on standing
Oxoxylopinine
A hydroxylactam
A hydroxylactam
LiAlH4
13β-Hydroxyxylopinine
13α-Hydroxyxylopinine

SCHEME 19.4

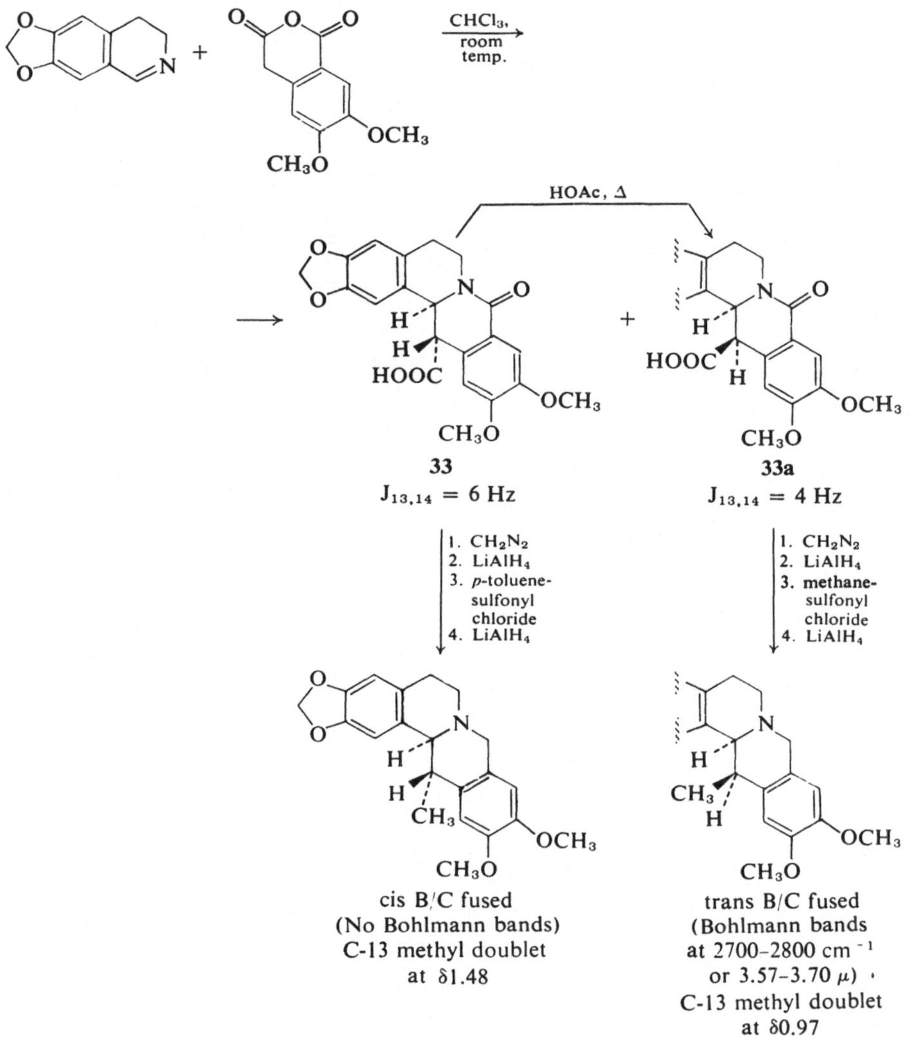

SCHEME 19.5

19.3. Reactions

19.3.1. Carbon–Nitrogen Bond Cleavage

Hofmann degradation of quaternary *N*-methyltetrahydroprotoberberine salts induced by methanolic alkali results in scission of ring B and, less often, in alternate cleavage of ring C to afford methine bases.[40] With the diphenolic coreximine methochloride however, the products were the secoberbine 34 (30%) and the novel tetracyclic base 35 (15%). Heating the secoberbine 34 with methanolic alkali for 24 hr gave a good yield of the tetracyclic base 35; and indeed, when coreximine methochloride was heated for 40 hr under these conditions, bases 34 and 35 were obtained in 10 and 60% yields, respectively. The structure of the tetracyclic base 35 was confirmed by the synthesis of its *O,O*-dimethyl derivative from laudanosine[41]:

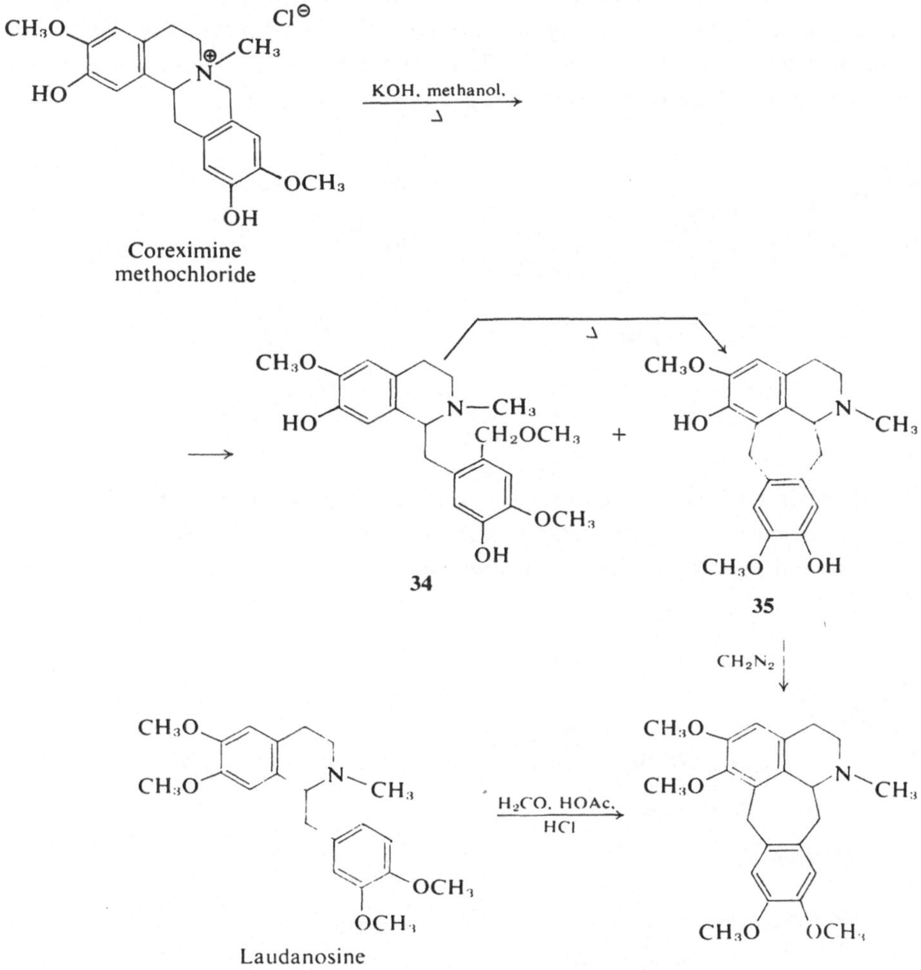

Coreximine
methochloride

34

35

Laudanosine

The monophenolic discretine methiodide, by contrast, under the same conditions, yielded only the normal methine base 36.[42]

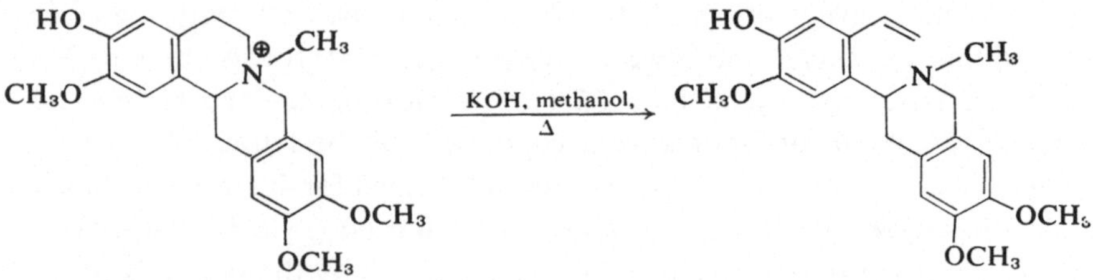

Discretine methiodide   36 (43%)

A requirement for cleavage of ring C is the presence of a phenolic function at C-9 or C-11 in the starting methiodide salt; in addition, for the formation of a tetracyclic base such as **27**, a phenol is needed at C-2. A plausible mechanism for these transformations is shown below.

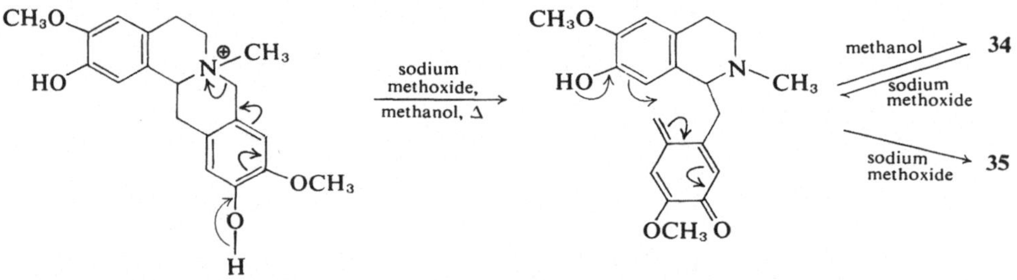

Similar secoberbine derivatives are obtained when nonphenolic tetrahydroprotoberberines are treated with acylating agents in the presence of sodium iodide. *N*-Methylsecotetrahydropalmatinol was prepared in such fashion[43]:

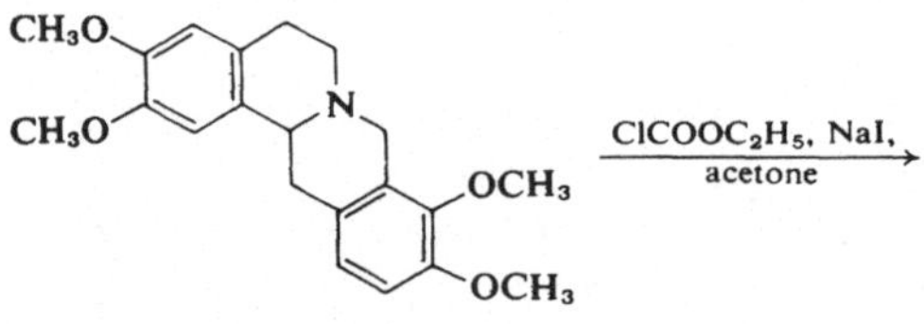

Tetrahydropalmatine

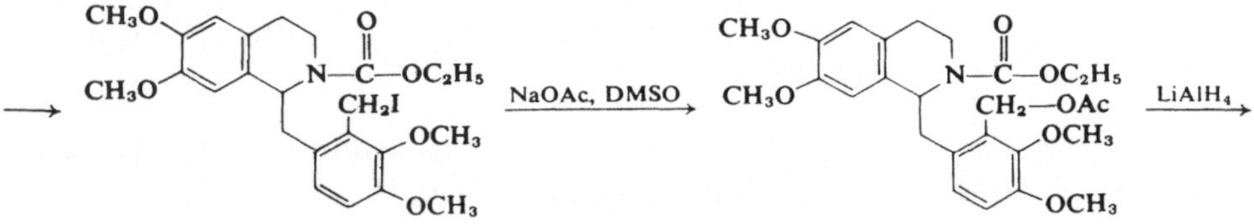

N-Methylsecotetrahydro-
palmatinol

Nagata and co-workers have reported the preferential (91%) cleavage of
N-methyltetrahydroprotoberberine salts at the N-7 to C-14 bond by Birch
reduction. The use of the cyclopropylmethyl ether protecting group in this
instance is to be noted.[44] The cyclopropylmethyl group can withstand sodium
borohydride, lithium aluminum hydride, and catalytic reduction using a palladium catalyst, as well as Bischler–Napieralski conditions, thus illustrating
its high utility as a phenol-protecting group.

Similarly, the Birch-type reduction of various tetrahydroisoquinoline
N-oxides has been reported to generate secondary amines in good yields
through cleavage of the central carbon to nitrogen bridge.[45]

The N-7 to C-8 bond of protoberberinium salts can be cleaved by use of
sodium acetate in acetic anhydride, leading directly to β-naphthol derivatives. In
the case of berberine, the major product (30%) undergoes keto–enol tautomer-

ization, while the minor product (24%) is a vinyl diacetate.[46] A mechanism to explain this transformation is offered below. Bridged intermediates related to **37** have been encountered in the Diels–Alder addition of ketene diethyl acetal to isoquinolinium and acridizinium salts[47]:

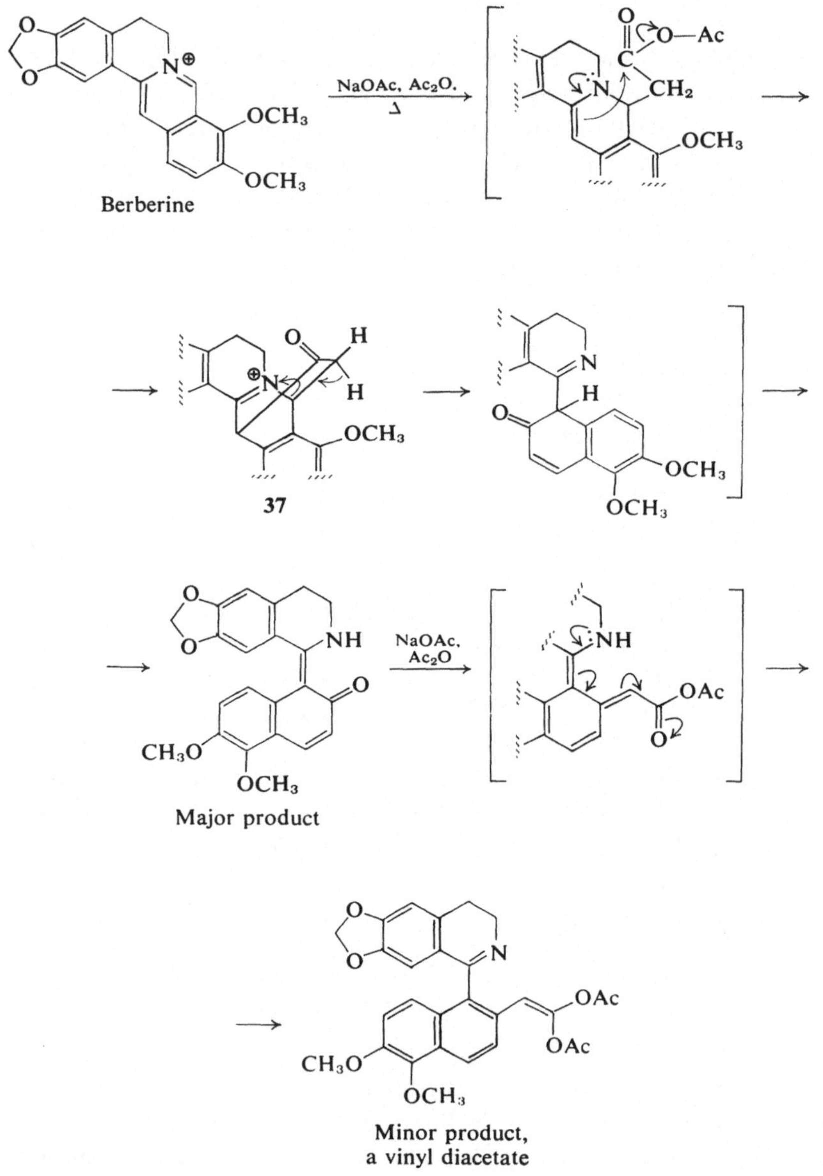

Berberine

37

Major product

Minor product,
a vinyl diacetate

19.3.2. Rearrangements

Pyrolysis of tetrahydroprotoberberine methiodides is reported to yield, in addition to the expected N-demethylation product, 13-methyltetrahydro-protoberberine via migration of the N-methyl group to C-13. This rearranged product possesses a *trans*-quinolizidine system with the 13-methyl group occupying the axial position.[48] (For the base-catalyzed rearrangement of N-methyltetrahydroprotoberberine salts to spirobenzylisoquinolines, see Sec. 25.2.7.)

(±)-Stylopine methiodide

19.3.3. O-Demethylation of Berberinium Salts

Conditions have been reported for a one-step O-demethylation-O-acylation. Berberine was thus converted to 9-O-acylberberrubine salts.[49]

A 9-O-acylberberrubine salt

19.3.4. Oxidation

The chemistry of protoberberine–acetone adducts has been reconsidered within the context of ease of alkylation and oxidation. Naruto and his students have reported that reactive alkyl halides such as methyl iodide react with berberine–acetone to yield, in addition to the expected berberine and 13-methylberberine, the bridged derivatives **38** and **39**.[50]

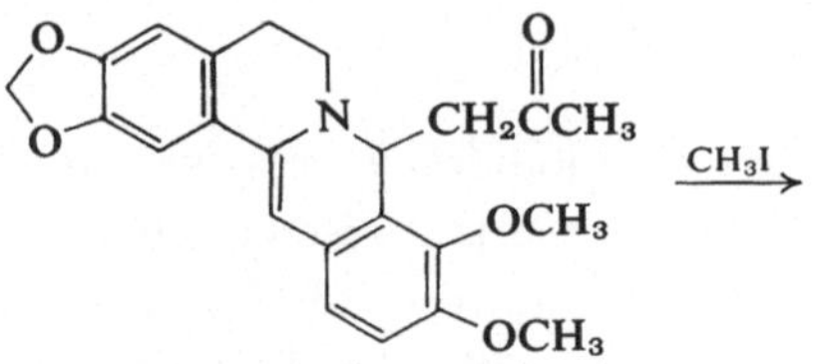

Berberine–acetone

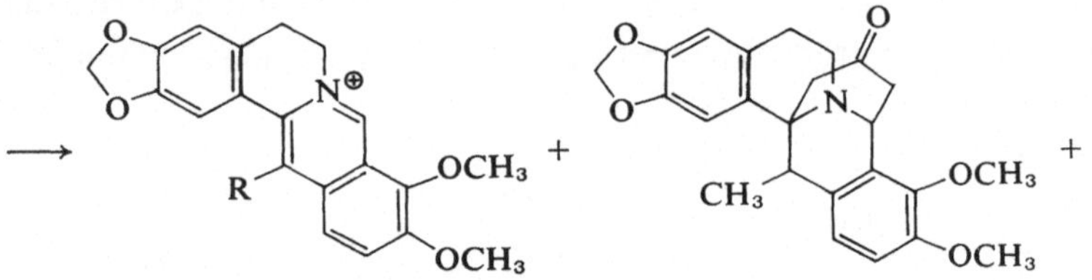

Berberine, R = H
13-Methylberberine, R = CH₃

38

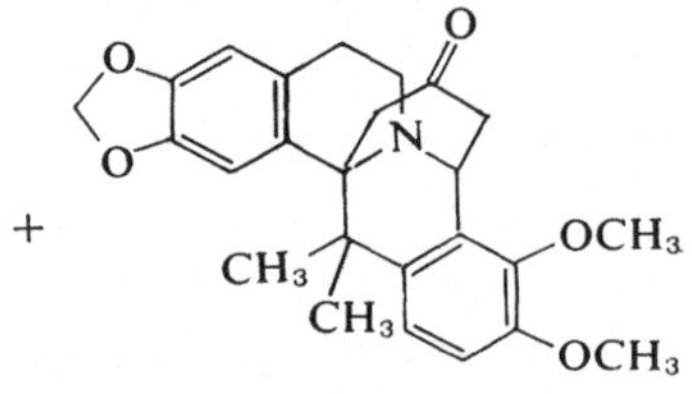

39

They also found that while palmatine–acetone oxidizes with permanganate to afford the expected 8,14-propano derivative **40**, similar oxidation of 13-methylpalmatine–acetone led to cleavage of the 13–14 bond to yield the amidoketone **41**[51]:

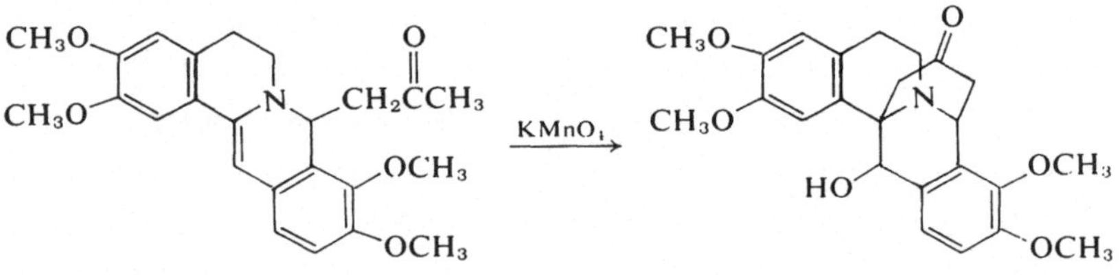

Palmatine–acetone

40

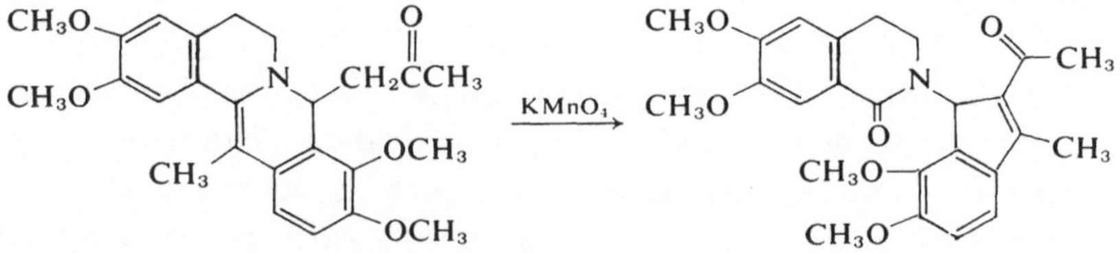

13-Methylpalmatine–acetone

41

8,14-Propano bridge formation must be fairly facile since it has been observed that the crude acetone adducts of a mixture of quaternary protoberberines were contaminated with small amounts of propano derivatives such as **42**, as a result of intramolecular condensation.[52]

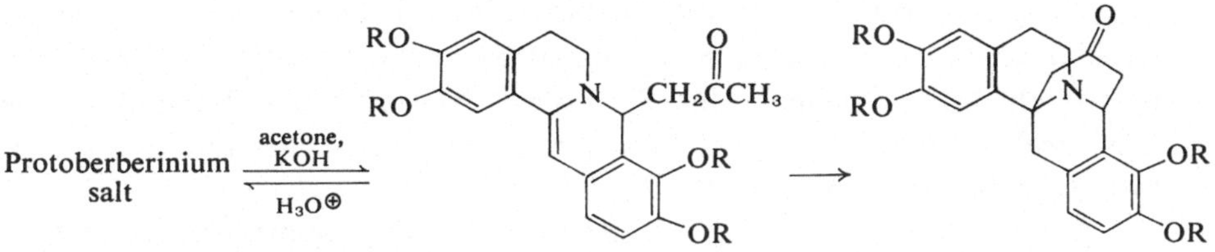

Berberine–acetone when subjected to mild permanganate oxidation is converted to neoxyberberine–acetone and to the lactamic acid **43** reminiscent of Perkin's berberal.[53]

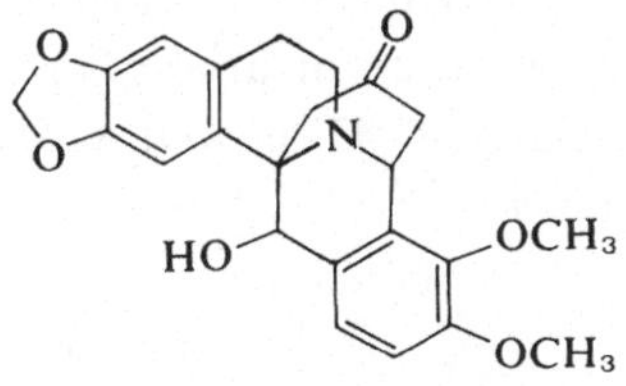

Neoxyberberine–acetone

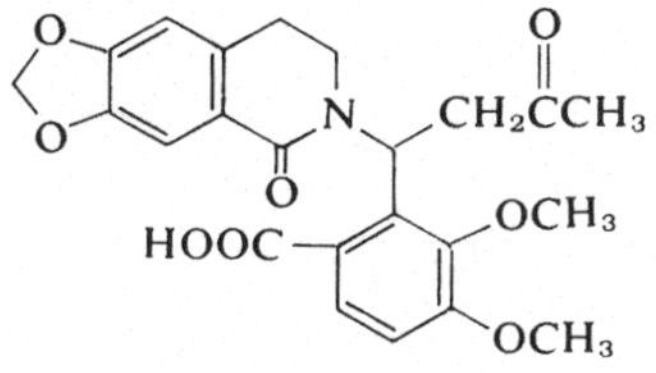

43

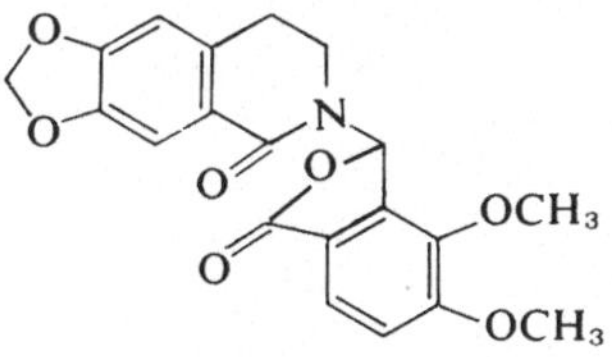

Berberal

19.3.5. The Chemistry of Oxybisberberine

Although at the time of publication the structure of oxybisberberine, the crystalline ferricyanide oxidation product of berberine, is still unknown in its details, the chemistry of this dimer has proved to be rich in its diversity. Oxybisberberine played a key role in the first known *in vitro* conversion of berberine to the phthalideisoquinoline alkaloid β-hydrastine.[54] This photosensitive dimer, which shows no carbonyl absorption in the IR, is stable to base, but cleaves rapidly and irreversibly in acid. In the simple and efficient conversion to hydrastine, oxybisberberine was cleaved with methanolic hydrogen chloride to produce berberine chloride and 8-methoxyberberinephenolbetaine. The phenolbetaine has the requisite oxygen function at C-13 as well as the potential carboxylic ester at C-8 for transformation to a phthalideisoquinoline. The unmasking of the C-8 carboxyl was achieved by simple hydration to furnish dehydronorhydrastine methyl ester. Subsequent *N*-alkylation with methyl iodide and borohydride reduction led to a mixture of ($\pm$)-α-hydrastine and ($\pm$)-β-hydrastine (see Scheme 19.6).[54]

Alternatively, treatment of oxybisberberine with pyridine hydrochloride in pyridine led to oxidative scission, and the nature of the products was dictated by the nucleophiles present and by the pH of the medium during the work-up. When the reaction was quenched with methanolic and aqueous acids, the products were the 8,13-dioxoberbines **44** and **45**, respectively. These compounds undergo rapid, reversible, interconversion in acidic media through an immonium quinoid species. On the other hand, neutral work-up with buffered systems gives rise to the aporhoeadane **46**, which is also produced rapidly but irreversibly on treatment of **45** with ammonium hydroxide. Alkaline work-up invariably produced as the major product Perkin's anhydroberberilic acid **47** as well as its solvolysis product noroxyhydrastinine (see Scheme 19.7).

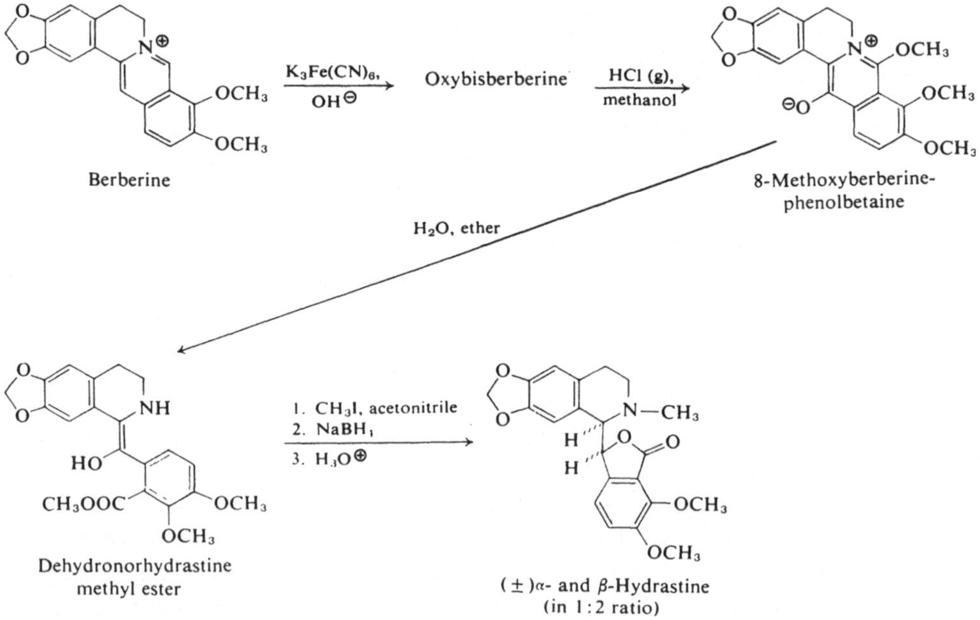

SCHEME 19.6

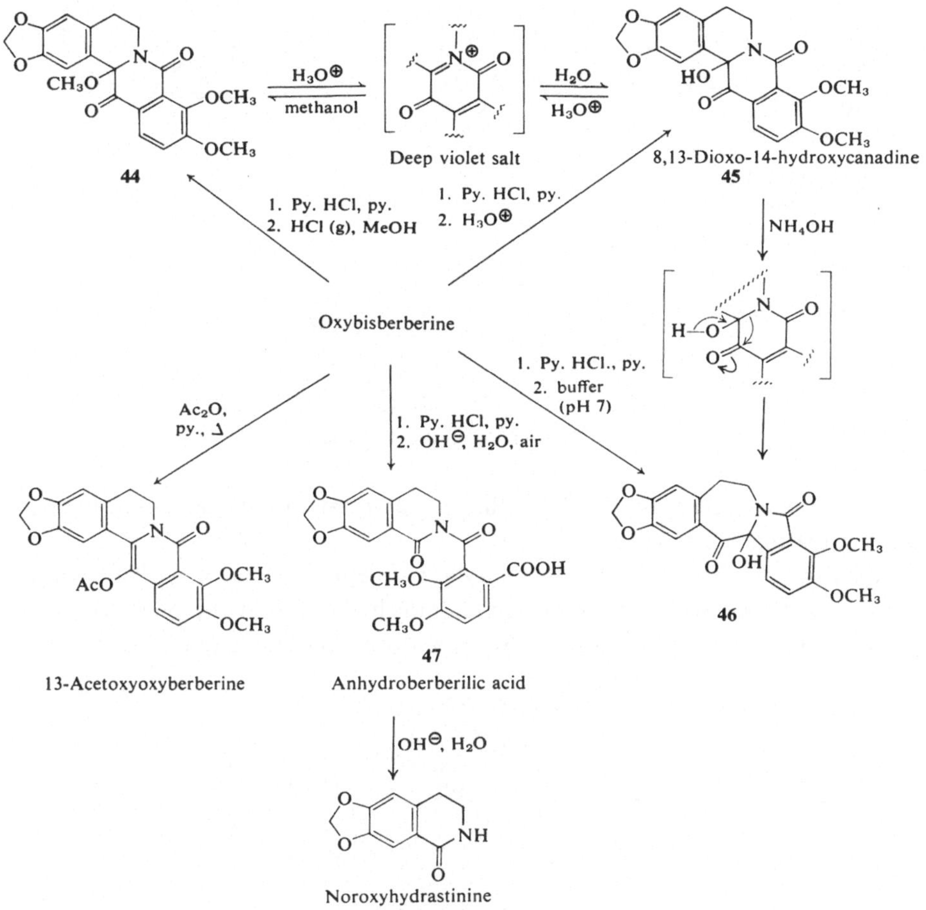

SCHEME 19.7

Direct acetylation of oxybisberberine with acetic anhydride in pyridine provided 13-acetoxyoxyberberine. Berberine is inevitably produced upon cleavage of oxybisberberine, and therefore accompanies such products as 8-methoxyberberinephenolbetaine or the dioxoberbines **44** and **45**.[54]

In contrast to the hydration of 8-methoxyberberinephenolbetaine, berberinephenolbetaine itself under identical hydration conditions leads to numerous highly oxidized products, including **45–47**.[54]

The Further Chemistry of Berberinephenolbetaine, 8-Methoxyberberinephenolbetaine, and 8,13-Dioxo-14-hydroxycanadine. Hydration of 8-methoxyberberinephenolbetaine in wet tetrahydrofuran, a solvent in which it is appreciably soluble, generates methyl isoanhydroberberilate in 80% yield by the

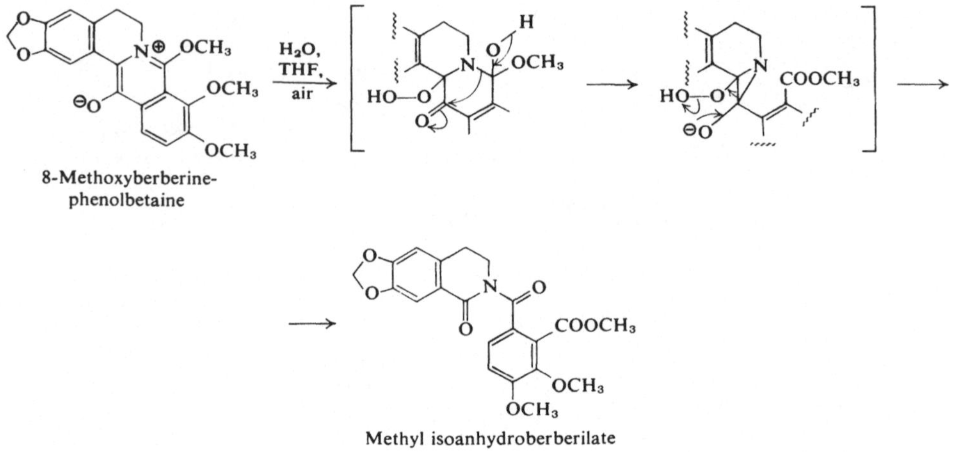

8-Methoxyberberine-
phenolbetaine

Methyl isoanhydroberberilate

oxidative mechanism indicated, so that an unusual overall carbon-to-nitrogen acyl migration takes place.[54a]

If, on the other hand, 8-methoxyberberinephenolbetaine is *O*-acetylated, and then treated with methanolic potassium hydroxide in the presence of air, no rearrangement occurs, and methyl anhydroberberilate is obtained.[54a]

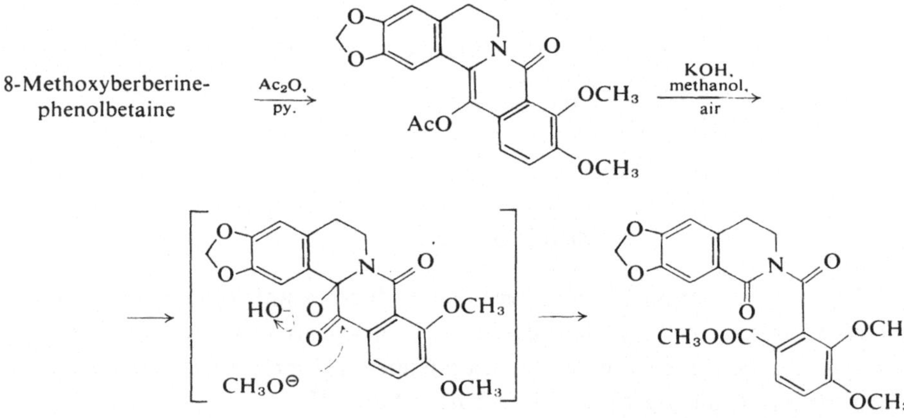

8-Methoxyberberine-
phenolbetaine

Methyl anhydroberberilate

An alternate preparation of 8-methoxyberberinephenolbetaine proceeds through irradiation of berberine in methanol containing rose bengal and sodium methoxide, with a stream of oxygen. A precipitate is thus obtained which upon recrystallization from methanol yields orange-colored crystals of 8-methoxy-berberinephenolbetaine.[54b] Borohydride reduction then affords mostly ophio-carpine and a much smaller yield of epiophiocarpine.[54b,54c]

Ophiocarpine   Epiophiocarpine

If 7,8-dihydroberberine is irradiated in the presence of rose bengal and oxygen, the known berberinephenolbetaine is obtained in good yield. Further irradiation of a concentrated methanolic solution of berberinephenolbetaine in the presence of oxygen gives a stable peroxide whose borohydride reduction again leads to mostly ophiocarpine and a little epiophiocarpine.[54d]

Berberinephenolbetaine   A stable peroxide

Irradiation of a methanolic solution of 8-methoxyberberinephenolbetaine in a stream of oxygen takes a different course and affords a spirobenzylisoquinoline, which upon *N*-acetylation and acid hydrolysis, gives rise to a diketospirobenzylisoquinoline.[54e]

8-Methoxyberberine-
phenolbetaine

A diketospiro-
benzylisoquinoline

A stereospecific conversion of berberine to β-hydrastine has been achieved through the intermediacy of 8,13-dioxo-14-hydroxycanadine (**45**). Treatment of this colorless compound with 25% aqueous sulfuric acid produces instantly the deep violet color characteristic of dioxoimmonium salt formation. Upon heating *in situ*, hydrolysis to a water-soluble yellow immonium keto acid takes place. Neutralization then supplies a colorless γ-lactol in 90% overall yield from **45**. *N*-Methylation of the γ-lactol, followed by borohydride reduction, generates essentially pure racemic β-hydrastine with none of the diastereomeric α-hydrastine.[54f]

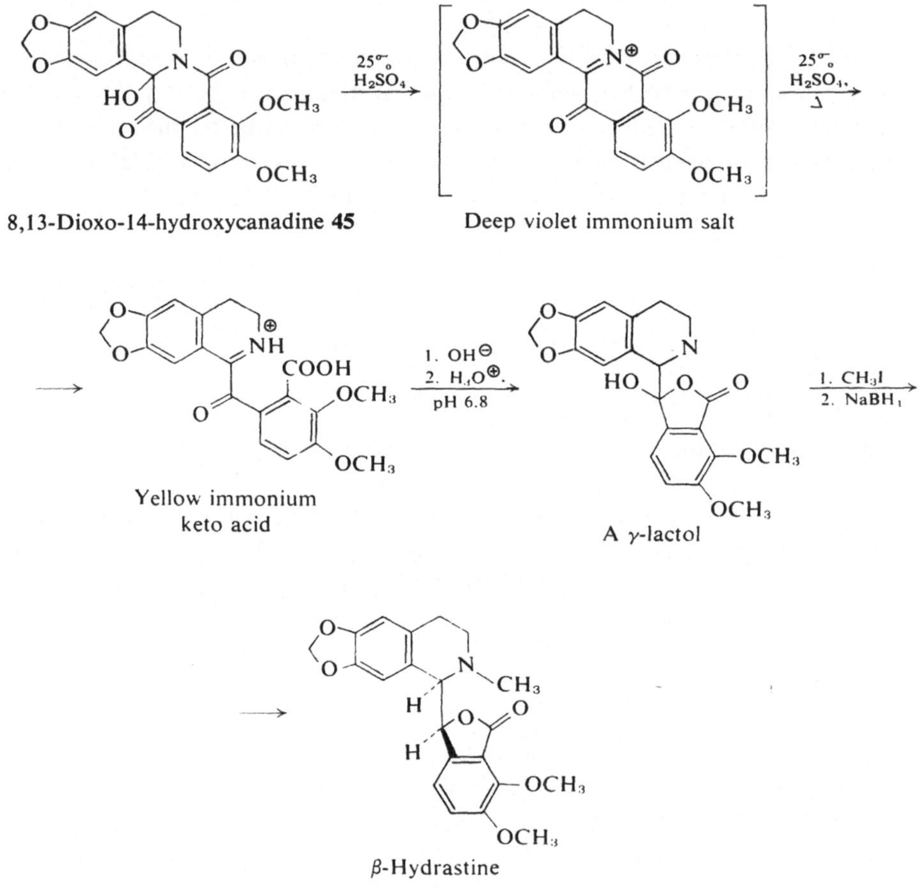

8,13-Dioxo-14-hydroxycanadine **45**   Deep violet immonium salt

Yellow immonium
keto acid

A γ-lactol

β-Hydrastine

19.3.6. Reactions with Alkylating and Acylating Agents

Dimeric protoberberinium salts have been obtained by reaction of dihydroprotoberberines with formaldehyde, most probably by the mechanism indicated below[55]:

The *N*-methyldihydroprotoberberine salt **49**, derived from the ditosylate **48** of narcotinediol, upon heating with sodium iodide and acetic anhydride afforded the quaternary salt **50**.[56] A mechanism is suggested below[57]:

19.3.7. Transformations of Dehydroprotoberberines

Dehydroprotoberberinium salts may be used as intermediates for the introduction of substituents at the C-5 position of tetrahydroprotoberberine bases.[58] Two-step reduction of dehydronorcoralydine leads to the enamine **51** which can be C-acylated as in **52**, or hydroborated and oxidized to give a separable mixture of diastereomeric C-5 hydroxylated tetrahydroprotoberbines:

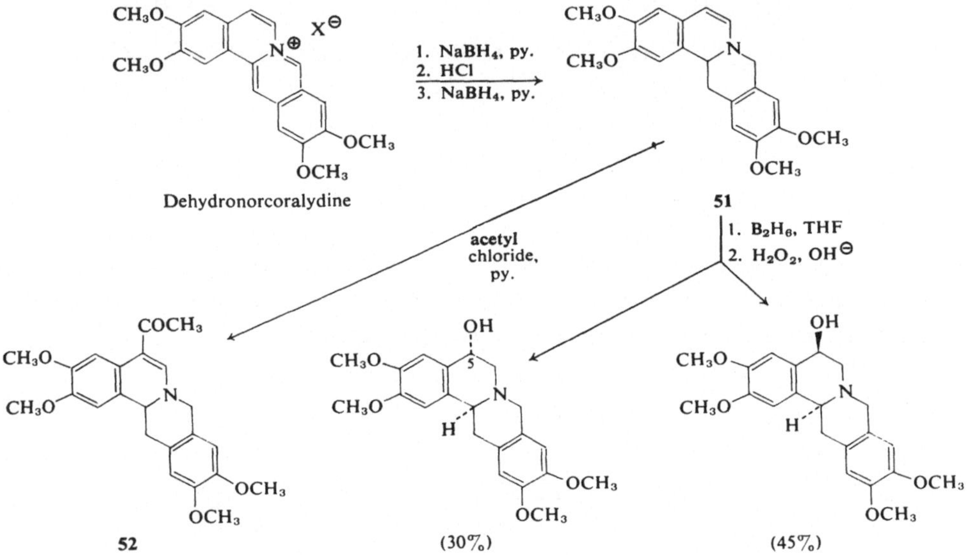

Yet another route to the secoberbines is by oxidation and saponification of dehydroprotoberberinium salts. Alkylation and borohydride reduction of the amino acid **53** yields a secoberbine methyl ester. Noteworthy is the finding that oxyberberine, which lacks the C-5,6 unsaturation, does not saponify in base to an analog of **53**[58]:

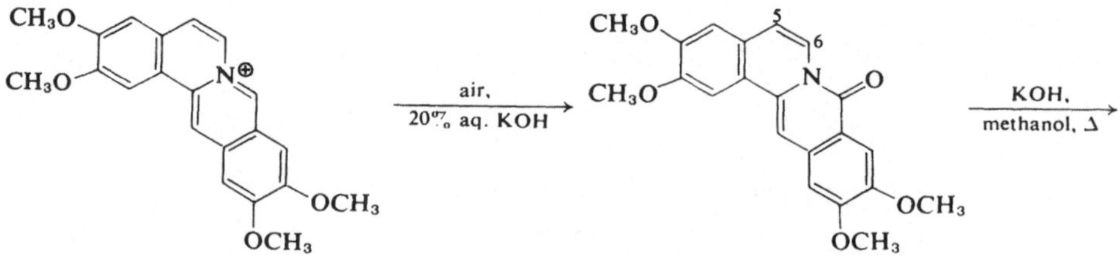

53

19.3.8. Smiles Rearrangement

Quaternary protoberberinium salts react with primary and secondary amines to generate aminated protoberberine salts.[59] Oxygenated substituents at C-9 or C-11 are efficiently replaced with nitrogen by this procedure. The use of hindered secondary amines leads to side products due to competing *O*-demethylation of the C-9 or C-11 methoxyls.

19.3.9. C-Acylation, a New Reaction of Benzylic N-Oxides

Known reactions of alkaloidal *N*-oxides have usually been limited to variants of the Polonovski reaction. Working with canadine *N*-oxide, it has now been demonstrated that as an alternative to that reaction, acylation at the C-8 benzylic site can occur when the reagents are acetic anhydride in pyridine, and the catalyst is 4-dimethylaminopyridine. The products are 8-acetyldihydroberberine (5%) and 8,13-diacetyldihydroberberine (43%). These can be oxidized with iodine to the hitherto unknown 8-acetylberberine and 13-acetylberberine, respectively.[59a]

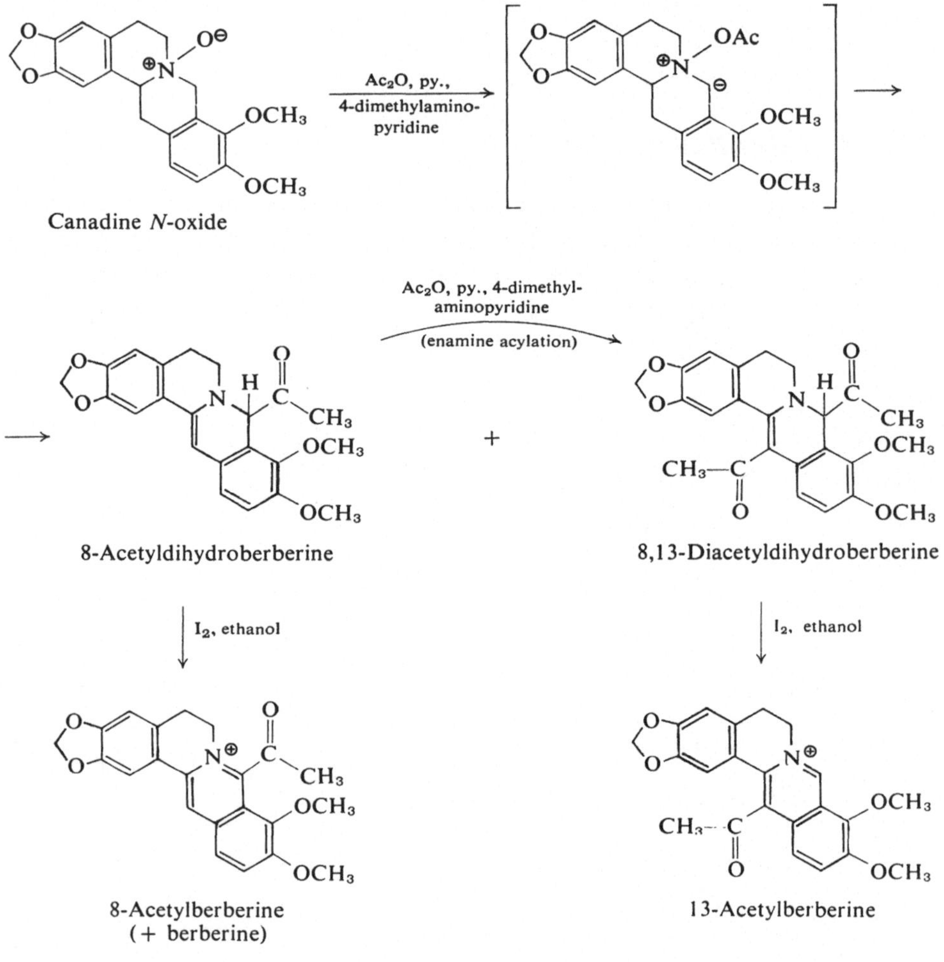

Canadine *N*-oxide

8-Acetyldihydroberberine

8,13-Diacetyldihydroberberine

8-Acetylberberine
(+ berberine)

13-Acetylberberine

19.4. Biogenesis

Incorporation studies of *Corydalis solida* Sw. (Fumariaceae) fed with [3-¹⁴C]-tyrosine and [methyl-¹⁴C]-methionine as well as [methyl-T,¹⁴C]-methionine indicated that methionine supplies the ▲ carbons of the protoberberine corydaline and the spirobenzylisoquinoline ochotensimine.[60]

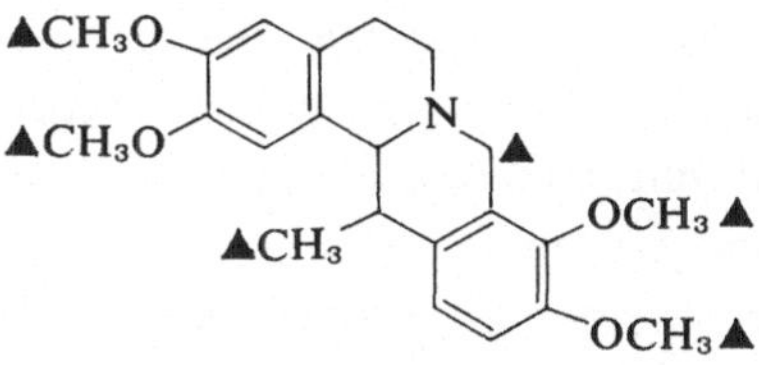

Corydaline

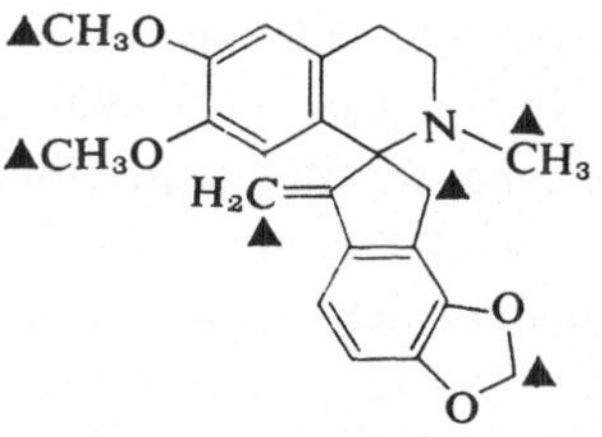

Ochotensimine

Rat liver homogenates have been reported to convert $(\pm)$-[N-^{14}CH$_3$]-reticuline into racemic coreximine with 11.7% incorporation of the label.[61] Also detected in the same experiments were labeled scoulerine and unlabeled norreticuline. Laudanosine was similarly biotransformed into xylopinine, tetrahydropalmatine, and norlaudanosine.[62]

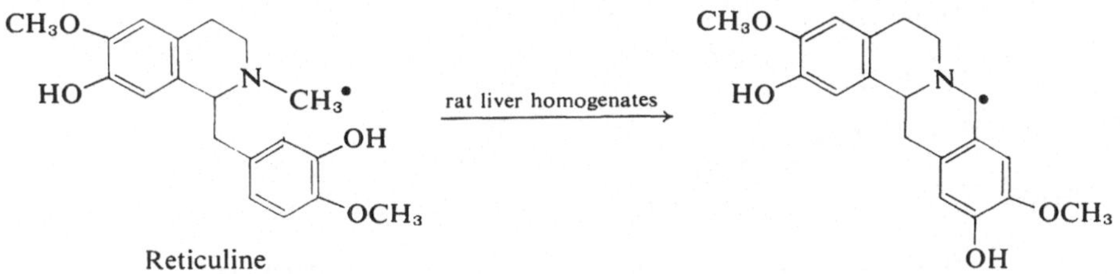

Reticuline

rat liver homogenates →

Coreximine

In a study of the biosynthesis of stylopine in *Chelidonium majus* L. (Papaveraceae), labeled $(+)$-reticuline was found to be converted first to $(-)$-scoulerine and then to $(-)$-stylopine; in addition, no loss of label occurred in the conversion to $(-)$-stylopine[63] when the reticuline was tritiated at the C-1 position.

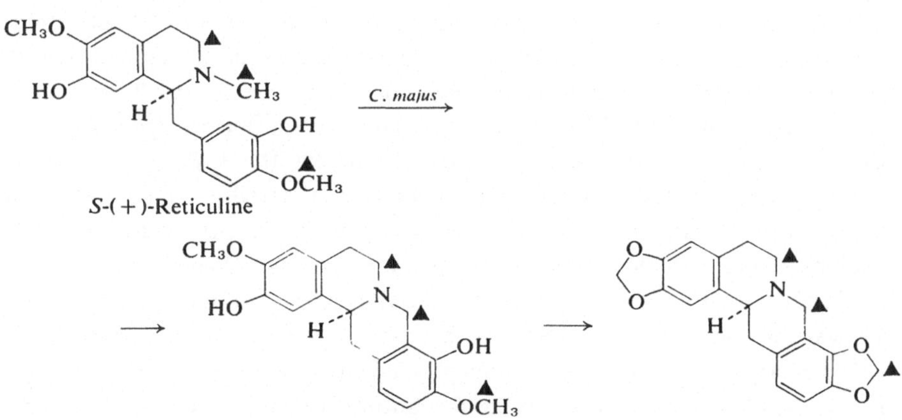

S-$(+)$-Reticuline

C. majus →

S-$(-)$-Scoulerine

→

S-$(-)$-Stylopine

Battersby and his group have recently reported the details of an extensive study of biogenetic interrelations among tetrahydroprotoberberines, benzophenanthridines, rhoeadines, and phthalideisoquinolines. A key step in this study of the stereospecificity of enzyme reactions was the use of horse liver alcohol dehydrogenase to prepare the complementary *R*- and *S*-benzyl alcohols shown.[64] These tritiated materials were needed for the synthesis of chirally labeled scoulerine.

In an effort to avoid loss of label, an interesting and unusual route to the starting tetrahydroprotoberberine was employed. The anion **55**, obtained from the imine **54**, was condensed with the isotopically labeled isomeric halides, and subsequent hydrolysis gave rise to amines of high isotopic and configurational purity. By successive reactions first the C ring and then the B ring was formed to furnish the desired scoulerines (see Scheme 19.8).

Plants fed with the 13*S* isomer of tritiated scoulerine yielded chelidonine which had lost 72% of the original label, whereas the 13*R* isomer afforded chelidonine which had retained 75% of the tritium activity. Thus, the bioconversion of the tetrahydroprotoberberines scoulerine and stylopine to chelidonine involves a stereospecific removal of the pro-*S* hydrogen.[65] When *Papaver rhoeas* L. (Papaveraceae) was treated with the same precursor tetrahydroprotoberberines, they were incorporated into rhoeadine, and the results again agreed with a stereospecific removal of the pro-*S* hydrogen in this bioconversion.[66] (For a detailed discussion of benzophenanthridine and rhoeadine biogenesis, see Secs. 21.5 and 26.6, respectively.)

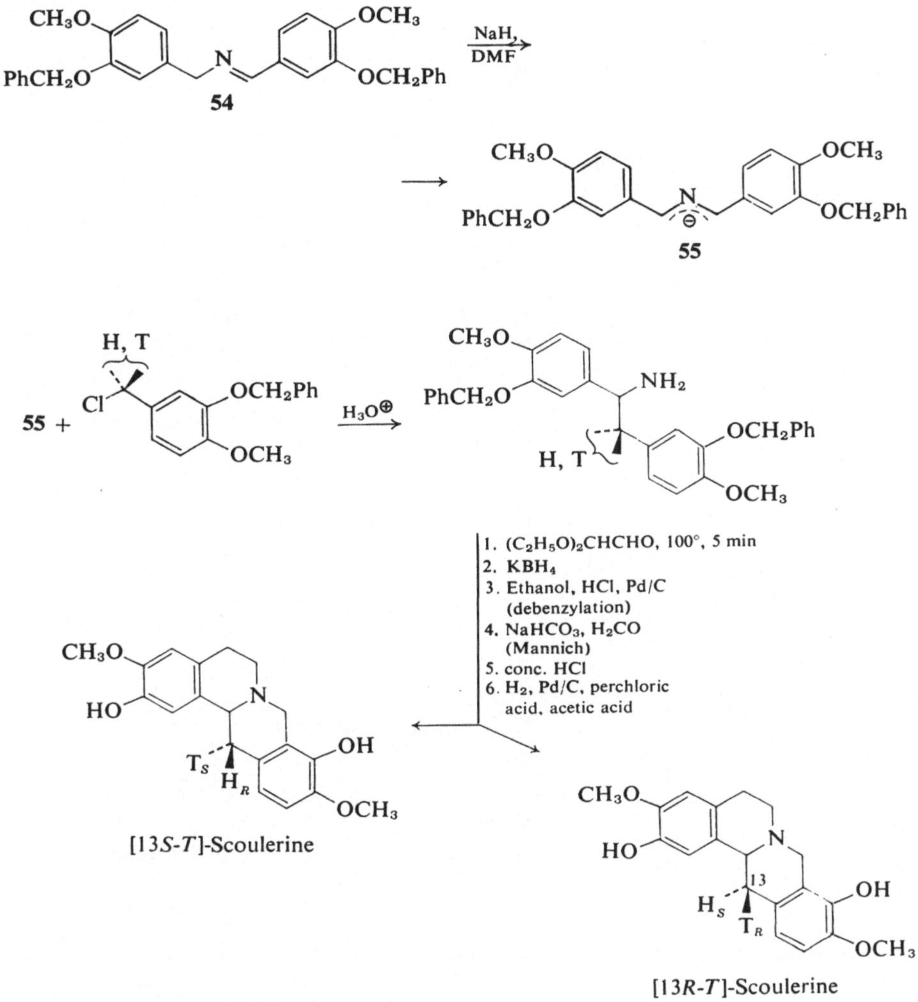

SCHEME 19.8

In a separate study by Jeffs and Scharver,[67] chirally labeled canadines were used in a survey of the biogenesis of ophiocarpine in *Corydalis ophiocarpa* Hook. f. & Thoms. (Fumariaceae). In this case, however, a simple transannular cyclization was applied to introduce the chiral labels in canadine. Thus the reaction of (±)-canadine with cyanogen bromide yielded the *trans*-dibenza-zecine **56** which, upon acid hydrolysis in a tritiated medium, gave rise to [13α-*T*]-canadine, via the highly stereoselective anti addition of the nitrogen lone pair to the double bond. The epimeric 13β-isomer was obtained by a related route:

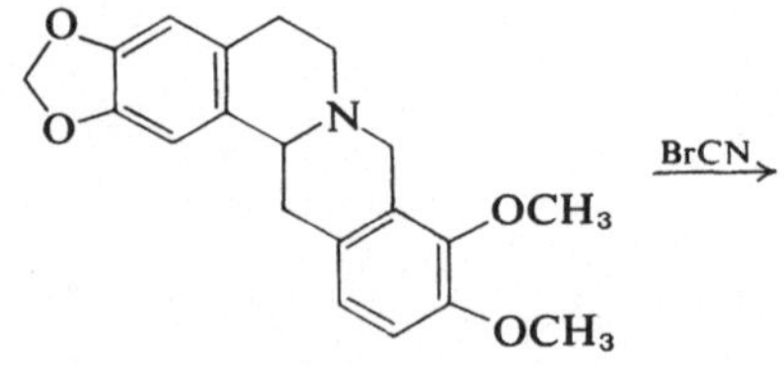

(±)-Canadine

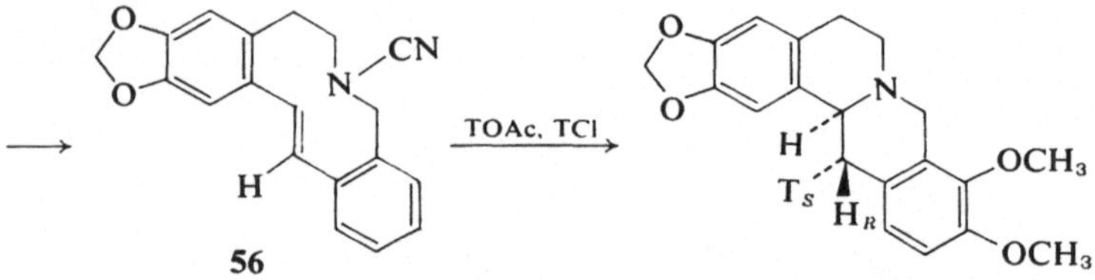

56

[13α-*T*]-Canadine

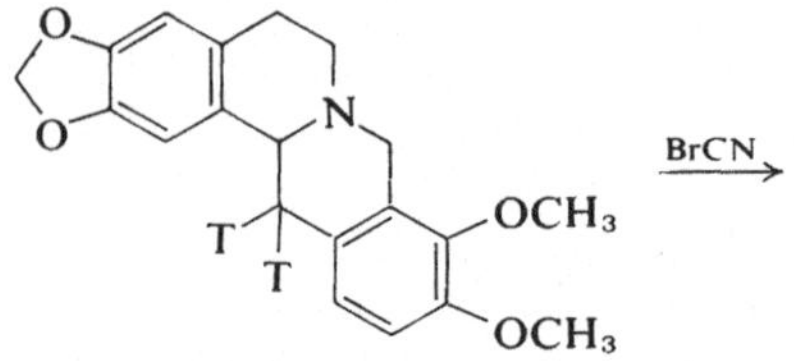

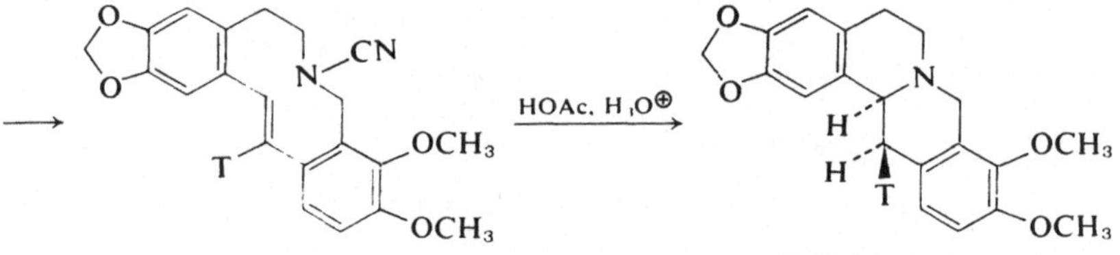

[13β-*T*]-Canadine

Incorporation experiments with the [13α-*T*]- and [13β-*T*]-canadines established that hydroxylation of the C-13 benzylic methylene, to afford ophiocarpine, proceeds with removal of the pro-13*R* hydrogen and corresponds to
an overall retention of configuration in the hydroxylation process.[67]

An observation based on the above studies is that the biogenetic oxidation
of tetrahydrobenzylisoquinolines at the C-13 benzylic carbon may proceed via

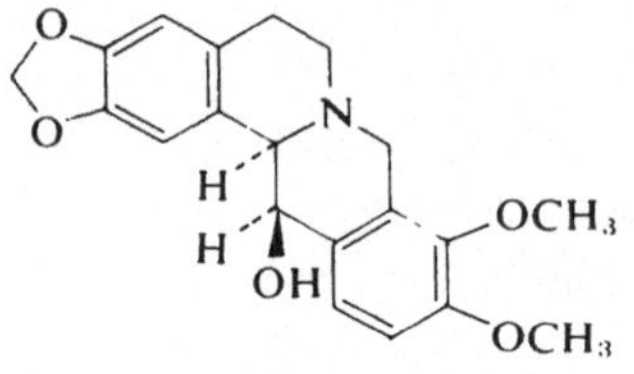

(−)-Ophiocarpine

two opposite, yet highly stereospecific pathways. The first, via pro-13S hydrogen removal, leads to 13α-hydroxylated tetrahydroprotoberberinoids, as yet unisolated from nature, which are highly predisposed for further structural modification to alkaloids of the rhoeadine, benzophenanthridine, and phthalide-isoquinoline types. The second pathway, via removal of the pro-13R hydrogen, gives rise to 13β-hydroxylated tetrahydroprotoberberines, which have been isolated from nature and may represent an anabolic endpoint.[68]

[N-[14]CH$_3$]-Tetrahydrocoptisine methiodide was incorporated (0.19%) into protopine by *Corydalis incisa* (Thunb.) Pers.; and by standard chemical degradation the label was found to be located exclusively at the N-methyl group.[69]

19.5. Pharmacology

The protoberberine alkaloids and their derivatives show at least three types of biological activity: antimicrobial, uterine, and antileukemic and antineoplastic.[11,70]

Berberrubine chloride has antimicrobial activity against *Mycobacterium smegmatis* at the 100 μg/ml level,[71] whereas berberine may be markedly valuable in the treatment of cutaneous leshmaniasis.[72] A summary of the antimicrobial effects of berberine has recently become available.[73]

Berberine chloride shows anthelmintic activity and is able to eliminate *Syphacia obvelata* from mice.[73a] A series of synthetic 8-substituted tetrahydroprotoberberines has demonstrated both antifungal and antiarrhythmic activity.[73b] Berberine sulfate and tetrahydropalmatine inhibit the respiratory chain by interfering with the action of NADH oxidase.[73c]

Neuropsychopharmacological studies using (−)-tetrahydrocoptisine with mice and rats have indicated this methylenedioxy-substituted tertiary base to possess antipsychotic and neuroleptic activity.[73d]

Quaternary protoberberine chlorides, including berberine, palmatine, jatrorrhizine, coptisine, and dehydrocorydaline, cause marked contraction of uterine muscle, but only weak spasmolytic activity on isolated mouse intestine.[74] Protoberberine alkaloid bearing plants which exhibit abortifacient and uterine stimulant properties have been reviewed by Farnsworth and co-workers.[75]

Oral administration of dehydrocorydaline prevented gastric and duodenal ulcers in guinea pigs and rats.[76] Intravenous injection of [^{14}C]-dehydrocorydaline in mice followed by macro- and microautoradiography revealed that the heterogeneous reticular distribution of radioactivity is due to perilobular localization of the label in the liver.[77]

The two norcoralydine methiodides have shown some neuromuscular

Dehydrocorydaline chloride

blocking activity.[78] 2,3,10,11-Substituted protoberberines and their derivatives have been claimed to possess analgesic, vasodilating, and antihypertensive properties (no data).[79] In addition, berberine–acetone derivatives have been claimed to possess antihistamine and antiinflammatory effects (no data).[80]

Protoberberines and their derivatives as potential anticancer agents have been reviewed.[81] In recent years a large number of derivatives of berberine, berberrubine, and palmatrubine have been prepared for investigation of their potential antileukemic and antineoplastic activities (no data).[82] One of these, 9-methoxyprotoberberinium bromide increased the survival time of mice infected with sarcoma 180 and leukemia L-1210 by 100 and 241% respectively.[83]

Coralyne chloride, an 8-methyldehydroprotoberberinium salt, possesses reproducible activity against leukemias P-388 and L-1210 in the mouse.[84] A structure–activity relationship study of coralyne salts and analogs revealed several features:[85]

1. The 8-ethyl analog was more active than the parent 8-methyl compound, whereas the 8-propyl analog was inactive.

2. Replacement of one or both dimethoxy groupings with methylenedioxy groups slightly decreased the antileukemic activity. However, the 8-methyl and ethyl bismethylenedioxy analogs showed KB activity in cell culture, whereas the corresponding bis(dimethoxy) compounds did not.

3. Analogs having fewer alkoxy substituents than coralyne had diminished but detectable antileukemic activity.

4. The planarity and rigidity of molecules of this type is critical to the activity.

5. Various salts of coralyne all had comparable activity.

Coralyne as the sulfoacetate salt undergoes a reversible hydrolysis to 6'-acetylpapaverine on standing in concentrated aqueous sodium hydroxide solution.[86] Spectral methods have shown that coralyne forms a stable complex with thymus DNA.[87]

Protoberberine alkaloids interact with horse liver alcoholdehydrogenase and prevent ethanol inhibitors competitive with ethanol from binding. Spectrometric, fluorometric, and kinetic measurements have shown that the site of the

Coralyne salt

6'-Acetylpapaverine

binding is the region of the enzyme designated the "active site pocket."[88] Among the protoberberines tested the most effective inhibitor of the enzyme was 13-ethylberberine.[89] Salts of palmatrubine esters have been reported to show specific xanthine oxidase inhibition by the Schardinger method.[90] Finally, coralyne salts were found to be more potent inhibitors of catechol *O*-methyltransferase than pyrogallol.[87]

19.6. Mass Spectroscopy

The mass spectral fragmentation of tetrahydroprotoberberine bases is usually easy to interpret, since the major cleavage results from retro-Diels–Alder fragmentation of ring C, to produce ions **57** (base) and **58**.

A recent study of the fragmentation of variously substituted tetrahydro-protoberberines revealed that compounds possessing a C-9 *O*-methyl group give rise to strong (15% of M$^+$) ions for the loss of methoxyl, whereas those compounds having methoxyl groups at other sites on the molecule show only weak (2–3% of M$^+$) ions for methoxyl loss.[91]

This generalization has recently been used to correct the structures of a variety of tetrahydroprotoberberine alkaloids. Aequaline was shown to be identical with discretamine, and coramine with coreximine. Schefferine is the same as kikemanine, and discretinine corresponds to corypalmine.[4]

Discretamine

Coreximine

Kikemanine, R = CH₃, R₁ = H
Corypalmine, R = H, R₁ = CH₃

19.7. NMR Spectroscopy

The PMR spectra of 19 quaternary protoberberine salts in trifluoroacetic acid (TFA) have been tabulated and assignments made wherever possible.[92]

The CMR spectra of a number of protoberberine[93] and tetrahydroprotoberberine alkaloids[94–96] have been published. On the basis of these data a correlation between the chemical shift of the C-6 carbon and the conformation of the quinolizidine system of tetrahydroprotoberberines has been drawn. Tetrahydroprotoberberines which possess a C-1 methoxyl and exist in the cis conformation give rise to a signal for the C-6 carbon at higher field than $\delta 48.3$, whereas those with hydrogen at C-1 show a signal for the C-6 carbon at or about $\delta 51.3$.[97] The C-8 methyl groups of the tetrahydroprotoberberines O-methylcorytenchirine and coralydine appear at $\delta 17.7$ and 21.3, respectively.[10] Some typical CMR chemical shifts of tetrahydroprotoberberines are given below.

The detection of the conformation of the quinolizidine system of the tetrahydroprotoberberines has been approached from several angles including IR Bohlmann bands,[98] rates of quaternization,[99] circular dichroism,[100] PMR

140.4
103.1 30.1
100.7
128.5 46.9
123.9
N
134.5 147.8 54.7 57.3
CH₃O 31.6 124.8 or 127.3
59.5 108.7
124.8 or 127.3 114.6 145.3
144.3 OCH₃ 56.1
OH

1,10-Dimethoxy-2,3- methylenedioxy-
11-hydroxytetrahydroprotoberberine
(not a natural product)[94]

147.5 127.0
111.5 29.2
56.2 CH₃O
51.5
53.7
56.2 CH₃O 121.4
109.1 59.4 141.6
147.5 129.9 36.5 OH
128.1
119.3 144.2
109.1 OCH₃ 56.2

Tetrahydropalmatrubine[94]

chemical shifts of the C-8 protons,[99] as well as the CMR chemical shift of the C-6 carbon signal.[97] Yet another method for determining the conformation of the quinolizidine system is inspection of the coupling constant (^{13}C–H) for the ring-juncture carbon. Conformations which have the cis configuration around the nitrogen lone pair and the C-14 proton exhibit a J (^{13}C–H) of between 6 to 12 Hz larger than for the trans-fused analogs.[101] CMR analysis of a series of tetrahydroprotoberberine free bases, N-metho salts and N-protonated salts, indicated that the amount of B/C cis form, in comparison to the B/C trans form, increased substantially when a C-13 methyl group was present and the C-13 and C-14 hydrogens were trans to each other.[101a] This conclusion is also supported by the recent IR spectral data described below.

19.8. IR, UV, and Fluorescence Spectroscopy

The UV and IR spectra of a series of quaternary, dihydro-, tetrahydro-, and oxoprotoberberines have been recorded.[102,103] The maxima for the fluorescence bands are between 540 and 570 nm for quaternary protoberberines, and between 320 and 330 nm for tetrahydroprotoberberines.[103]

A careful study of the Bohlmann bands of tetrahydroprotoberberines has confirmed that[104]:

(a) If C-1 and C-13 are unsubstituted, the trans B/C quinolizidine conformation prevails.

(b) When no substituent is present at C-1, but a substituent is present at C-13 so that the hydrogens at C-13 and C-14 are cis to each other, the trans B/C quinolizidine conformation is again overwhelmingly favored.

(c) In the variation of case (b) above, where the C-13 and C-14 hydrogens are trans to each other, the cis quinolizidine conformation predominates.

(d) If a substituent is present at C-1, but C-13 is unsubstituted, intermediate conformations or mixtures of cis and trans conformations will exist.[104]

19.9 Solidaline, a Modified Protoberberine Alkaloid

A reinvestigation of the alkaloids of *Corydalis solida* (L.) Swartz (Fumariaceae) has yielded the unusual base solidaline, $C_{23}H_{27}NO_6$, λ_{max}^{EtOH} 233, 281, 316, and 364 nm (4.03, 3.51, 3.42, and 3.33), which possesses a nonphenolic hydroxyl, a tertiary C-methyl group, four methoxyls, and two tetrasubstituted aromatic rings. The CMR chemical shifts for the core atoms of the molecule are listed below; the three most intense peaks in the mass spectrum were at m/e 207 (100), 206 (89), and 191 (56).[105]

Solidaline[105]

$e \pm$ impact

m/e 207,
$C_{11}H_{13}NO_3$

m/e 206,
$C_{12}H_{14}O_3$

m/e 191,
$C_{11}H_{11}O_3$

The structural assignment was also supported by PMR spectral data including NOE and spin decoupling studies. In particular, irradiation of the C-13 methyl singlet at δ1.73 resulted in 21–24% NOE enhancement of the signals for H-1 and H-12, indicating that the C-13 methyl group is equatorial and close to H-1 and H-12.[105]

It should be stated, however, that the alternate structure **51** for solidaline still cannot be completely excluded from consideration.

Addition of acid at room temperature caused a reversible pseudobase–immonium equilibrium, noticeable in the UV spectrum. Furthermore, upon heating in acid, an irreversible change was observed consonant with formation of the protoberberinium salt **52**.[105] Salt **52** could be formed from either structures for solidaline.

51   **52**

References and Notes

1. S.-T. Lu, T.-L. Su, T. Kametani, A. Ujiie, M. Ihara, and K. Fukumoto, *Heterocycles*, **3**, 459 (1975); and *J. Chem. Soc. Perkin I*, 63 (1976).
2. R. H. F. Manske, R. Rodrigo, D. B. MacLean, and L. Baczynskyj, *An. Quim.*, **68**, 689 (1972).
3. P. W. Jeffs and J. D. Scharver, *J. Org. Chem.*, **40**, 644 (1975).
4. E. Brochmann-Hanssen and H.-C. Chiang, *J. Org. Chem.*, **42**, 3588 (1977).
5. C. Tani, N. Nagakura, and S. Hattori, *Chem. Pharm. Bull., Tokyo*, **23**, 313 (1975).
6. C. Tani, N. Nagakura, S. Hattori, and N. Masaki, *Chem. Lett.*, 1081 (1975).
7. V. Novák and J. Slavík, *Collect. Czech. Chem. Commun.*, 883 (1974).
8. A. Shafiee, I. Lalezari, P. Nasseri-Nouri, and R. Asgharian, *J. Pharm. Sci.*, **64**, 1570 (1975).
9. M. Shamma, *The Isoquinoline Alkaloids*, Academic Press, New York (1972), p. 268.
10. T. Kametani, M. Ihara, and T. Honda, *Heterocycles*, **4**, 483 (1976); B. R. Pai, K. Nagarajan, H. Suguna, and S. Natarajan, *Heterocycles*, **6**, 1377 (1977).
11. Y. Kondo, *Heterocycles*, **4**, 197 (1976).
12. T. Kametani, *The Chemistry of the Isoquinoline Alkaloids*, Vol. 2, The Sendai Institute of Heterocyclic Chemistry, Sendai, Japan, pp. 191–209 (1974).
13. F. Šantavý in *The Alkaloids*, R. H. F. Manske, ed., Vol. 12, Academic Press, New York (1970), p. 383. For the synthesis of 13-hydroxylated tetrahydroprotoberberines by the Mannich cyclization, see P. Osei-Gyimah, J. W. Fowble, D. R. Feller, and D. D. Miller, *J. Org. Chem.*, in press.
14. T. Kametani, E. Taguchi, Y. Yamaki, A. Kozuka, and T. Terui, *J. Pharm. Soc. Japan*, **93**, 529 (1973). See also Ref. 4 above.
15. T. Kametani, A. Ujiie, and K. Fukumoto, *Heterocycles*, **2**, 55 (1975); and *J. Chem. Soc. Perkin I*, 1954 (1974).
16. T. Kametani, A. Ujiie, M. Ihara, and K. Fukumoto, *Heterocycles*, **3**, 143 (1975); and *J. Chem. Soc. Perkin I*, 1822 (1975).
17. D. Berney and T. Jauner, *Helv. Chim. Acta*, **59**, 623 (1976).
18. T. Kamegaya, Japan. Pat. 74 20,800; through *Chem. Abstr.*, **82**, 73291y (1975).
18a. S. Natarajan, B. R. Pai, R. Rajaraman, C. S. Swaminathan, K. Nagarajan, V. Sundarsanam, D. Rogers, and A. Quick, *Tetrahedron Lett.*, 3573 (1975).

19. H. Bruderer, J. Metzger, and A. Brossi, *Helv. Chim. Acta*, **58**, 1719 (1975). H. Bruderer, J. Metzger, A. Brossi, and J. J. Daly, *Helv. Chim. Acta*, **59**, 2793 (1976).
19a. T. Kametani, H. Matsumoto, Y. Satoh, H. Nemoto, and K. Fukumoto, *J. Chem. Soc. Perkin I*, 376 (1977).
20. C. Tani, S. Takao, H. Endo, and E. Oda, *J. Pharm. Soc. Japan*, **93**, 268 (1973).
21. H. Suguna and B. R. Pai, *Collect. Chem. Czech. Commun.*, **41**, 1219 (1976).
22. H. G. Peer, *Rec. Trav. Chim. Pays-Bas*, **79**, 825 (1960).
23. W. Nagata, H. Itazaki, K. Okada, T. Wakabayashi, K. Shibata, and N. Tokutake, *Chem. Pharm. Bull., Tokyo*, **23**, 2867 (1975).
24. For a recent review of this reaction, see I. Ninomiya, *Heterocycles*, **2**, 105 (1974).
25. G. R. Lenz, *J. Org. Chem.*, **39**, 2839 (1974).
26. G. R. Lenz, *J. Org. Chem.*, **41**, 2201 (1976).
27. G. R. Lenz, *J. Org. Chem.*, **39**, 2846 (1974).
28. For an alternative enamide synthesis of the alkaloid xylopinine, see T. Kametani, T. Honda, T. Sugai, and K. Fukumoto, *Heterocycles*, **4**, 927 (1976).
29. I. Ninomiya, H. Takasugi, and T. Naito, *Heterocycles*, **1**, 17 (1973); and *J. Chem. Soc. Perkin I*, 1791 (1975).
30. I. Ninomiya, T. Naito, and H. Takasugi, *J. Chem. Soc. Perkin I*, 1720 (1975).
30a. G. R. Lenz, *J. Org. Chem.*, **42**, 1117 (1977).
30b. T. Kametani, T. Sugai, Y. Shoji, T. Honda, F. Satoh, and K. Fukumoto, *J. Chem. Soc. Perkin I*, 1977 (1977). Treatment of the bromoenamide with sodium amide in liquid ammonia furnishes the corresponding 8-oxoberbine in low yield; see T. Kametani, T. Honda, T. Sugai, and K. Fukumoto, *Heterocycles*, **4**, 927 (1976).
30c. T. Kametani, A. Ujiie, M. Ihara, K. Fukumoto, and S.-T. Lu, *J. Chem. Soc. Perkin I*, 1218 (1976).
31. For a recent review of the benzocyclobutene thermolysis reaction in alkaloid synthesis, see T. Kametani and K. Fukumoto, *Heterocycles*, **3**, 29 (1975).
32. T. Kametani, K. Ogasawara, and T. Takahashi, *Chem. Commun.*, 675 (1972); and *Tetrahedron*, **29**, 73 (1973).
33. T. Kametani, Y. Hirai, F. Satoh, K. Ogasawara, and K. Fukumoto, *Chem. Pharm. Bull., Tokyo*, **21**, 907 (1973).
34. T. Kametani, M. Takemura, K. Ogasawara, and K. Fukumoto, *J. Heterocycl. Chem.*, **11**, 179 (1974).
35. T. Kametani, T. Kato, and K. Fukumoto, *Tetrahedron*, **30**, 1043 (1974); T. Kametani, C. Ohtsuka, H. Nemoto, and K. Fukumoto, *Chem. Pharm. Bull., Tokyo*, **24**, 2525 (1976).
36. S. F. Dyke and E. P. Tiley, *Tetrahedron*, **31**, 561 (1975).
37. R. T. Dean, H. C. Padgett, and H. Rapoport, *J. Am. Chem. Soc.*, **98**, 7448 (1976).
38. H. Irie, K. Akagi, S. Tani, K. Yabusaki, and H. Yamane, *Chem. Pharm. Bull., Tokyo*, **21**, 855 (1973).
38a. D. Greenslade and R. Ramage, *Tetrahedron*, **33**, 927 (1977).
39. M. Shamma and V. St. Georgiev, *Tetrahedron Lett.*, 2339 (1974).
39a. M. Cushman, J. Gentry, and F. W. Dekow, *J. Org. Chem.*, **42**, 1111 (1977); and M. A. Haimova, N. M. Mollov, S. C. Ivanova, A. I. Dimitrova, and V. I. Ognyanov, *Tetrahedron*, **33**, 331 (1977).
40. T. Kametani, M. Takemura, K. Fukumoto, T. Terui, and A. Kozuka, *Heterocycles*, **2**, 433 (1974).
41. T. Kametani, M. Takemura, K. Fukumoto, T. Terui, and A. Kozuka, *J. Chem. Soc. Perkin I*, 2678 (1974).
42. T. Kametani, M. Takemura, K. Takahashi, M. Takeshita, M. Ihara, and K. Fukumoto, *Heterocycles*, **2**, 653 (1974); and *J. Chem. Soc. Perkin I*, 1012 (1975).
43. Professor H. Rönsch, private communication.

44. W. Nagata, K. Okada, H. Itazaki, and S. Uyeo, *Chem. Pharm. Bull.*, *Tokyo*, 23, 2878 (1975).
45. J. P. Yardley, *Synthesis*, 543 (1973).
46. M. Shamma, L. A. Smeltz, J. L. Moniot, and L. Töke, *Tetrahedron Lett.*, 3803 (1975); and M. Shamma, J. L. Moniot, L. A. Smeltz, W. A. Shores, and L. Töke, *Tetrahedron*, 33. 2907 (1977).
47. D. L. Fields, T. H. Regan, and J. C. Dignan, *J. Org. Chem.*, 33, 390 (1968).
48. C. Tani, S. Takao, and K. Tagahara, *J. Pharm. Soc. Japan*, 93, 197 (1974); and C. Tani, Y. Suzuta, and K. Tagahara, *J. Pharm. Soc. Japan*, 97, 591 (1977).
49. T. Shimada, T. Ikegawa, T. Endo, H. Kuroda, Y. Ikeda, K. Tachibana, and Y. Okazaki, *Chem. Abstr.*, 80, 146399k (1974).
50. S. Naruto, H. Nishimura, and H. Kaneko, *Chem. Pharm. Bull.*, *Tokyo*, 23, 1271 (1975).
51. S. Naruto, H. Nishimura, and H. Kaneko, *Chem. Pharm. Bull.*, *Tokyo*, 23, 1276 (1975).
52. S. Naruto, H. Nishimura, and H. Kaneko, *Chem. Pharm. Bull.*, *Tokyo*, 23, 1565 (1975).
53. Y. Kondo and T. Takemoto, *Chem. Pharm. Bull.*, *Tokyo*, 20, 2134 (1972).
54. J. L. Moniot and M. Shamma, *J. Am. Chem. Soc.*, 98, 6714 (1976); and M. Shamma, J. L. Moniot, and D. M. Hindenlang, *Tetrahedron Lett.*, 4273 (1977).
54a. J. L. Moniot, A. H. Abd el Rahman, and M. Shamma, *Tetrahedron Lett.*, 3787 (1977).
54b. M. Hanaoka, C. Mukai, and Y. Arata, *Heterocycles*, 6, 895 (1977).
54c. M. Shamma and J. L. Moniot, unpublished results.
54d. Y. Kondo, M. Inoue, and J. Imai, *Heterocycles*, 6, 953 (1977).
54e. M. Hanaoka and C. Mukai, *Heterocycles*, in press.
54f. M. Shamma, D. M. Hindenlang, T.-T. Wu, and J. L. Moniot, *Tetrahedron Lett.*, 4285 (1977).
55. S. Naruto, H. Mizuta, A. Kagemoto, and H. Nishimura, *J. Pharm. Soc. Japan*, 95, 1400 (1975).
56. V. Šimánek and A. Klásek, *Collect. Czech. Chem. Commun.*, 38, 1614 (1973).
57. The mechanism presented here is different from that which appears in the original paper.
58. M. Shamma and L. A. Smeltz, *Tetrahedron Lett.*, 1415 (1976).
59. S. Naruto, H. Mizuta, and H. Nishimura, *Tetrahedron Lett.*, 1595 (1976); *ibid.*, 1597 (1976).
59a. M. Shamma and P. Chinnasamy, unpublished results.
60. H. L. Holland, M. Castillo, D. B. MacLean, and I. D. Spenser, *Can. J. Chem.*, 52, 2818 (1974).
61. T. Kametani, M. Takemura, K. Takahashi, M. Ihara, and K. Fukumoto, *Heterocycles*, 3, 139 (1975); and T. Kametani, Y. Ohta, M. Takemura, M. Ihara, and K. Fukumoto, *Heterocycles*, 6, 415 (1977).
62. T. Kametani, M. Takemura, M. Ihara, K. Takahashi, and K. Fukumoto, *J. Am. Chem. Soc.*, 98, 1956 (1976).
63. A. R. Battersby, R. J. Francis, M. Hirst, E. A. Ruveda, and J. Staunton, *J. Chem. Soc. Perkin I*, 1140 (1975).
64. A. R. Battersby, J. Staunton, and H. R. Wiltshire, *J. Chem. Soc. Perkin I*, 1156 (1975).
65. A. R. Battersby, J. Staunton, H. R. Wiltshire, B. J. Bircher, and C. Fuganti, *J. Chem. Soc. Perkin I*, 1162 (1975).
66. A. R. Battersby and J. Staunton, *Tetrahedron*, 30, 1707 (1974).
67. P. W. Jeffs and J. D. Scharver, *J. Am. Chem. Soc.*, 98, 4301 (1976).
68. M. Shamma and J. L. Moniot, unpublished observations.

69. C. Tani and K. Tagahara, *Chem. Pharm. Bull.*, *Tokyo*, **22**, 2457 (1974).
70. V. Preininger, in *The Alkaloids*, R. H. F. Manske, ed., Vol. 15, Academic Press, New York (1975), p. 231.
71. S. A. Gharbo, J. L. Beal, R. W. Doskotch, and L. A. Mitscher, *Lloydia*, **36**, 349 (1973).
72. E. A. Steck in *Progress in Drug Research*, E. Jucker, ed., Vol. 18, Birkhauser Verlag, Basel (1974), p. 289.
73. F. E. Hahn and J. Ciak, *Antibiotics*, D. Gottlieb, P. D. Shaw, and J. W. Cocoran, eds., Vol. 3, Springer, New York (1975), p. 577.
73a. K. C. Singhal, *Indian J. Exp. Biol.*, **14**, 345 (1976).
73b. G. R. Lenz, U.S. Appl. pat. 4,013,666; through *Chem. Abstr.*, **87**, 23587g (1977).
73c. T. Schewe and W. Mueller, *Acta Biol. Med. Ger.*, **35**, 1019 (1976); through *Chem. Abstr.*, 171590b (1976).
73d. S. K. Bhattacharya, V. B. Pandey, A. B. Ray, and B. Dasgupta, *Arzneim.-Forsch.*, **26**, 2187 (1976); through *Chem. Abstr.*, **86**, 83687a (1977).
74. Y. Kitabatake, K. Ito, and M. Tajima, *J. Pharm. Soc. Japan*, **84**, 73 (1964).
75. N. R. Farnsworth, A. S. Bingel, G. A. Cordell, F. A. Crane, and H. H. S. Fong, *J. Pharm. Sci.*, **64**, 535 (1975).
76. Y. Soli, K. Kawashima, and M. Shimizu, *Folia Pharmacol. Japan*, **70**, 425 (1974).
77. H.. Miyazaki, T. Fujii, K. Nambu, and M. Hashimoto, *Chem. Pharm. Bull.*, *Tokyo*, **23**, 2182 (1975).
78. J. B. Stenlake, W. D. Williams, N. C. Dhar, R. D. Waigh, and I. G. Marshall, *Eur. J. Med. Chem.-Chim. Ther.*, **9**, 243 (1974).
79. T. Kametani, Chemipha Co., Ltd., Japan. Pats.; through *Chem. Abstr.*, **81**, 63839y, 15248w (1974); **83**, 193577p (1975); **84**, 31284j, 44513h, 105883c, 150824v, 150825w, 165092d, 180457v (1975).
80. S. Noguchi and M. Imanishi, Japan. Pat., through *Chem. Abstr.*, **80**, 121178z, 121180u (1974).
81. G. A. Cordell and N. R. Farnsworth, *Heterocycles*, **4**, 393 (1976).
82. Kanebo Co., Ltd., Japan. pats.; through *Chem. Abstr.*, **81**, 120848b (1974); **82**, 171275r, 171276s (1974); **83**, 28424d, 28425e, 28426f, 28427g, 28428h, 28429j, 43562r, 43563s, 43564t, 43565u, 43566v, 43567w, 43568x, 43569y, 43570s, 43571t, 43572u, 43574w, 164407c, 164408d, 164409e, 164410y (1975); **84**, 59833m, 122110r (1975).
83. Y. Sawa, Kanebo Co., Ltd., Japan. Pat.; through *Chem. Abstr.*, **84**, 90384q (1975).
84. K. Y. Zee-Cheng and C. C. Cheng, *J. Pharm. Sci.*, **61**; 969 (1972).
85. K. Y. Zee-Cheng, K. D. Paul, and C. C. Cheng, *J. Med. Chem.*, **17**, 347 (1974).
86. M. J. Cho, A. J. Repta, C. C. Cheng, K. Y. Zee-Cheng, T. Higuchi, and I. H. Pitman, *J. Pharm. Sci.*, **64**, 1825 (1975). For a detailed study of pseudobase formation, see V. Šimanek and V. Preininger, *Heterocycles*, **6**, 475 (1977). Dihydrocoralyne is readily oxidized to 13-hydroxycoralyne which can be converted to 6'-acetylpapaveraldine, see J. Imai and Y. Kondo, *Heterocycles*, **5**, 153 (1976).
87. K. Y. Zee-Cheng and C. C. Cheng, *J. Pharm. Sci.*, **62**, 1572 (1973); W. D. Wilson, A. N. Gough, J. J. Doyle, and M. W. Davidson, *J. Med. Chem.*, **19**, 1261 (1976).
88. J. Kovář and S. Pavelka, *Collect. Czech. Chem. Commun.*, **41**, 1081 (1976).
89. S. Pavelka and J. Kovář, *Collect. Czech. Chem. Commun.*, **40**, 753 (1975).
90. T. Shimada, T. Ikegawa, S. Daibo, Y. Okazaki, K. Tachibana, T. Endo, T. Kono, H. Kuroda, and Y. Ikeda, Kanebo Co., Ltd., Japan. Pat.; through *Chem. Abstr.*, **83**, 114716g (1975).
91. W. J. Richter and E. Brochmann-Hanssen, *Helv. Chem. Acta*, **58**, 203 (1975). See also H.-C. Chiang and E. Brochmann-Hanssen, *J. Org. Chem.*, **42**, 3190 (1977).
92. K. Jewers, A. H. Manchanda, and P. N. Jenkins, *J. Chem. Soc. Perkin II*, 1393 (1972).

93. K. Yoshikawa, I. Morishima, J. Kunitomo, M. Juichi, and Y. Yoshida, *Chem. Lett.*, 961 (1975).

94. T. Kametani, A. Ujiie, M. Ihara, K. Fukumoto, and H. Koizumi, *Heterocycles*, 3, 371 (1975).

95. For a discussion of methods for the synthesis of protoberberines, see T. Kametani, M. Ihara, and T. Honda, *Heterocycles*, 4, 483 (1976).

96. D. W. Hughes, H. L. Holland, and D. B. MacLean, *Can. J. Chem.*, 54, 2252 (1976).

97. T. Kametani, K. Fukumoto, M. Ihara, A. Ujiie, and H. Koizumi, *J. Org. Chem.*, 40, 3280 (1975).

98. S. Wolfe, A. B. Schlegel, M.-H. Whangbo, and F. Bernardi, *Can. J. Chem.*, 52, 3787 (1974).

99. M. Shamma, C. D. Jones, and J. A. Weiss, *Tetrahedron*, 25, 4347 (1969).

100. G. Snatzke, J. Hrbek, L. Hruban, A. Horeau, and F. Šantavý, *Tetrahedron*, 26, 5013 (1970).

101. G. Van Binst and D. Tourwe, *Heterocycles*, 1, 257 (1973).

101a. N. Takao, K. Iwasa, M. Kamigauchi, and M. Sugiura, *Chem. Pharm. Bull.*, *Tokyo*, 25, 1426 (1977).

102. S. Pavelka and J. Kovář, *Collect. Czech. Chem. Commun.*, 41, 3654 (1976).

103. S. Pavelka and E. Smekal, *Collect. Czech. Chem. Commun.*, 41, 3157 (1976).

104. N. Takao and K. Iwasa, *Chem. Pharm. Bull.*, *Tokyo*, 24, 3185 (1976).

105. R. H. F. Manske, R. Rodrigo, H. L. Holland, D. W. Hughes, D. B. MacLean, and J. K. Saunders, *Can. J. Chem.*, 56, 383 (1978).

THE SECOBERBINES

Occurrence: Fumariaceae, Papaveraceae, and Ranunculaceae

Structures and sources of the six naturally occurring secoberbines are shown in Scheme 20.1.

20.1. Structural Elucidation and Synthesis

20.1.1. Canadaline, Aobamine, and Corydalisol

The structures of canadaline[1] and aobamine[2] were established through analysis of physical data, especially their IR (1680 cm^{-1}, 5.95 μ) and PMR spectra which clearly indicated the presence of a substituted benzaldehyde

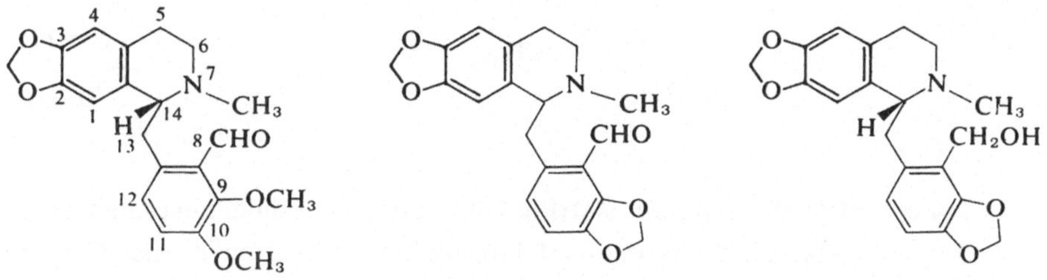

(+)-Canadaline[1]
Hydrastis canadensis L.

Aobamine[2]
Corydalis ochotensis Turcz.

(+)-Corydalisol[3]
Corydalis incisa
(Thunb.) Pers.

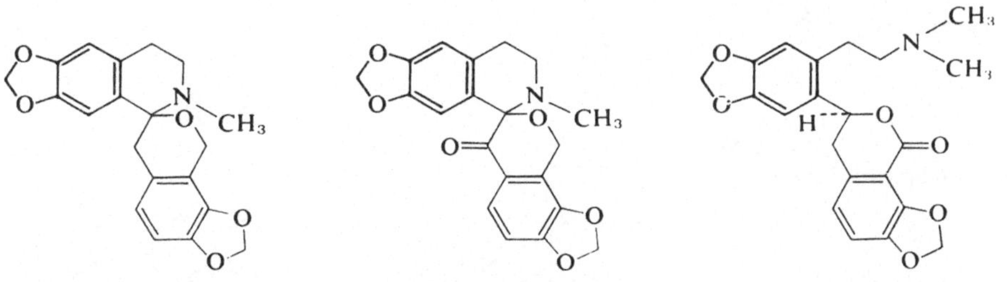

(±)-Hypercorine[1]
Hypecoum erectum L.

(±)-Hypecorinine
Pteridophyllum racemosum
Sieb. & Zucc.[5]
Hypecoum erectum L.[4]
Corydalis incisa[3]

(−)-Peshawarine[6]
Hypecoum parviflorum
Kar. & Kir.

SCHEME 20.1

261

moiety, while their mass spectra pointed to a benzylisoquinoline skeleton. The structures of canadaline and aobamine and also that of corydalisol,[3] which corresponds to the alcohol derived from aobamine, were confirmed by syntheses from protoberberine salts.

Berberine chloride was degraded by the method of Freund and Fleischer[7] to the styrylbenzylisoquinoline **1** which upon Lemieux–Johnson oxidation afforded canadaline in 67% overall yield from berberine.[6]

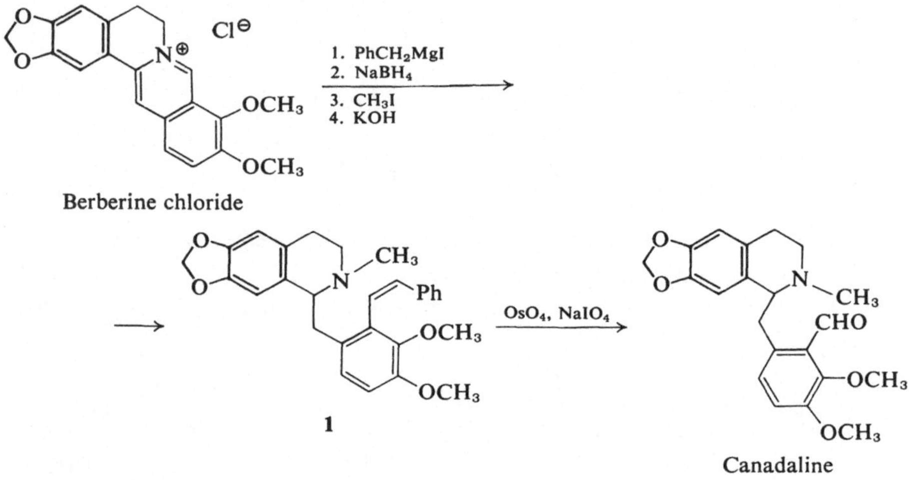

Berberine chloride

1

Canadaline

Application of this method, starting with coptisine iodide, generated aobamine in comparable yield. Subsequent borohydride reduction furnished corydalisol.[8]

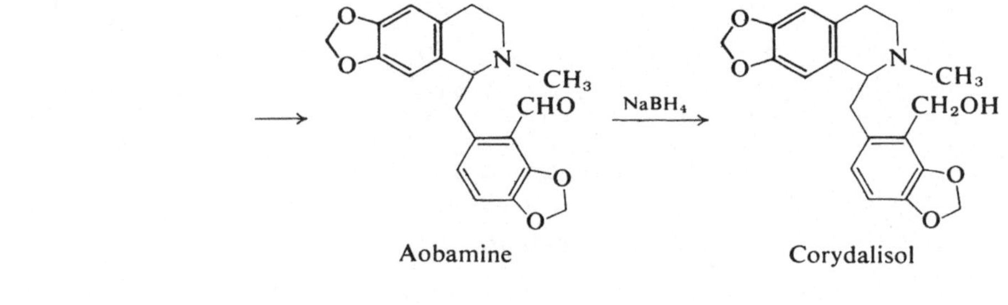

Coptisine iodide

Aobamine   Corydalisol

20.1.2. Hypecorine

The optically inactive base hypecorine, $C_{20}H_{19}NO_5$, exhibits a PMR spectrum suggestive of a bismethylenedioxybenzylisoquinoline structure. The natural product is acid unstable, and upon treatment with acetic anhydride in chloroform is transformed to the N,O-diacetyl derivative 2, thus confirming the presence in hypecorine of an aminoketal grouping. Permanganate oxidation of hypecorine yields 3,4-methylenedioxyphthalide which must originate from ring D of the natural product.[4]

Hypecorine

3,4-Methylenedioxyphthalide

2

Sodium borohydride reduction of the open (pseudohypecorine) form of the natural product, prepared by simple treatment of the alkaloid with hydrochloric acid, afforded the hydroxymethylbenzylisoquinoline 3. The latter compound, although not so stated specifically in the literature, should correspond to (±)-corydalisol. Base treatment of pseudohypecorine regenerates the amino-ketal function, while no formation of the corresponding enamine 4 was mentioned[9]:

Hypecorine

Pseudohypecorine

(±)-Corydalisol 3

4

20.1.3. Hypecorinine

The ketonic alkaloid hypecorinine, $C_{20}H_{17}O_6N$, ν_{max} 1690 and 1630 cm^{-1} (5.92 and 6.14 μ) was obtained from several sources, including *Corydalis incisa* Pers. (Fumariaceae). The characteristic instability of the molecule toward acid was indicative of the presence of an aminoketal function. Lithium aluminum hydride reduction of hypecorinine gave rise to a pair of diastereomeric diols **5** and **6** identified as bicucullinediol and adlumidinediol, respectively. This chemical interrelation, coupled with the absence of hydroxyl absorption in the IR spectrum of the alkaloid confirmed the structure assigned[3]:

($\pm$)-Hypercorinine

($\pm$)-Bicucullinediol **5**   ($\pm$)-Adlumidinediol **6**

20.1.4. Peshawarine

($-$)-Peshawarine, $C_{21}H_{21}NO_6$, exhibits an IR absorption band at 1725 cm^{-1} (5.80 μ) indicative of a conjugated δ-lactone. A significant feature of the PMR spectrum is a one-proton absorption appearing as two doublets at $\delta5.55$ and 5.73 for the hydrogen alpha to the lactone oxygen at C-14. Hydrogenolysis of peshawarine yielded the amino acid **7**, identical with the product obtained by similar hydrogenolysis of authentic ($-$)-bicuculline methochloride. In contrast, lithium aluminum hydride reduction of peshawarine supplied the diol **8**, different from that obtained from similar reduction of bicuculline methine. The mass spectrum of **8** which shows ions at m/e 222 and 165, each of which possesses three oxygen atoms, further pointed to the presence of a δ- rather than a γ-lactone in peshawarine[6]:

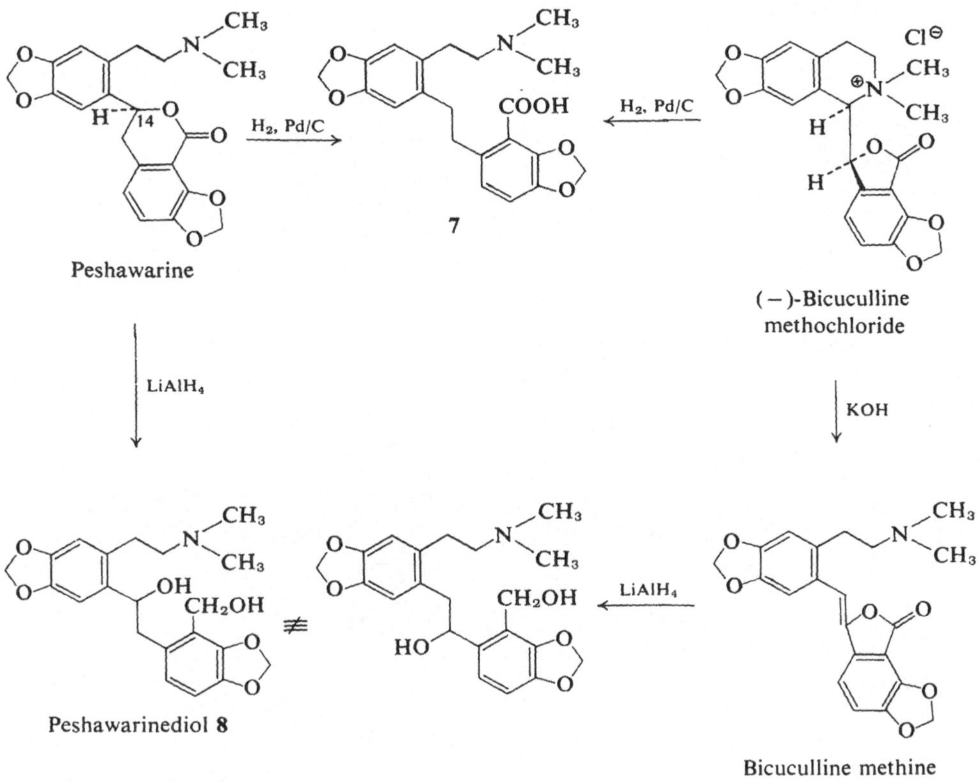

Peshawarine

7

(−)-Bicuculline methochloride

LiAlH₄

KOH

Peshawarinediol **8**

LiAlH₄

Bicuculline methine

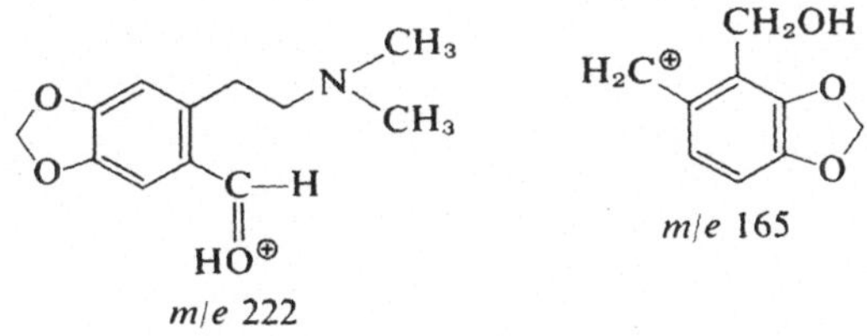

m/e 222

m/e 165

The structure proposed for peshawarine was confirmed by its synthesis from aobamine, in turn derived from coptisine (Sec. 20.1.1).[8] The transformation of isoquinoline to benzopyran was induced by treatment of aobamine with ethyl chloroformate under Schotten–Baumann conditions. This reaction most probably proceeds by an S_N2 displacement,[10] and leads to isolation of the hemiacetal **9** in high yield.[8] The acetal **10**, derived from trimethyl orthoformate treatment of **9**, was reduced to the *N,N*-dimethyl derivative **11**. Subsequent hydrolysis to the hemiacetal and Jones oxidation furnished peshawarine in 80% yield from **11**[8]:

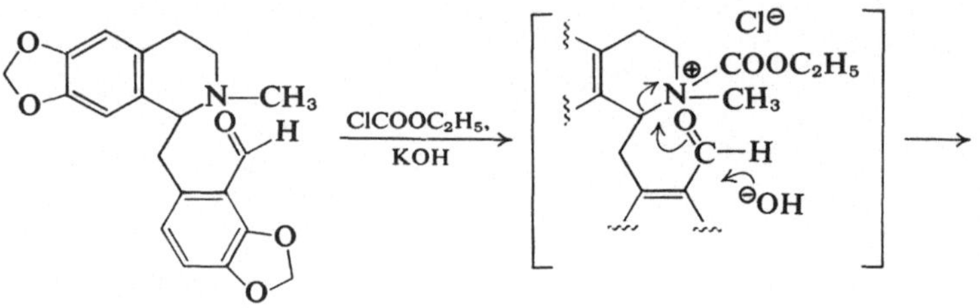

Aobamine

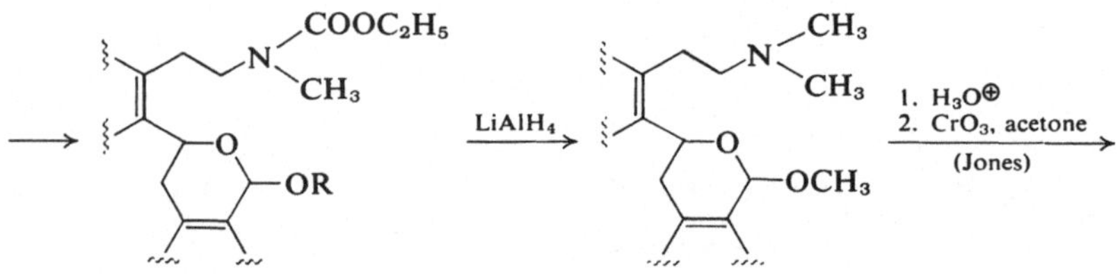

9, R = H
10, R = CH₃

11

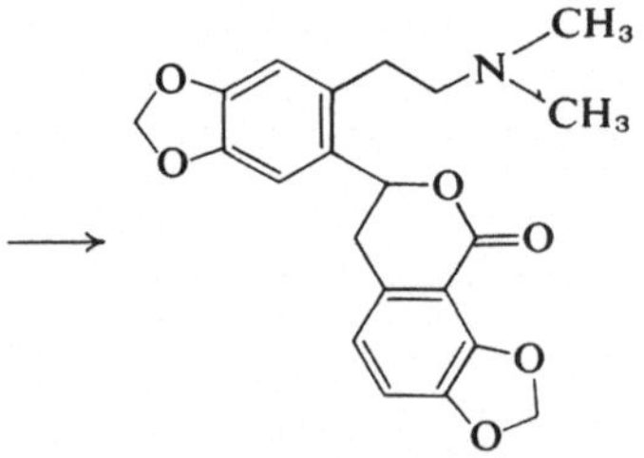

Peshawarine

Alternatively, (±)-peshawarine has been obtained from (+)-rhoeadine methiodide through Emde reduction followed by acid hydrolysis which provided a racemic hemiacetal. Jones oxidation of this material furnished the secoberberine alkaloid as a racemate.[10a]

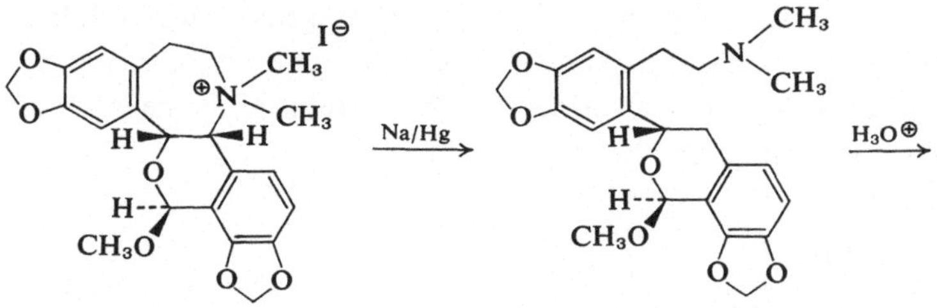

Rhoeadine methiodide   Optically active (30%)

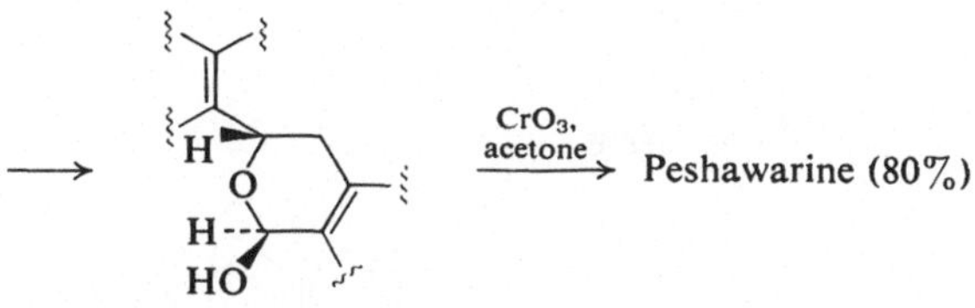

Racemic hemiacetal
(74%)

20.2. Absolute Configuration

The absolute configuration of (+)-corydalisol was established by its chemical interrelation with (+)-stylopine of known chirality.[3] Canadaline and its corresponding alcohol, canadalisol, are weakly dextrorotatory.[1] By analogy, they were assigned the same absolute configuration as (+)-corydalisol.

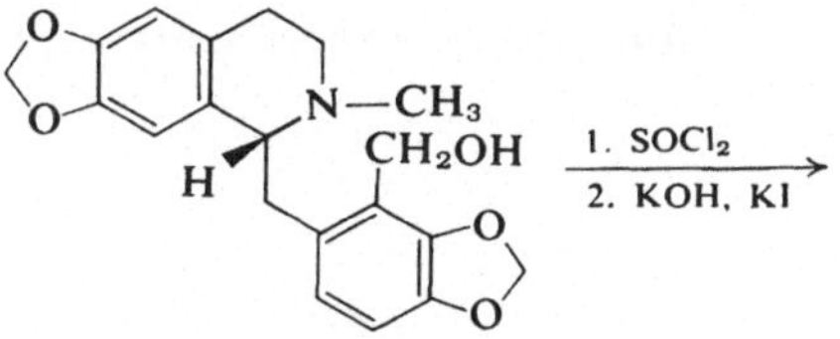

(+)-Corydalisol

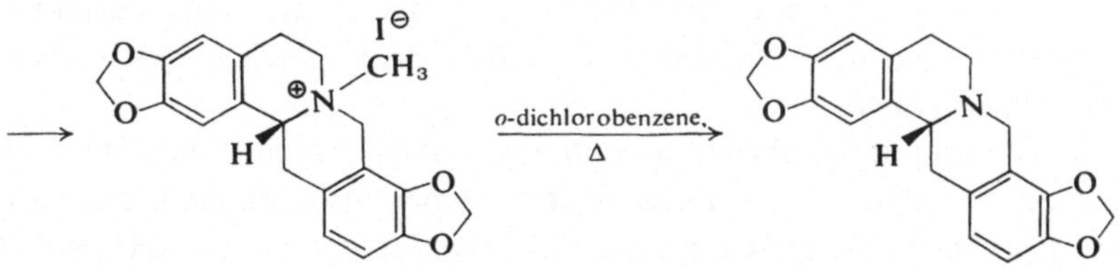

(+)-Stylopine

The Emde degradation product **12** from (+)-rhoeagenine methiodide of known stereochemistry[11,12] has been shown to be identical in terms of mp, IR, PMR, and CD curves with peshawarinediol, thus establishing the absolute configuration of (−)-peshawarine[13]:

(−)-Peshawarine

(+)-Peshawarinediol **12**

(+)-Rhoeagenine methiodide

20.3. Biogenesis

Secoberbines, as implied by their name, must be catabolic products from protoberberine salts or tetrahydroprotoberberines (berbines). But proof of the biogenesis of the secoberbines can come only from *in vivo* experiments with labeled precursors.

20.4. Mass Spectroscopy

The mass spectra of the open-form secoberbines, such as canadaline, show typically weak molecular ions, and a base peak which corresponds to the isoquinoline moiety.

The isoquinolinobenzopyran bases such as hypecorinine, on the other hand, exhibit ions due to retro-Diels–Alder fragmentation of the pyran ring. In addition, peshawarine gives rise to a base peak at *m/e* 58, for $CH_2N(CH_3)_2^{\oplus}$, due to facile cleavage of the dimethylaminoethyl side chain.

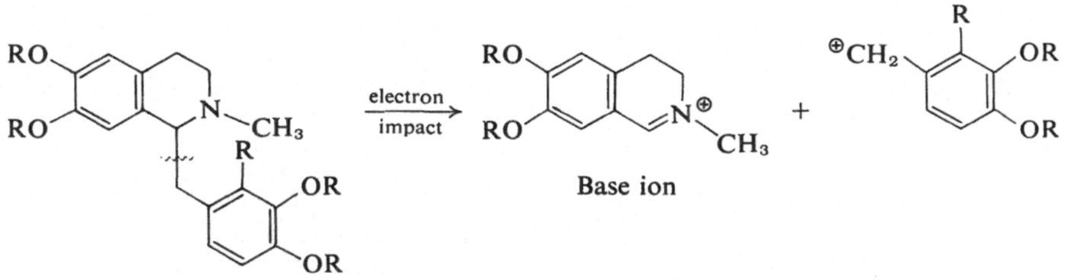

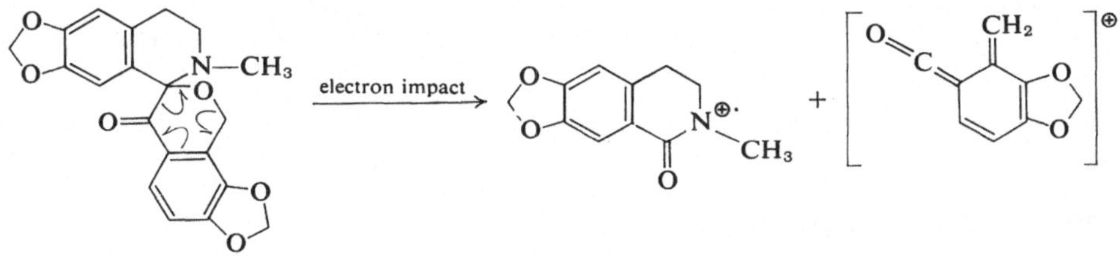

Hypecorinine

20.5. PMR Spectroscopy

The following PMR spectral values have been collected:

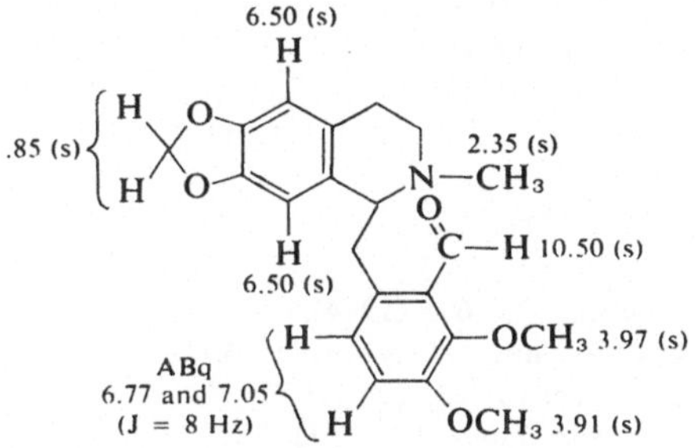

Canadaline[1]

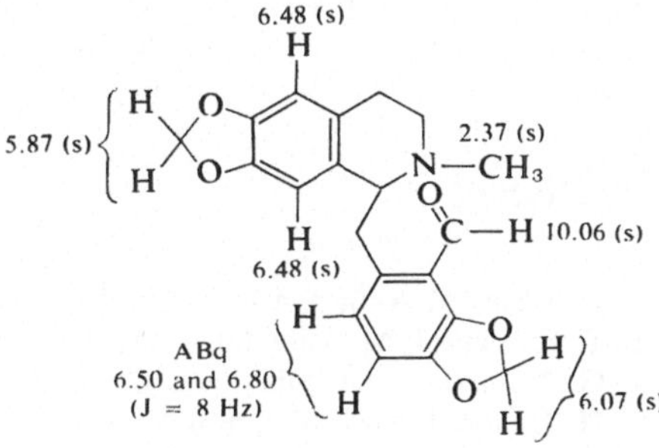

Aobamine[2]

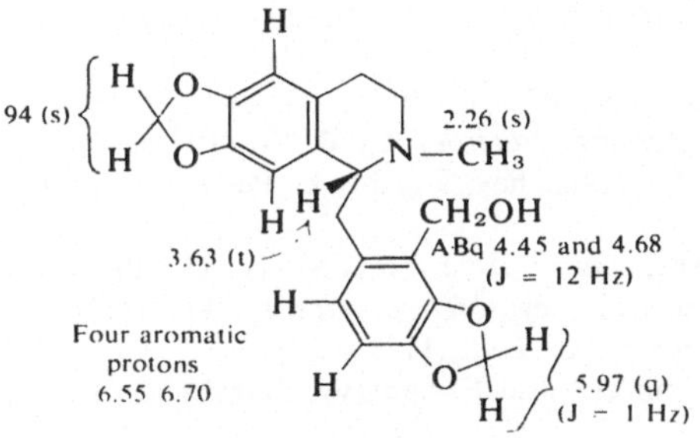

Corydalisol[3]

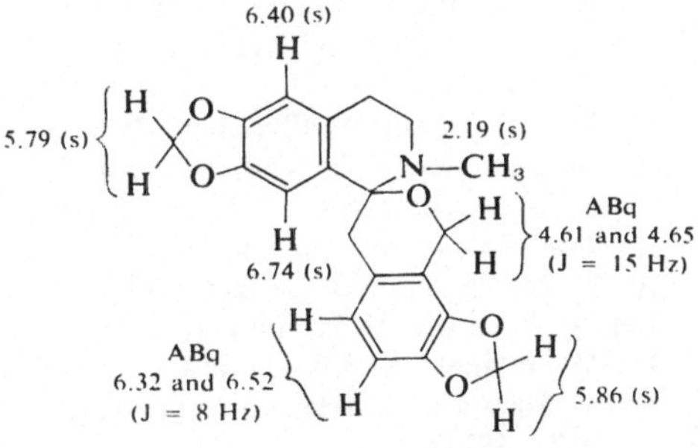

Hypecorine (in CCl₄)[4]

Hypecorinine[3,4]   Peshawarine[6]

20.6. UV Spectroscopy

Canadaline	λ_{max}^{MeOH}	288 nm (4.89)
Corydalisol	λ_{max}^{MeOH}	240 and 295 nm (3.75 and 3.83)
Hypecorinine	λ_{max}^{MeOH}	240, 292, and 320 nm (4.38, 4.10, and 3.95)
	$\lambda_{max}^{MeOH-HCl}$	247, 300, 343, and 382 sh
Hypecorine	λ_{max}^{MeOH}	236 and 290 nm (3.93 and 3.87)
Peshawarine	λ_{max}^{EtOH}	228, 245, 293, and 333 nm
		(4.96, 4.32, 3.78, and 3.69)

References

1. J. Gleye, A. Ahond, and E. Stanislas, *Phytochemistry*, **13**, 675 (1974).
2. T. Kametani, M. Takemura, M. Ihara, and K. Fukumoto, *Heterocycles*, **4**, 723 (1976).
3. G. Nonaka and I. Nishioka, *Chem. Pharm. Bull., Tokyo*, **23**, 294 (1975); G. Nonaka, H. Okabe, I. Nishioka, and N. Takao, *J. Pharm. Soc. Japan*, **93**, 87 (1973).
4. L. D. Yakhontova, M. N. Komarova, M. E. Perel'son, K. F. Blinova, and O. N. Tolkachev, *Khim. Prir. Soedin.*, 624 (1972); *Chem. Nat. Compds.*, 592 (1972).
5. A. Ikuta and H. Itokawa, *Phytochemistry*, **15**, 577 (1976).
6. M. Shamma, A. S. Rothenberg, G. S. Jayatilake, and S. F. Hussain, *Heterocycles*, **5**, 41 (1976).
7. M. Freund and K. Fleischer, *Ann. Chem., Liebigs*, **397**, 30 (1913).
8. M. Shamma, A. S. Rothenberg, and S. F. Hussain, *Heterocycles*, **6**, 707 (1977).
9. L. D. Yakhontova, M. N. Komarova, O. N. Tolkachev, and M. E. Perel'son, *Khim. Prir. Soedin.*, 491 (1976).
10. H. M. Stephan, G. Langer, and W. Wiegrebe, *Pharm. Acta Helv.*, **51**, 164 (1976).
10a. V. Šimánek, V. Preininger, F. Šantavý, and L. Dolejš, *Heterocycles*, **6**, 711 (1977).
11. V. Šimánek, A. Klásek, and F. Šantavý, *Tetrahedron Lett.*, 1779 (1973).
12. V. Šimánek, A. Klásek, L. Hruban, V. Preininger, and F. Šantavý, *Tetrahedron Lett.*, 2171 (1974).
13. M. Shamma, A. S. Rothenberg, and S. F. Hussain, *Tetrahedron*, in press.

THE BENZOPHENANTHRIDINES

21.1. Introduction

Some new benzophenanthridine alkaloids and their sources are shown below*:

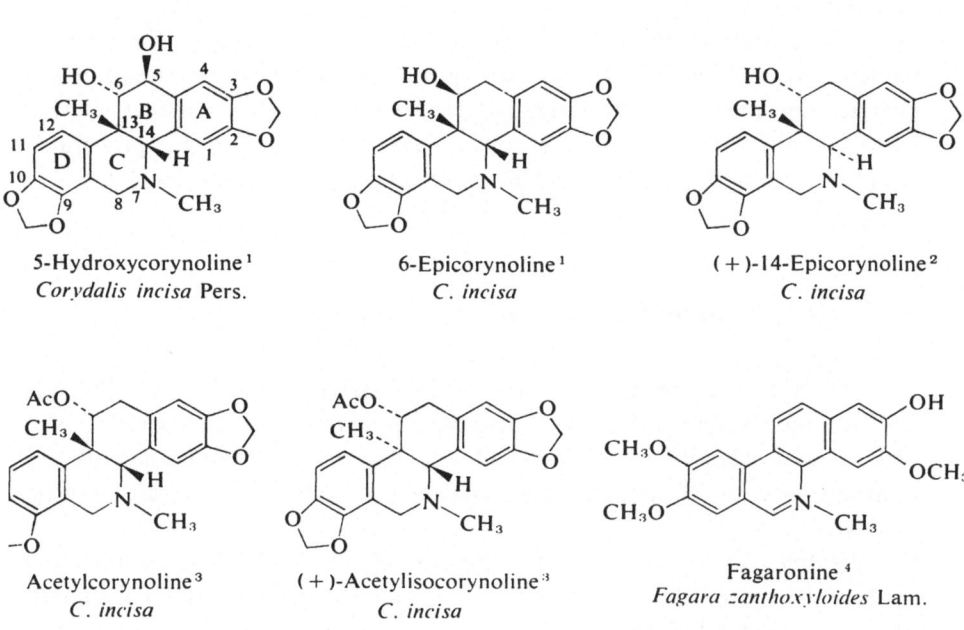

5-Hydroxycorynoline[1]
Corydalis incisa Pers.

6-Epicorynoline[1]
C. incisa

(+)-14-Epicorynoline[2]
C. incisa

Acetylcorynoline[3]
C. incisa

(+)-Acetylisocorynoline[3]
C. incisa

Fagaronine[4]
Fagara zanthoxyloides Lam.

Arnottianamide, isoarnottianamide, and iwamide are three unusual amidic alkaloids present in *Zanthoxylum* spp. (Rutaceae). They are clearly related to the benzophenanthridines, but possess an open ring C.

---

* There is no generally accepted numbering system for the benzophenanthridines, so that 5-hydroxycorynoline and 6-epicorynoline are also called 12-hydroxycorynoline and 11-epicorynoline. The numbering system adopted here is based on biogenetic considerations; see M. Shamma, *The Isoquinoline Alkaloids*, Academic Press, New York (1972), p. 315; and Ref. 30 below.

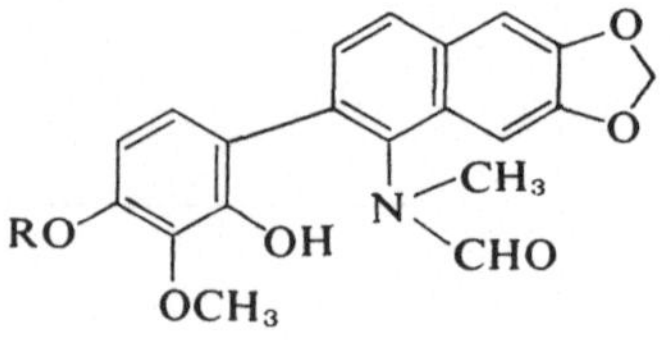

Arnottianamide,[5] R = CH$_3$
Iwamide,[6] R = H

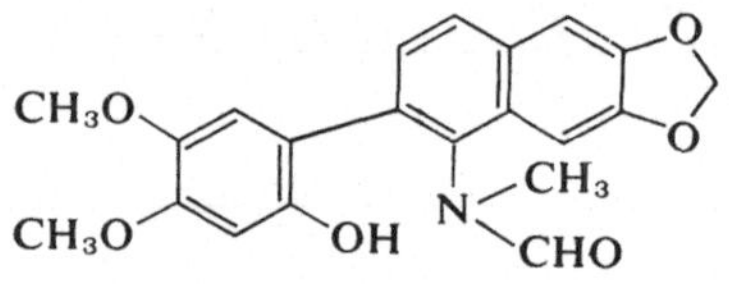

Isoarnottianamide[5]

Bocconine has been found to be identical with chelirubine,[7] so that the older name chelirubine will be used in the following discussion of the structural elucidation of this alkaloid.

A large-scale purification of commercially available sanguinarine involves reduction to the dihydro derivative using sodium borohydride, extraction with pentane, and reoxidation to sanguinarine using palladium on carbon followed by silver nitrate.[8] Dihydrochelerythrine has been obtained by preparative liquid chromatography of the crude extracts from a *Fagara* species.[8a]

21.2. Arnottianamide and Isoarnottianamide

The Rutaceae are a common source of benzophenanthridine alkaloids. Recent investigations of some Formosan and Japanese *Zanthoxylum* spp. (Rutaceae) yielded two colorless and phenolic amides, arnottianamide and isoarnottianamide. Both compounds analyze for $C_{21}H_{19}NO_6$ and are optically inactive.

Treatment of arnottianamide with lithium aluminum hydride furnished the base deoxoarnottianamide which was O-methylated using Rodionow conditions to afford methyl deoxoarnottianamide. Subsequently, in a transformation with biogenetic implications, reaction of the known benzophenanthridine alkaloid chelerythrine chloride with *m*-chloroperbenzoic acid in HMPA at 40° gave arnottianamide in 70% yield (see Scheme 21.1).[5]

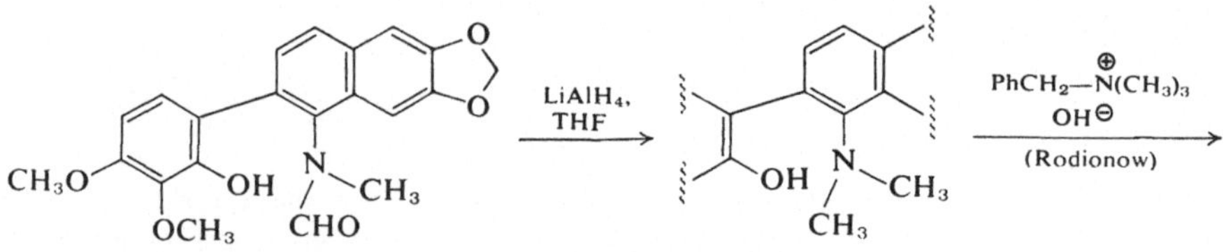

Arnottianamide   Deoxoarnottianamide

SCHEME 21.1

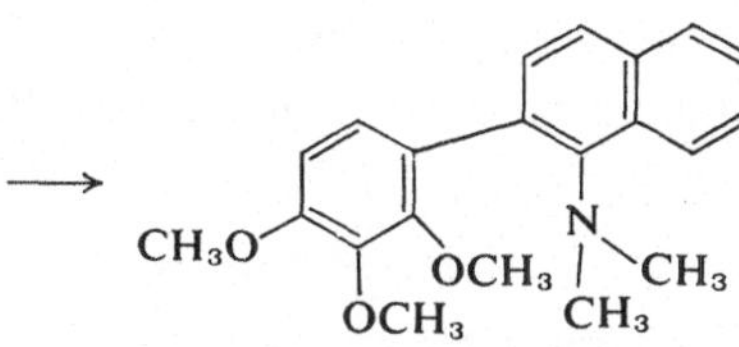

Methyl deoxoarnottianamide

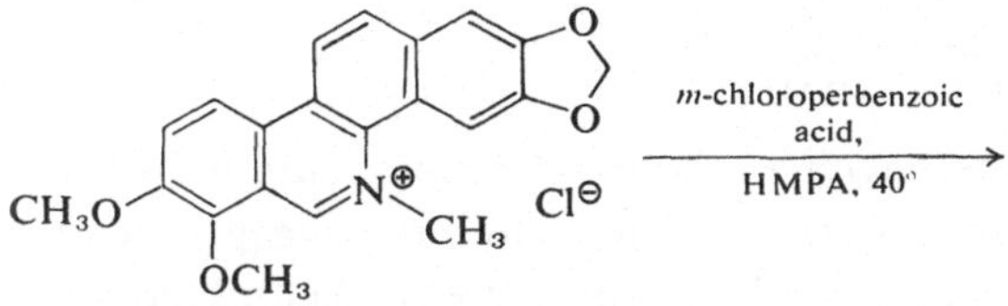

Chelerythrine chloride

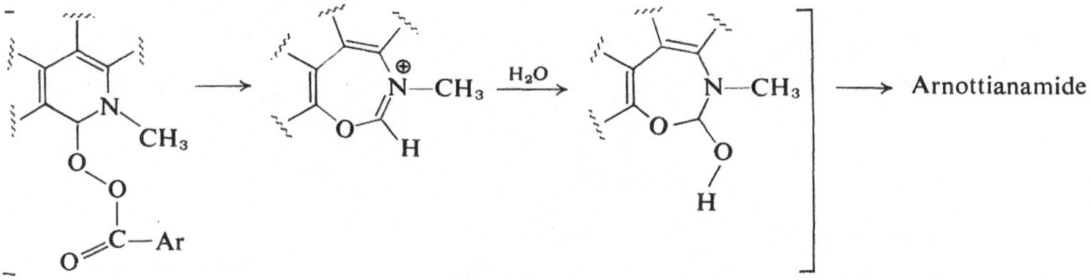

m-chloroperbenzoic acid, HMPA, 40°

$\longrightarrow$ Arnottianamide

SCHEME 21.1 (Continued)

In isoarnottianamide, the two methoxyl substituents are meta and para to the phenolic function, rather than ortho and meta as in arnottianamide.[5]

21.3. Synthesis

21.3.1. Photocyclization[8b]

Photocyclization of an enamide bearing an *o*-methoxyl group takes place at the site occupied by the methoxyl to afford a substituted pyridone after work-up.[9]

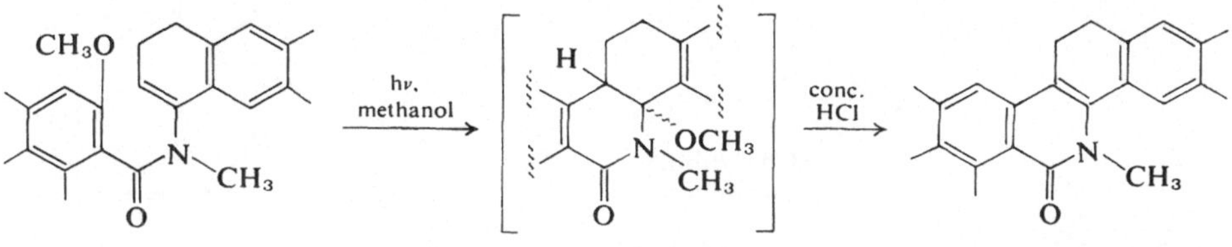

hν, methanol

conc. HCl

A substituted pyridone

This approach has been used by Ninomiya, Ishii, and their co-workers in the synthesis of such aromatic benzophenanthridine alkaloids as dihydronitidine and dihydroavicine.[10,11]

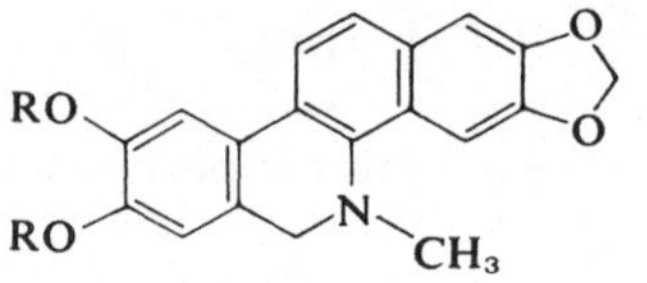

Dihydronitidine, R = CH₃
Dihydroavicine, R + R = CH₂

An interesting extension of the enamide photocyclization route is in the synthesis of oxychelirubine (oxybocconine) by Ishii and his colleagues (see Scheme 21.2). This work[12] proved conclusively the structure of chelirubine.

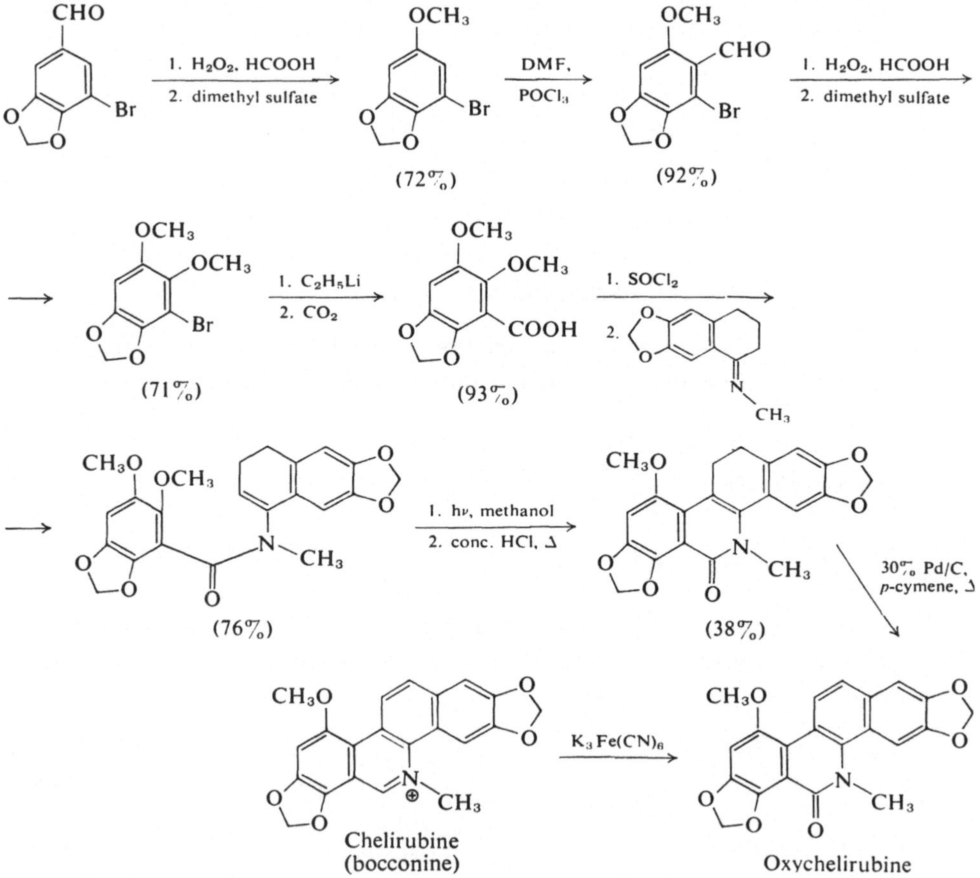

SCHEME 21.2

A corollary is that chelilutine, sanguirubine, and sanguilutine, which are analogs of chelirubine, also incorporate a trioxygenated ring D, and indeed possess the structures indicated below.[12]

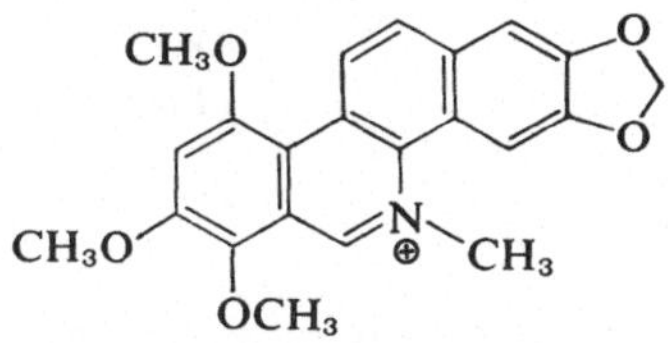

Chelilutine

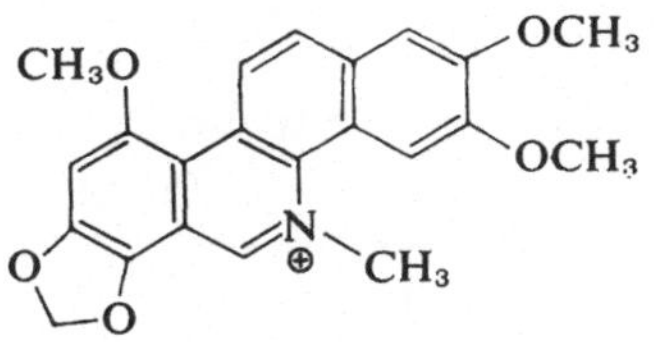

Sanguirubine

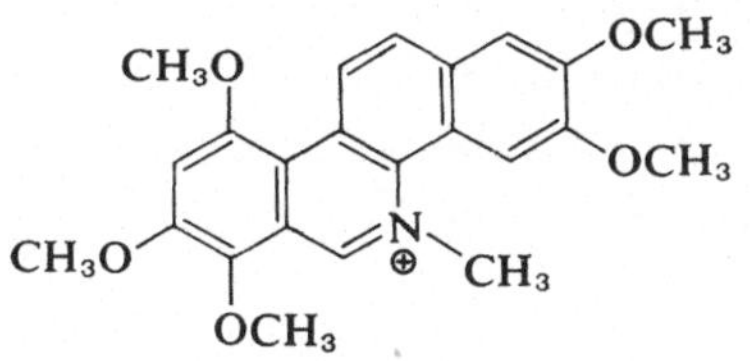

Sanguilutine

Ninomiya and co-workers have recently succeeded in a total synthesis of corynoline and 5-hydroxycorynoline using enamide photocyclizations. The main challenge was to obtain a cis B/C ring fusion, and this was achieved by immediate hydrogenolysis of the crude photocyclization product **2**, which was obtained in 20% yield from the enamide **1**[13]:

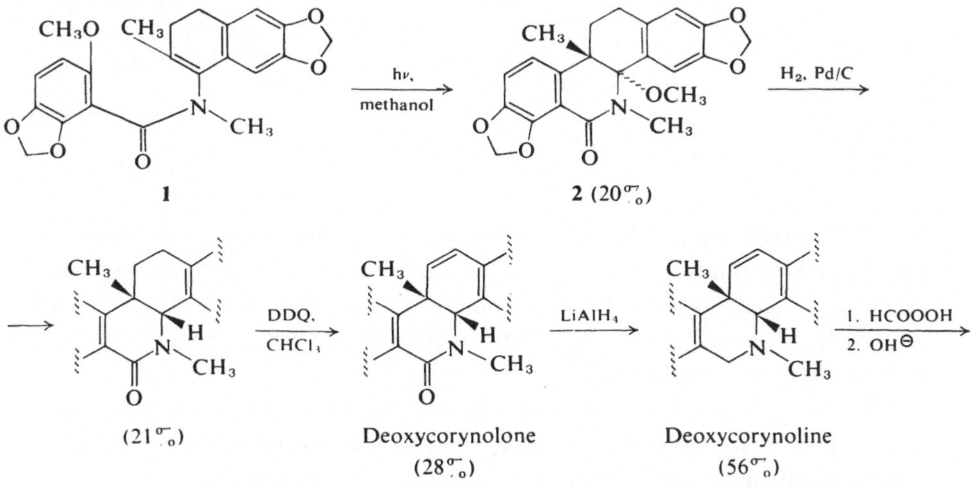

1

2 (20%)

(21%)

Deoxycorynolone
(28%)

Deoxycorynoline
(56%)

5-Hydroxycorynoline
(91%)

Corynoline
(35%)

The above sequence was extended to a preparation of 6-epicorynoline. Performic acid oxidation of deoxycorynolone, followed by treatment with alkali, afforded a mixture of diols 3, epimeric at C-5. Lithium aluminum hydride reduction of diols 3, followed by palladium hydrogenolysis of the benzylic C-5 hydroxyl, gave rise to 6-epicorynoline. It will be noted that in this instance performic acid oxidation, followed by hydrolysis, furnished C-6 beta hydroxyl derivatives.[14]

Deoxycorynolone

3

6-Epicorynoline

N-Benzoylenamides lacking an *o*-methoxyl group undergo photocyclization to yield a trans B/C fused product. However, if an electron-withdrawing substituent, such as cyano or carbomethoxyl, is present in the *N*-benzoyl moiety, the original trans fused product isomerizes to the cis analog upon further irradiation in methanol.[15]

Kessar and co-workers have shown that the photocyclization of aromatic amides possessing an *o*-bromo group represents another efficient avenue to aromatic benzophenanthridines such as nitidine and avicine.[16]

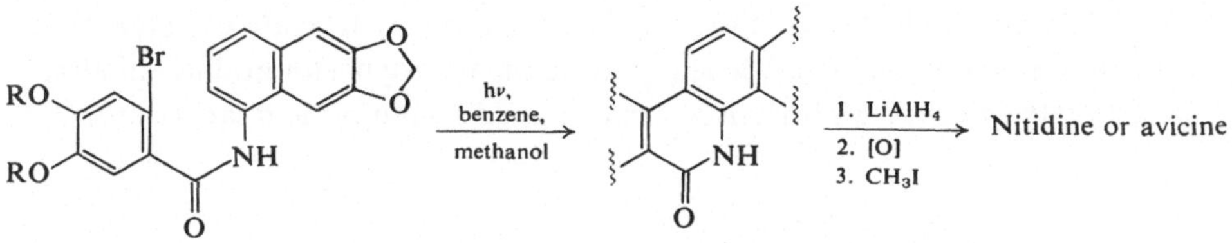

R = CH₃
or R + R = CH₂

(70%)

Also, the ketonic isocarbostyryl **4** was converted by Onda *et al.* into a mixture of Z- and E-enol acetates by reaction with acetic anhydride and potassium acetate. Furthermore, the Z-isomer could be isomerized to the E-isomer by irradiation. The E-isomer was desired since its further irradiation in the presence of iodine was found to yield aromatic benzophenanthridines **5** and **5a**[17]:

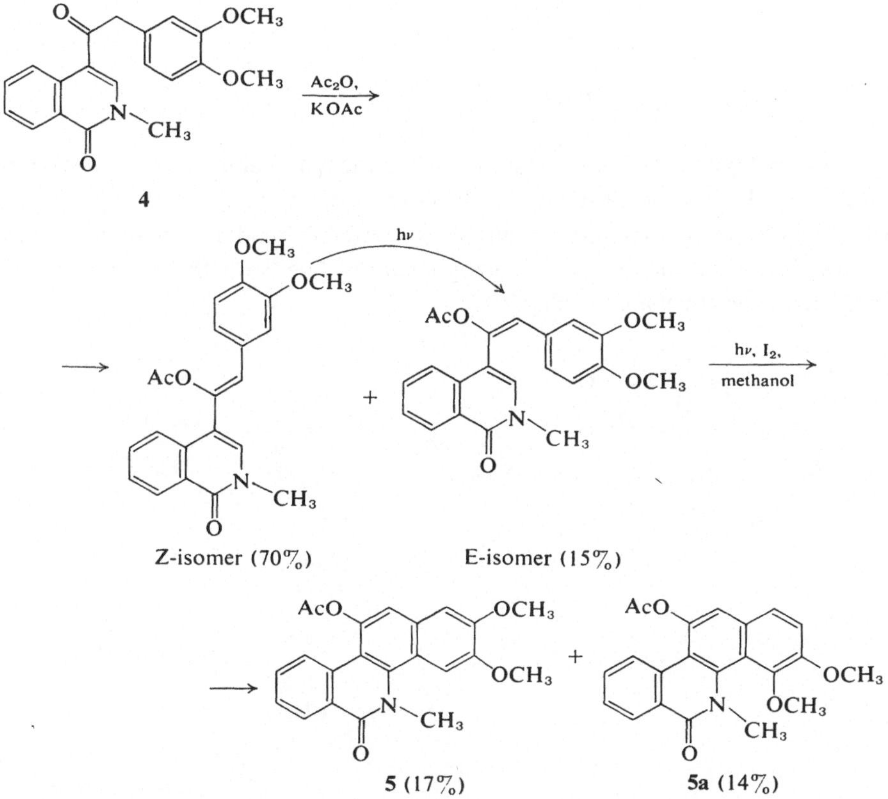

Z-isomer (70%)   E-isomer (15%)

5 (17%)   **5a** (14%)

Syntheses of chelilutine and sanguilutine have very recently been reported. Irradiation of amines **5b** and **5c** led to aromatic benzophenanthridines in about 50% isolated yield. *N*-Methylation then furnished chelilutine and sanguilutine.[17a]

5b, R + R = CH₂
5c, R = CH₃

Chelilutine (R + R = CH₂)
Sanguilutine (R = CH₃)

21.3.2. *Benzyne Intermediates*

The benzyne route to aromatic benzophenanthridines was investigated initially by Kessar. Treatment of the *N*-(*o*-bromobenzyl)-1-naphthylamine **6** with potassium amide in liquid ammonia, followed by oxidation with manganese dioxide, furnished the aromatic benzophenanthridine **7** which was readily converted to chelerythrine chloride.[18,19]

6   **7 (80%)**

Chelerythrine chloride

21.3.3. From Protopines and Protoberberines

Onda has given complete details[20] on his conversion of the protopinoid alkaloid allocryptopine to chelerythrine (see Scheme 21.3). The critical step was the photolysis of anhydroallocryptopine to yield an unstable base, probably **8**, which readily isomerized to tetrahydrochelerythrine. An alternate route to the required starting material, dihydroberberine methochloride, involves sodium borohydride in pyridine reduction of berberine to dihydroberberine, followed by quaternization with dimethyl sulfate.[20]

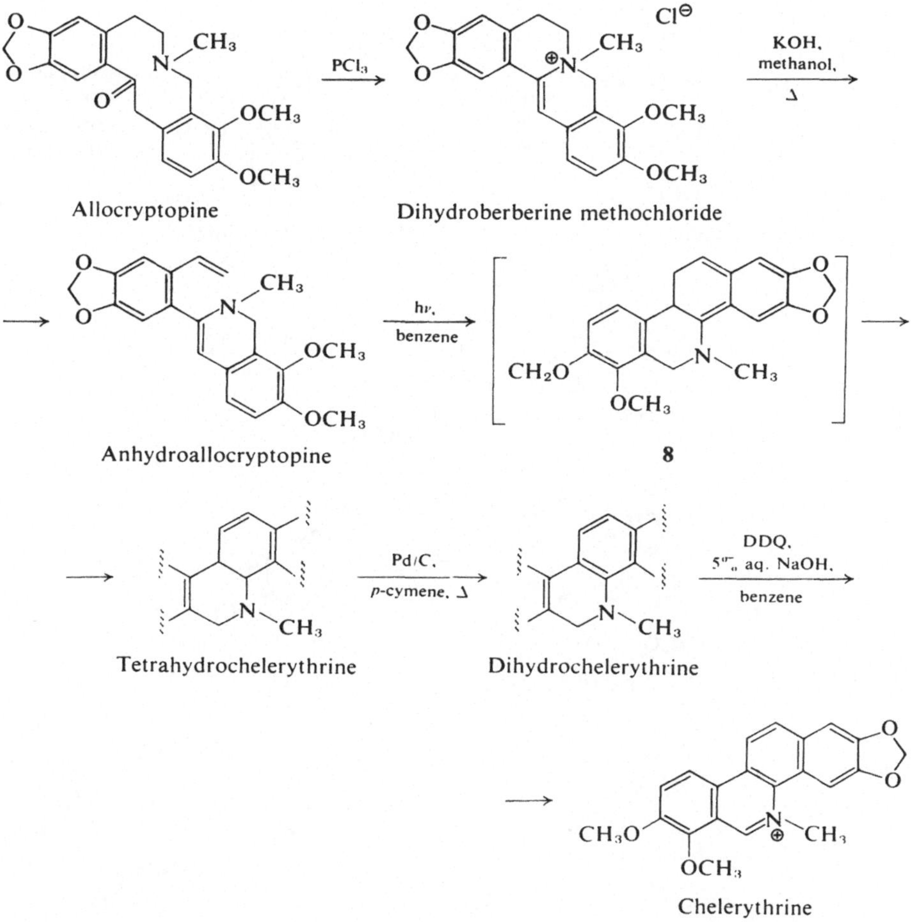

Allocryptopine

Dihydroberberine methochloride

Anhydroallocryptopine

8

Tetrahydrochelerythrine

Dihydrochelerythrine

Chelerythrine

SCHEME 21.3

The above sequence has been adapted to a synthesis of the deoxy analog **12** of corynoline. Berberine–acetone, easily prepared from berberine, was converted to the enamine **9** whose photolysis yielded two pseudobases, **10** and **11**. These could be reduced with sodium borohydride to the angularly methylated benzophenanthridine **12** (see Scheme 21.4).[21]

In order to obtain a close analog of corynoline, pseudobase **10** was treated with sodium cyanide to furnish pseudocyanide **13**. This material could be readily oxidized to the lactam **14**. The critical step in the sequence was the oxidation of **14** to the keto amide **15** which possesses a double bond adjacent to the angular methyl group, thus allowing introduction of the required alcoholic function as indicated in Scheme 21.5.[22] It should be pointed out that the cis B/C

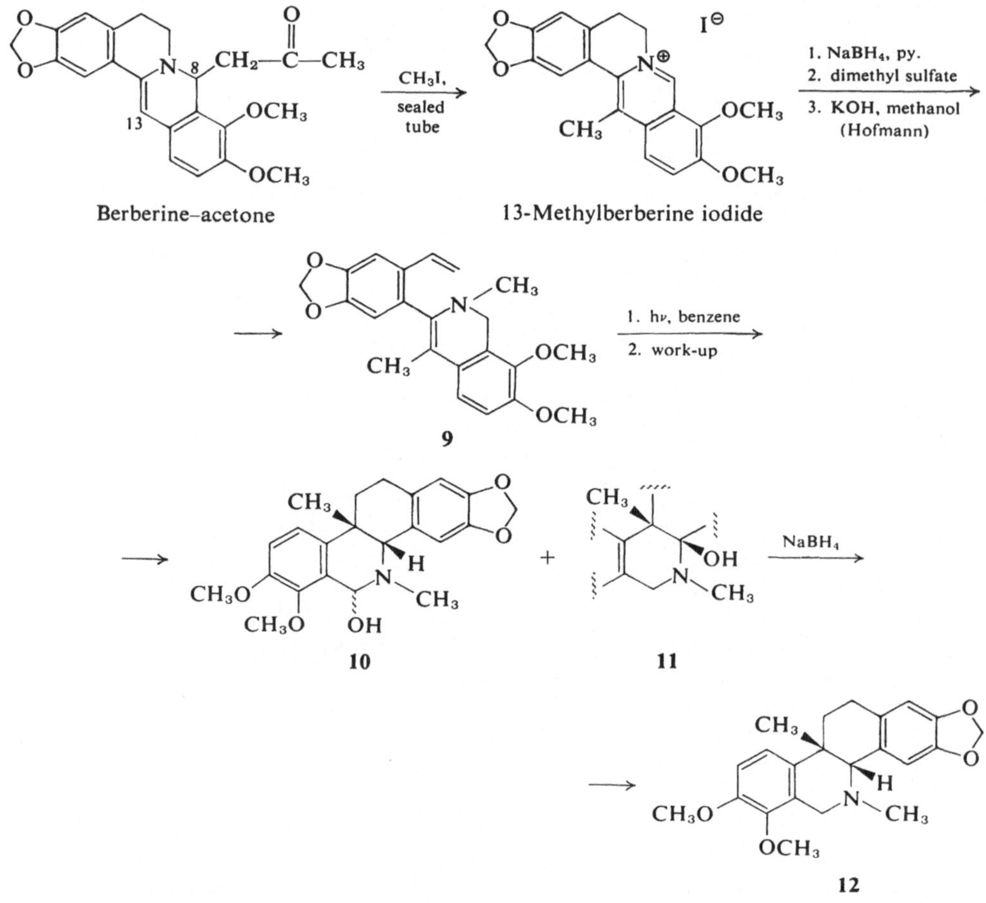

SCHEME 21.4

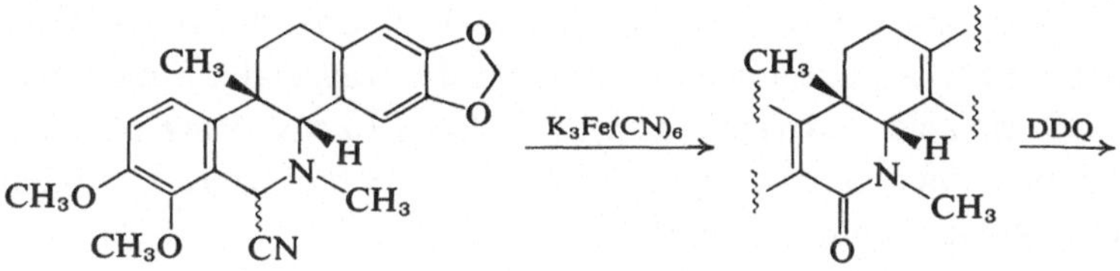

13 **14**

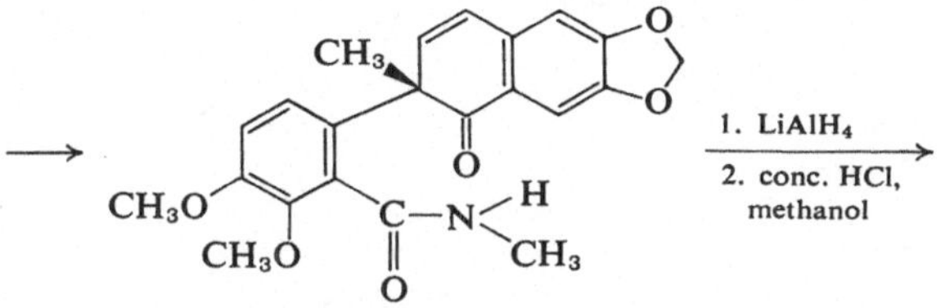

15

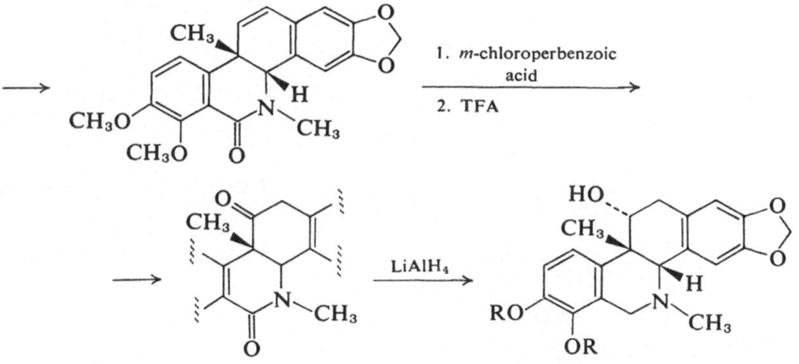

Corynoline analog, R = CH_3
Corynoline, R + R = CH_2

SCHEME 21.5

ring fusion in the angularly methylated benzophenanthridine series is probably the thermodynamically more stable arrangement. The aforementioned synthetic sequence can potentially be adapted to a preparation of corynoline itself.

Oxidation of pseudobase **10** with oxygen leads to the stable peroxide **16** which furnishes **12** upon borohydride reduction.[23]

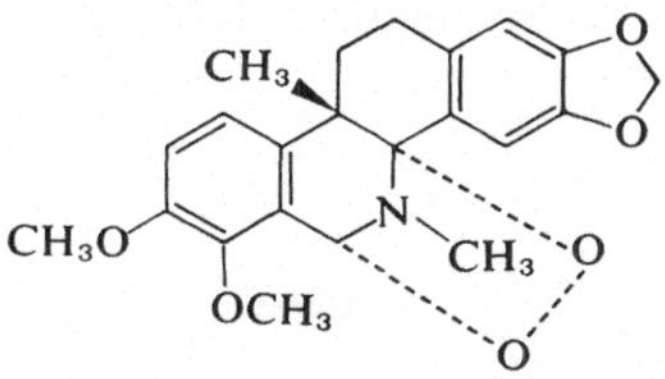

16

Finally, the Hofmann methine base from a C-10 hydroxylated protoberberine can be oxidized to a quinol acetate using lead tetraacetate. Subsequent treatment with sulfuric acid and acetic anhydride leads to a C-5 acetoxylated benzophenanthridine which can be aromatized in ring B by further treatment with hydrochloric acid in boiling ethanol[24]:

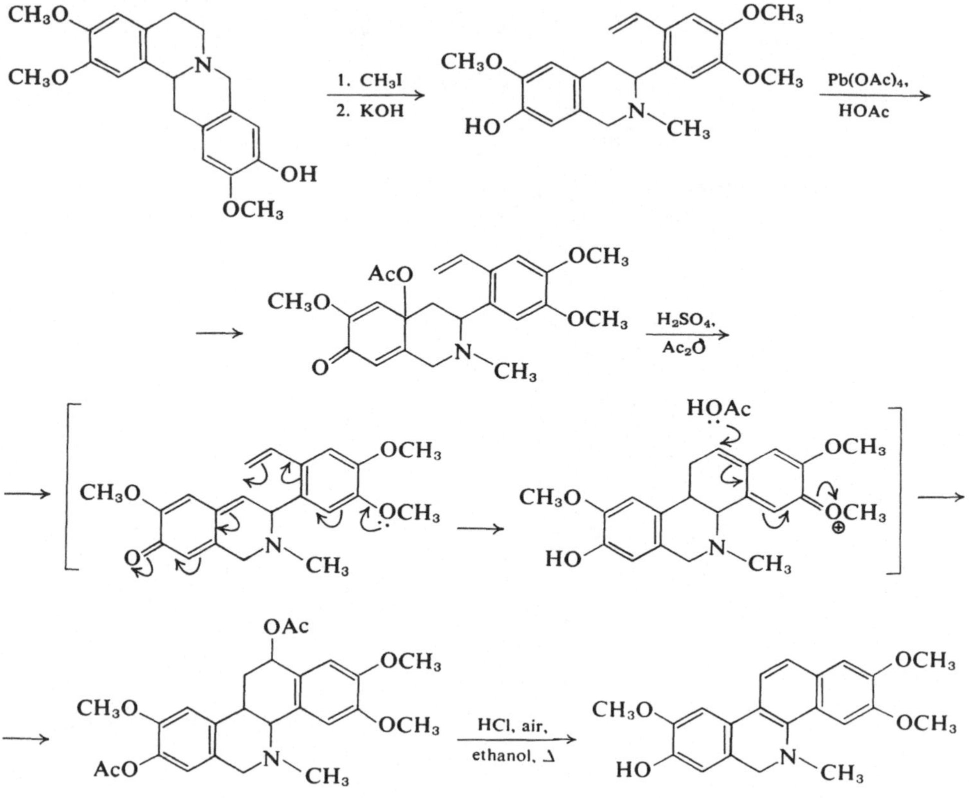

21.3.4. Mannich Cyclization

A short synthesis of nitidine involves Mannich cyclization of the cis amine **17** as indicated below[25,26]:

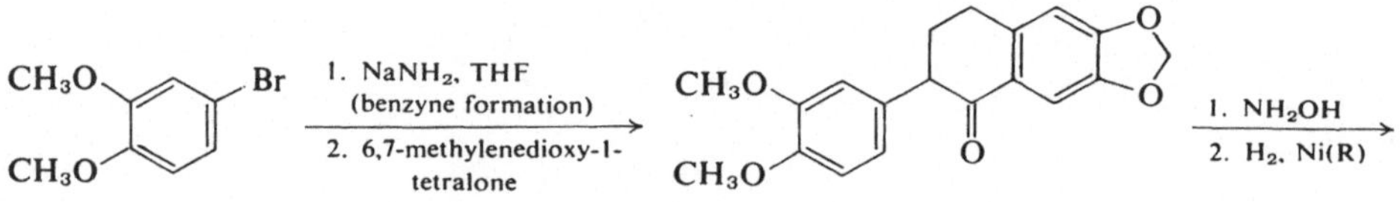

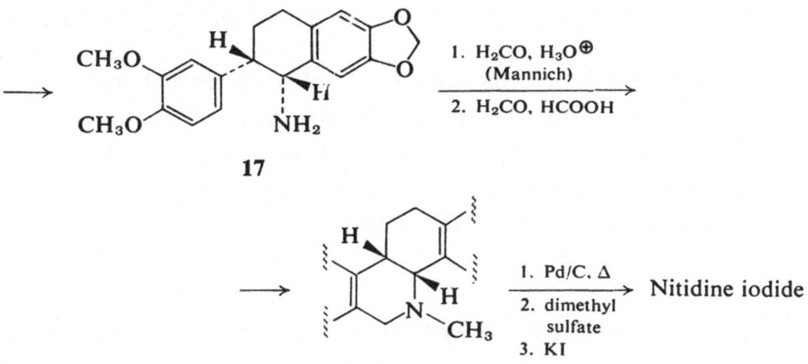

21.3.5. Friedel–Crafts Alkylation Using an Immonium Salt

The 4-substituted isocarbostyryl derivative **18**, obtained as indicated in Scheme 21.6, was reduced with lithium aluminum hydride to give the two cis fused benzophenanthridines **20** and **21**, through the intermediacy of the immonium salt **19**. The exact position of the phenolic function in **21** was not established with certainty.[27]

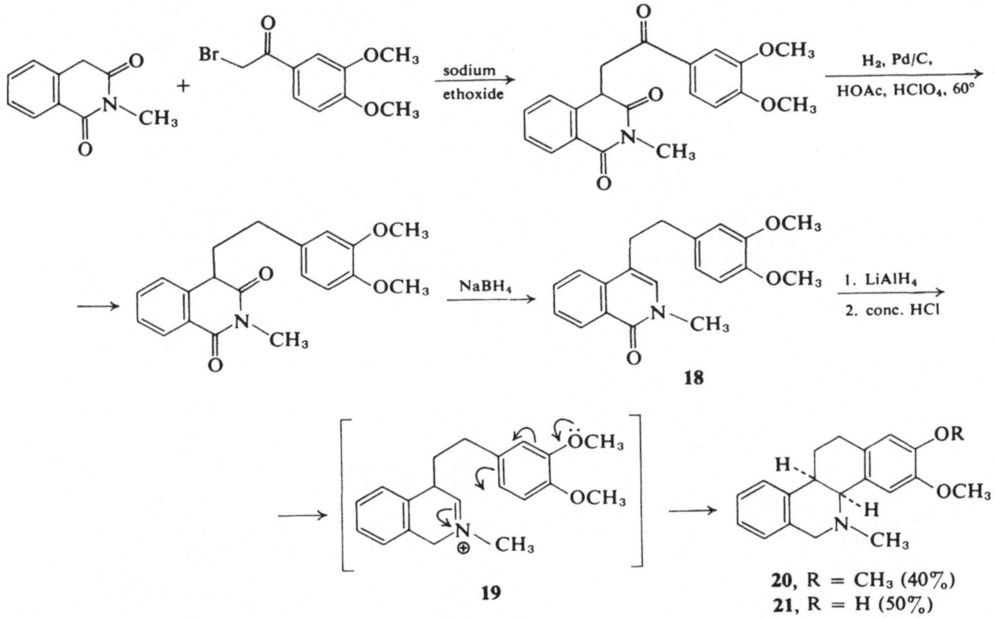

SCHEME 21.6

21.3.6. Condensation of an Imine with a Homophthalic Ester

Condensation of the diester **22** with the methyl imine of piperonal furnished the trans-fused lactam amide **23** which was homologated to **24**. Saponification and Friedel–Crafts acylation then provided the crystalline keto lactam **25** which was converted to oxysanguinarine (see Scheme 21.7).[27a]

An independent synthesis of nitidine has used a nearly parallel route.[27b]

21.4. Conversion of Isoquinolinium Salts into Pseudobases

The ease of formation of pseudobases (carbinolamines) from a variety of isoquinolinium salts has been investigated. Increasing pH facilitates pseudobase formation. Nitidine forms a pseudobase more readily than berberine, while 6,7-dimethoxy-2-methylisoquinoline did not form a pseudobase even at pH 13.[28] Sanguinarine and chelerythrine acetates exist in an equilibrium mix-

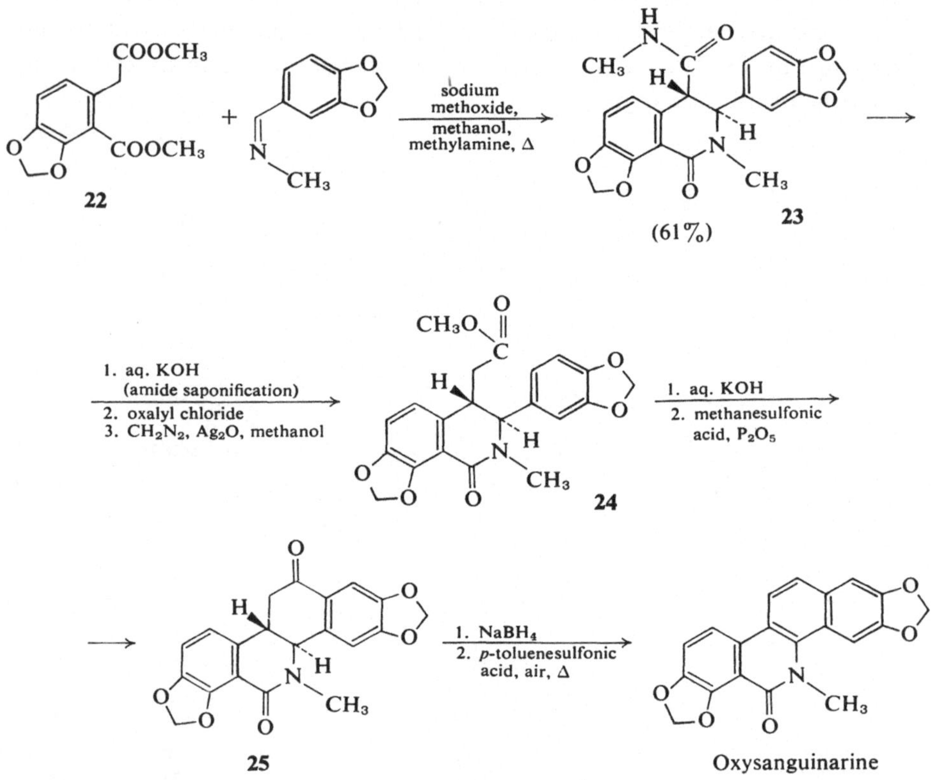

SCHEME 21.7

ture of the immonium salt and the pseudobase acetate, the ratio between these two species depending upon the polarity of the medium.[29]

21.5. Biogenesis

In a series of papers, Battersby and co-workers have described the origins of chelidonine and sanguinarine in *Chelidonium majus* L. (Papaveraceae).[30,31] Scheme 21.8 can be written for the biogenesis of these two benzophenanthridine alkaloids. The conclusions were supported by careful and thorough studies with labeled precursors.

Some of the relevant findings are[30]:

1. (+)-Reticuline is an effective precursor for chelidonine, whereas its enantiomer is not.

2. (−)-Scoulerine is well incorporated into chelidonine, but its enantiomer is ineffective. (−)-Scoulerine is also converted into sanguinarine and chelerythrine.

3. (−)-Stylopine is a good precursor of chelidonine.

4. If the chiral center of (−)-scoulerine is labeled with tritium, there is complete loss of tritium as the conversion to chelidonine takes place. This result supports the intermediacy of the enamine **26**.

5. (±)-Scoulerine is well incorporated into both chelidonine and stylopine, supporting the sequence scoulerine → stylopine → chelidonine.

6. Nandinine, which is isomeric with cheilanthifoline and corresponds to 2,3-methylenedioxy-9-hydroxy-10-methoxyberbine, is not effective as a precursor of either stylopine or chelidonine. These last two alkaloids must, therefore, be derived from cheilanthifoline and not from nandinine.

7. Stylopine methochloride is incorporated into the biogenetic scheme.

8. Chelidonine is formed by stereospecific removal of the pro-*S* hydrogen atom from (−)-stylopine, itself derived from (−)-scoulerine.[31]

9. Sanguinarine is not derived from chelidonine. Rather, both alkaloids have a common precursor represented by structure **26**.

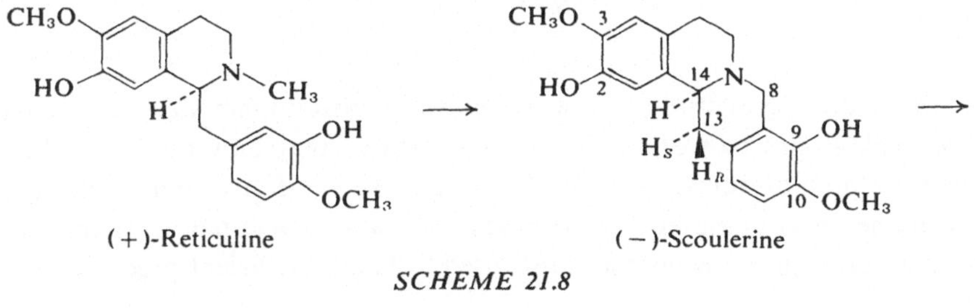

(+)-Reticuline   (−)-Scoulerine

SCHEME 21.8

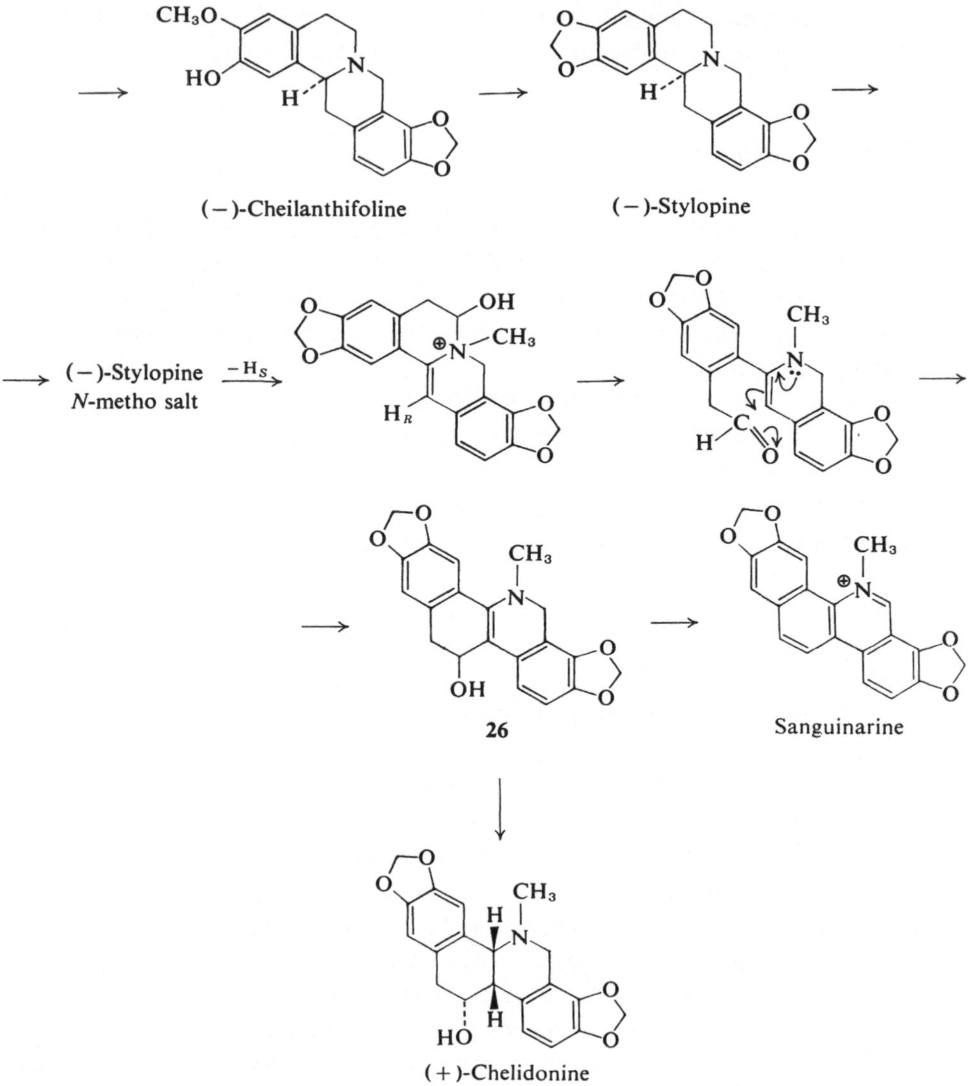
(−)-Cheilanthifoline
(−)-Stylopine
(−)-Stylopine N-metho salt
26
Sanguinarine
(+)-Chelidonine

SCHEME 21.8 (*Continued*)

In a more recent study, and working with *Corydalis incisa* and *Chelidonium majus*, Takao and co-workers have demonstrated that only the cis B/C fused *N*-metho salts of tetrahydroprotoberberines such as stylopine, tetrahydrocorysamine, and mesotetrahydrocorysamine, are converted into the corresponding protopines, while the trans fused salts are inefficient precursors.[31a]

cis-Stylopine methochloride

cis-Tetrahydrocorysamine
methochloride

cis-Mesotetrahydrocorysamine
methochloride

In that same study, it was found that when *C. majus* was treated with labeled protopine, labeled sanguinarine and chelidonine were produced.[31a] There is, therefore, a possibility that the sequence cis-B/C tetrahydroprotoberberine salt → protopine → N-metho-7,8-dihydroprotoberberine salt → enamino aldehyde → enamine (e.g., **26**) → benzophenanthridine may occur in nature.

The formation of saturated benzophenanthridines angularly methylated at C-13, such as corynoline, clearly involves the intermediacy of C-13 methylated protoberberines.[1,31a,32a] Labeled methionine supplies the *C*- and *N*-methyl groups, as well as the methylenedioxy carbon of corynoline in *C. incisa*.[1]

The benzophenanthridines chelirubine, chelilutine, sanguirubine, and sanguilutine all bear as a biogenetic scar the unusual 9,10,12-trioxygenation pattern in ring D. Significantly, the ring C opened alkaloid isoarnottianamide also possesses this same ring D trioxygenation pattern. It is probable, therefore, that the 9,10,12-trioxygenated benzophenanthridines of the chelirubine type are derived biogenetically by recyclization of isoarnottianamide-type compounds.[32b]

21.6. Pharmacology

The most important benzophenanthridines from a pharmacological standpoint are nitidine[32] and fagaronine.[4,33] Nitidine is active in the P388 lymphocytic test and is strongly cytotoxic. Fagaronine is as active as nitidine in the

P388 system, but is devoid of cytotoxicity. Preclinical toxicity studies have so far been limited to nitidine.[34]

27   Allonitidine

28

The methoxy derivative **27** of nitidine is active in the P388 test and the more resistant L1210 mouse leukemia test.[32]

As a further extension of the above results, it has been found that salts of nitidine, allonitidine and the tetramethoxy analog **28**, all possess activity against both leukemias L1210 and P388, while nitidine chloride and **28** show curative activity against Lewis lung carcinoma.[26c] Often, a 10,11-dimethoxy substitution pattern results in highly active (L1210 and P388 test) benzophenanthridinium salts.[19b]

Chelidimerine, which is *meso*-1,3-bis(8-hydrosanguinarinyl)acetone is cytotoxic, but sanguidimerine, which is diastereomeric with chelidimerine and corresponds to (+)-1,3-bis(8-hydrosanguinarinyl)acetone is inactive in the 9KB cytotoxicity assay as well as in the P388 test.[34]

Chelidimerine and sanguidimerine

21.7. Spectral Characteristics

A substantial amount of NMR and UV spectral data is now available for the benzophenanthridines, and some numerical values are cited below.

N-Demethylfagaronine[4]
Three methoxyl peaks at δ4.07, 4.17, and 4.18
λ_{max}^{ROH} 227, 272, 280, and 315 sh nm
(4.30, 4.67, 4.67 and 4.04)

5-Hydroxycorynoline[1]
λ_{max}^{MeOH} 240 and 289 nm
(3.84 and 3.75)

6-Epicorynoline[1]
λ_{max}^{MeOH} 238 sh and 290 nm (4.03 and 3.95)

Arnottianamide (NMR in TFA)[5]
λ_{max}^{EtOH} 236, 280 sh, 321 sh, 324, and 332 nm
(4.73, 4.01, 3.63, 3.65, and 3.81)

21.8. X-Ray Crystallography

X-Ray diffraction studies on corynoline p-bromobenzoate[35] and (+)-14-epicorynoline bromoacetate[36] have confirmed that the B/C fusion is cis in the former ester, and trans in the latter.

References and Notes

1. G. Nonaka and I. Nishioka, *Chem. Pharm. Bull., Tokyo*, **23**, 521 (1975). See also C. Tani and K. Tagahara, *J. Pharm. Soc. Japan*, **97**, 87 (1977).
2. N. Takao, H.-W. Bersch, and S. Takao, *Chem. Pharm. Bull., Tokyo*, **21**, 1096 (1973).
3. G. Nonaka, H. Okabe, I. Nishioka, and N. Takao, *J. Pharm. Soc. Japan*, **93**, 87 (1973).
4. M. Tin-Wa, C. L. Bell, C. Bevelle, H. H. S. Fong, and N. R. Farnsworth, *J. Pharm. Sci.*, **63**, 1476 (1974).
5. H. Ishii, T. Ishikawa, S.-T. Lu, and I.-S. Chen, *Tetrahedron Lett.*, 1203 (1976).
6. T. Ishikawa and H. Ishii, *Heterocycles*, **5**, 275 (1976).
7. J. Slavík and F. Šantavý, *Collect. Czech. Chem. Commun.*, **37**, 2804 (1972).
8. R. D. Shipanovic, C. R. Howell, and A. A. Bell, *J. Heterocycl. Chem.*, **9**, 1453 (1972).
8a. K. Hostettmann, M. J. Pettei, I. Kubo, and K. Nakanishi, *Helv. Chim. Acta*, **60**, 670 (1977).
8b. For a review on the photochemistry of imines to obtain phenanthridines and benzophenanthridines, see A. C. Pratt, *Chem. Soc. Rev.*, **6**, 63 (1973).
9. For a review on the applications of the enamide photocyclization to the synthesis of natural products, see I. Ninomiya, *Heterocycles*, **2**, 105 (1974). See also I. Ninomiya, T. Naito, T. Kiguchi, and T. Mori, *J. Chem. Soc. Perkin I*, 1696 (1973); and I. Ninomiya, T. Naito, and T. Kiguchi, *J. Chem. Soc. Perkin I*, 2257 (1973).
10. I. Ninomiya, T. Naito, H. Ishii, T. Ishida, M. Ueda, and K. Harada, *J. Chem. Soc. Perkin I*, 762 (1975).
11. H. Ishii, H. Ohida, and J. Haginawa, *J. Pharm. Soc. Japan*, **92**, 118 (1972).

12. H. Ishii, K. Harada, T. Ishida, E. Ueda, and K. Nakajima, *Tetrahedron Lett.*, 319 (1975).
13. I. Ninomiya, O. Yamamoto, and T. Naito, *Chem. Commun.*, 437 (1976); and *Heterocyclces*, 4, 743 (1976). *Trans*-benzophenanthridinones undergo thermal or photochemical isomerization to the cis analogs; see I. Ninomiya, O. Yamamoto, T. Kiguchi, T. Naito, and H. Ishii, *Heterocycles*, 6, 1730 (1977).
14. I. Ninomiya, O. Yamamoto, and T. Naito, *Heterocycles*, 5, 67 (1976).
15. I. Ninomiya, T. Kiguchi, O. Yamamoto, and T. Naito, *Heterocycles*, 4, 467 (1976).
16. S. V. Kessar, G. Singh, and P. Balakrishnan, *Tetrahedron Lett.*, 2269 (1974). See also I. Ninomiya and T. Naito, *Heterocycles*, 3, 307 (1975), for a synthesis of *N*-demethylfagaronine.
17. M. Onda, Y. Harigaya, and T. Suzuki, *Heterocycles*, 4, 1669 (1976). For related work, see U.K. Pandit, *Heterocycles*, 6, 1520 (1977).
17a. S. V. Kessar, Y. P. Gupta, K. Dhingra, G. S. Sharma, and S. Narula, *Tetrahedron Lett.*, 1459 (1977).
18. S. V. Kessar, M. Singh, and P. Balakrishnan, *Indian J. Chem.*, 12, 323 (1974).
19. This approach has also been used in a synthesis of fagaronine and other benzophenanthridinium salts by (a) J. P. Gillespie, L. G. Amoros, and F. R. Stermitz, *J. Org. Chem.*, 39, 3239 (1974); and (b) F. R. Stermitz, J. P. Gillespie, L. G. Amoros, R. Romero, T. A. Stermitz, K. A. Larson, and J. E. Ogg, *J. Med. Chem.*, 18, 708 (1975).
20. M. Onda, K. Yonezawa, and K. Abe, *Chem. Pharm. Bull.*, *Tokyo*, 19, 31 (1971). For related work, see also M. Onda and K. Kawakami, *Chem. Pharm. Bull.*, *Tokyo*, 20, 1484 (1972).
21. M. Onda, K. Yuasa, J. Okada, K. Kataoka, and K. Abe, *Chem. Pharm. Bull.*, *Tokyo*, 21, 1333 (1973).
22. M. Onda, K. Yuasa, and J. Okada, *Chem. Pharm. Bull.*, *Tokyo*, 22, 2365 (1974).
23. M. Onda, M. Gotoh, and J. Okada, *Chem. Pharm. Bull.*, *Tokyo*, 23, 1561 (1975).
24. T. Kametani, M. Takemura, M. Ihara, and K. Fukumoto, *Heterocycles*, 6, 99 (1977).
25. T. Kametani, K. Kigasawa, M. Hiiragi, and O. Kusama, *J. Heterocycl. Chem.*, 10, 31 (1973).
26. For syntheses of nitidine and other aromatic benzophenanthridines by a useful modification of the Robinson–Arthur approach, see (a) K.-Y. Zee-Cheng and C. C. Cheng, *J. Heterocycl. Chem.*, 10, 85 (1973); (b) *ibid.*, 867 (1973); and (c) K.-Y. Zee-Cheng and C. C. Cheng, *J. Med. Chem.*, 18, 66 (1975).
27. H. Iida, K. Takahashi, and T. Kikuchi, *Heterocycles*, 4, 1497 (1976). For an alternate synthesis of an enamine–immonium system analogous to 19, see W. J. Gensler, S. F. Lawless, A. L. Bluhm, and H. Dertouzos, *J. Org. Chem.*, 40, 733 (1975).
27a. M. Shamma and H. H. Tomlinson, *J. Org. Chem.*, in press.
27b. M. Cushman and L. Cheng, 174th ACS Meeting, Chicago, Ill., Aug. 1977, Item 119.
28. V. Šimánek, *Khim. Rast. Veshchestv.*, 95 (1972); through *Chem. Abstr.*, **78**, 136478e (1973). For a detailed study of pseudobase formation from isoquinolinium salts, see V. Šimánek and V. Preininger, *Heterocycles*, 6, 475 (1977). Also O. N. Tolkachev, O. E. Lasskaya, and G. A. Maslova, *Khim. Prirod. Soedin.*, 615 (1975); and *Chem. Natural Compds.*, 645 (1976).
29. O. N. Tolkachev and O. E. Lasskaya, *Khim. Prir. Soedin.*, 741 (1974); through *Chem. Abstr.*, **82**, 171254h (1975); and through *Chem. Nat. Compds.*, 762 (1976).
30. A. R. Battersby, J. Staunton, H. R. Wiltshire, R. J. Francis, and R. Southgate, *J. Chem. Soc. Perkin I*, 1147 (1975). See also A. R. Battersby and J. Staunton, *Tetrahedron*, 30, 1707 (1974).
31. A. R. Battersby, J. Staunton, H. R. Wiltshire, B. J. Bircher, and C. Fuganti, *J. Chem. Soc. Perkin I*, 1162 (1975).

31a. N. Takao, K. Iwasa, M. Kamigauchi, and M. Sugiura, *Chem. Pharm. Bull.*, *Tokyo*, **24**, 2859 (1976).

32. M. E. Wall, M. C. Wani, and H. L. ʟaylor, 162nd National ACS Meeting, Washington, D.C., Sept. 1971, Abstract MEDI 34.

32a. A. Yagi, G. Nonaka, S. Nakayama, and I. Nishioka, *Phytochemistry*, **16**, 1197 (1977).

32b. H. Ishii, Y. Murakami, and T. Ishikawa, *Heterocycles*, **6**, 1686 (1977).

33. W. M. Messmer, M. Tin-Wa, H. H. S. Fong, C. Bevelle, N. R. Farnsworth, D. J. Abraham, and J. Trojánek, *J. Pharm. Sci.*, **61**, 1858 (1972).

34. For excellent reviews on potential anticancer plant principles, see G. A. Cordell and N. R. Farnsworth, *Heterocycles*, **4**, 393 (1976); and *Lloydia*, **40**, 1 (1977).

35. T. Kametani, T. Honda, M. Ihara, H. Shimanouchi, and Y. Sasada, *Tetrahedron Lett.*, 3729 (1972).

36. N. Takao, M. Kamigauchi, and K. Iwasa, *Tetrahedron Lett.*, 805 (1974).

22

THE 3-ARYLISOQUINOLINES

Occurrence: Fumariaceae

Structures:

(+)-Corydalic acid methyl ester [1]
Corydalis incisa (Thunb.) Pers.

Corydamine,[2] R = H
N-Formylcorydamine,[2] R = CHO

22.1. Introduction

In this small group of alkaloids, the 3-arylisoquinolines, as in the benzo-phenanthridines (see Sec. 21.1) to which they are clearly related, the nitrogen heterocycle can be either fully aromatic or saturated. All three members of the 3-arylisoquinolines have been isolated from *Corydalis incisa*, a plant also known to produce benzophenanthridines.

22.2. Structural Elucidation and Synthesis

22.2.1. Corydalic Acid Methyl Ester

Corydalic acid methyl ester, $C_{22}H_{33}NO_6$, possesses in addition to the ester function two methylenedioxy groups, a secondary *C*-methyl, and a tertiary *N*-methyl group. Reduction of the ester grouping with lithium aluminum hydride afforded the alcohol 1 which upon treatment with thionyl chloride in benzene followed by neutralization furnished the quaternary salt 2. Refluxing in *o*-dichlorobenzene led to *N*-demethylation of 2 to (+)-mesotetrahydro-corysamine, a material spectrally and chromatographically identical with synthetic (±)-mesotetrahydrocorysamine (3). As further proof of structure, the

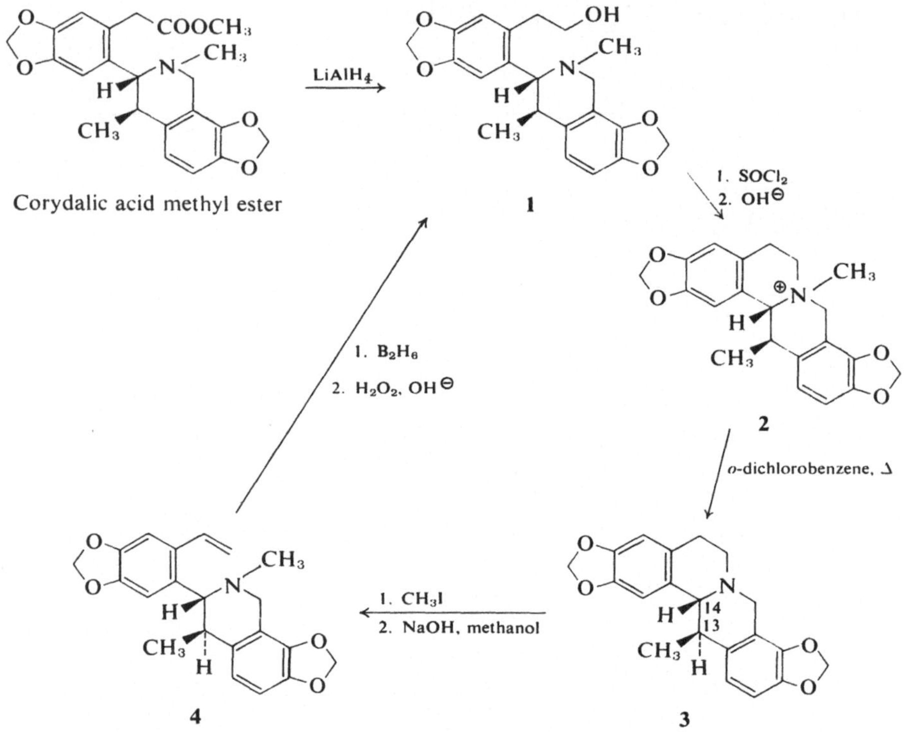

SCHEME 22.1

styrene **4**, prepared by Hofmann degradation of the methiodide of synthetic **3**, upon hydroboration and basic oxidation, gave rise to an alcohol identical to **1**. The stereochemical assignment for corydalic acid methyl ester was based on rates of methiodide formation, the 13-methyl group shift in the PMR, and the absence of Bohlmann bands in the IR spectrum of the protoberberinoid derivative **3** derived from the natural product (see Scheme 22.1).

22.2.2. Corydamine and N-Formylcorydamine

Corydamine, $C_{20}H_{18}N_2O_4$, incorporates a bis(methylenedioxy)-3-aryl-isoquinoline system with an *N*-methylaminoethyl side chain.[3] Reductive methylation of corydamine led to the tertiary base **5** which was converted by quaternization and Hofmann elimination to the isoquinolinostyrene **6**. *N*-Methylation of **6** followed by borohydride reduction gave rise to the known anhydroprotopine-B. Alternatively, when the methiodide of **5** was pyrolyzed in dimethylformamide, the protoberberine salt coptisine iodide was obtained.[2]

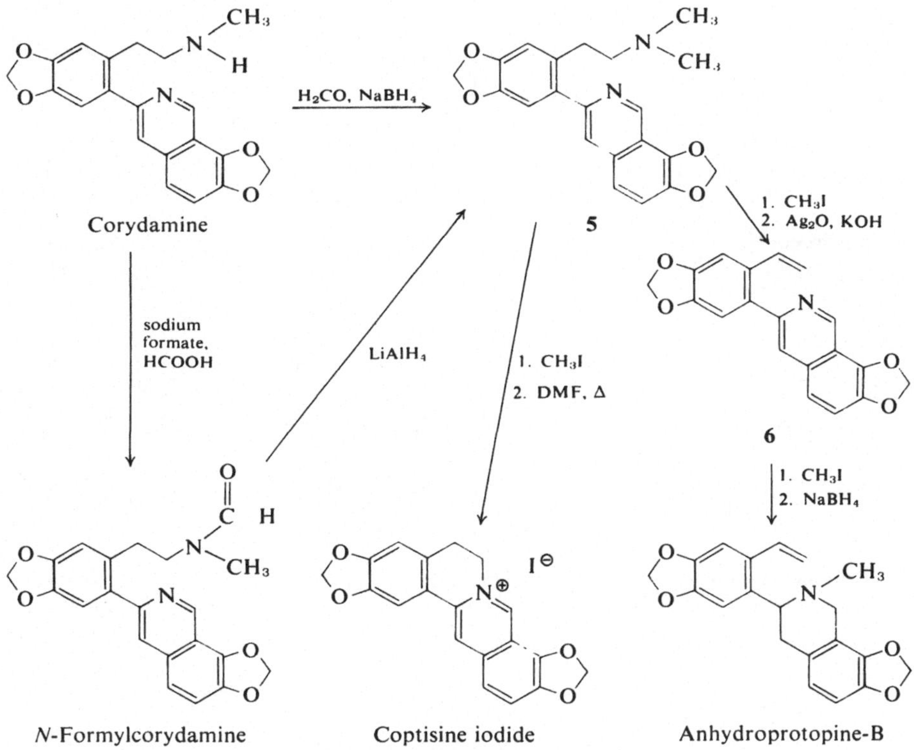

SCHEME 22.2

Lithium aluminum hydride reduction of *N*-formylcorydamine afforded the tertiary base **5**, and sodium formate–formic acid formylation of corydamine produced material identical with *N*-formylcorydamine, thus completing the structural elucidation of this amidic alkaloid (Scheme 22.2).[2]

22.3. Biogenesis

By feeding to *Corydalis incisa* ($\pm$)-tetrahydrocoptisine labeled with tritium at C-8 and C-14, it has been demonstrated that this tetrahydroprotoberberine can be converted to corynoline, corydalic acid methyl ester, and corydamine. Additionally, tetrahydrocorysamine acts as a precursor for corynoline and corydalic acid methyl ester by oxidative fission of the C-6 to N-7 bond (Scheme 22.3).[3a]

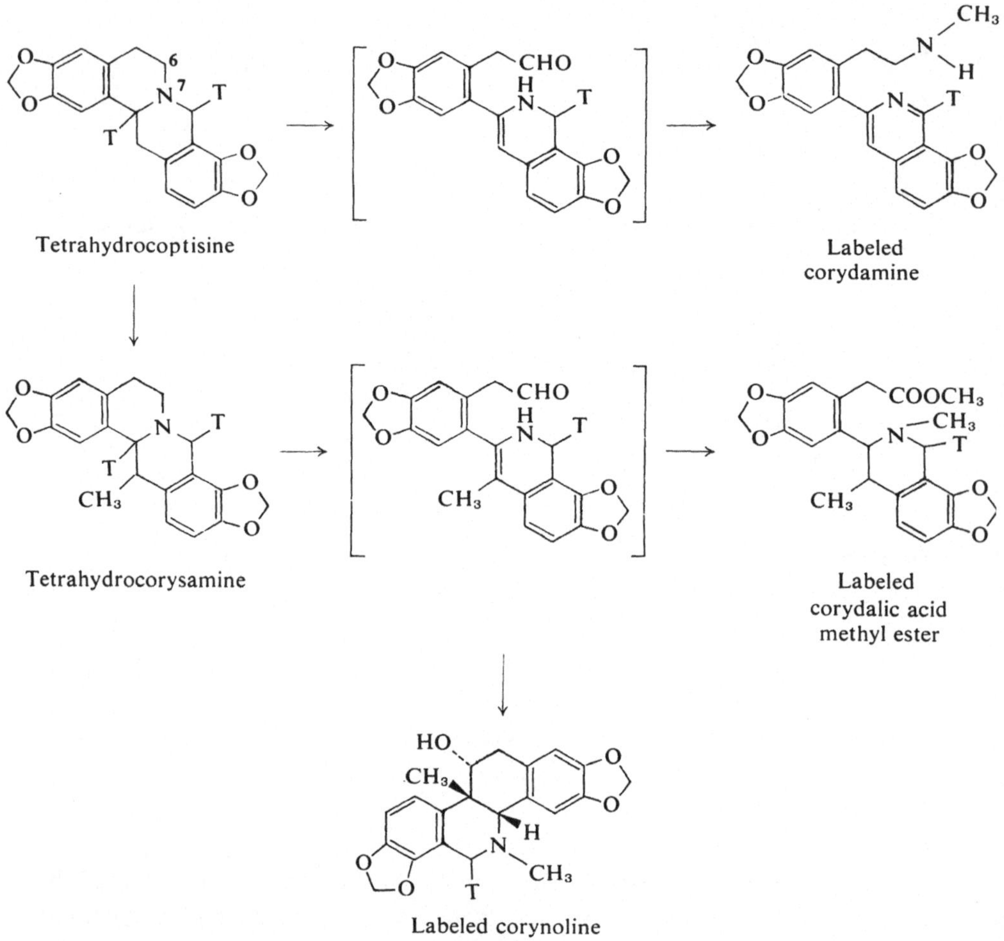

Tetrahydrocoptisine

Labeled corydamine

Tetrahydrocorysamine

Labeled corydalic acid methyl ester

Labeled corynoline

SCHEME 22.3

22.4. Pharmacology

Tetrahydrocorydamine has been reported to possess antiulcer activity (no data).[4]

22.5. Mass Spectroscopy

The aromatic 3-arylisoquinolines corydamine and *N*-formylcorydamine show a strong peak for the loss of the benzylic nitrogen-containing side chain.[2]

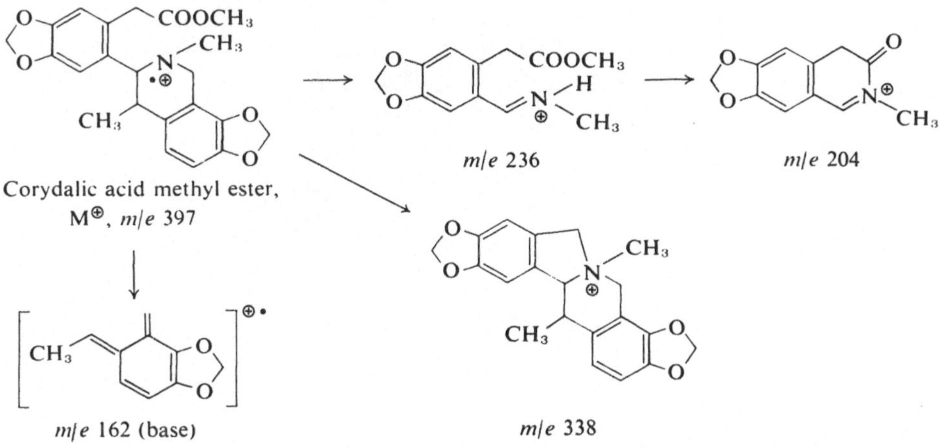

SCHEME 22.4

Corydalic acid methyl ester, which incorporates a tetrahydroisoquinoline unit, shows peaks due to the expected retro-Diels–Alder fragmentation, loss of the carbomethoxy group, and interaction of the ester and amine functions (Scheme 22.4).[1]

22.6. PMR Spectroscopy

The PMR chemical shifts for two 3-arylisoquinolines are indicated below.

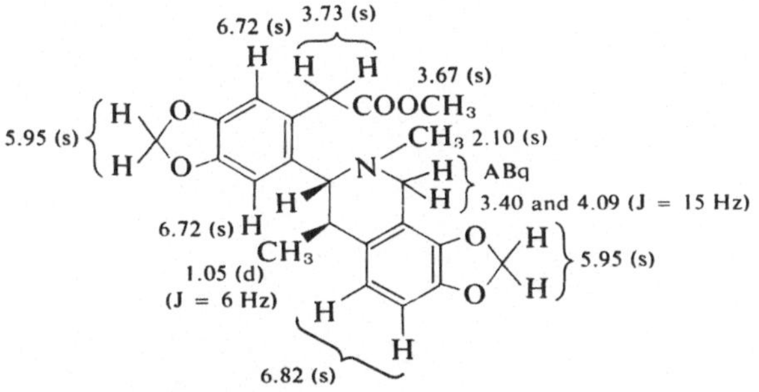

Corydalic acid methyl ester[1]

2.24–2.64 (m)

6.80 (s) H   H H   N—CH₃   2.11 (s)
5.93 (s) { H, O   CH₃
         { H O H   H 2.64–3.04 (m)
                N   H 9.35 (s)
6.88 (s)H
7.57 (s) H   O   H   6.19 (s)
        H   O   H
           H
7.33 (s)

N-Methylcorydamine²

22.7. UV Spectroscopy

Corydalic acid methyl ester¹: λ^{MeOH}_{max} 240 and 289 nm
(4.07 and 3.96)

Corydamine hydrochloride²: λ^{MeOH}_{max} 245, 312, and 380 nm
(4.51, 4.18, and 3.59)

References and Note

1. G. Nonaka, Y. Kodera, and I. Nishioka, *Chem. Pharm. Bull., Tokyo,* **21**, 1020 (1973).
2. G. Nonaka and I. Nishioka, *Chem. Pharm. Bull., Tokyo,* **21**, 1410 (1973).
3. The simultaneous independent isolation and characterization of corydamine has been reported by N. Takao and K. Iwasa, *Chem. Pharm. Bull., Tokyo,* **21**, 1587 (1973).
3a. A. Yagi, G. Nonaka, S. Nakayama, and I. Nishioka, *Phytochemistry,* **16**, 1197 (1977).
4. I. Nishioka, N. Takao, G. Nonaka, and K. Iwasa, Japan. Pat. 74,125,398; through *Chem. Abstr.,* **83**, 28432e (1975).

THE PROTOPINES

23.1. Syntheses and Interconversions

23.1.1. The Transformation of a 13-Oxoprotoberberinium N-Metho Salt to a Protopine Analog

When the berberinium salt **1**, derived from berberinephenolbetaine, was treated with sodium hydride in the presence of air and potassium iodide, it was smoothly converted into 13-oxoallocryptopine. This transformation represents a simple

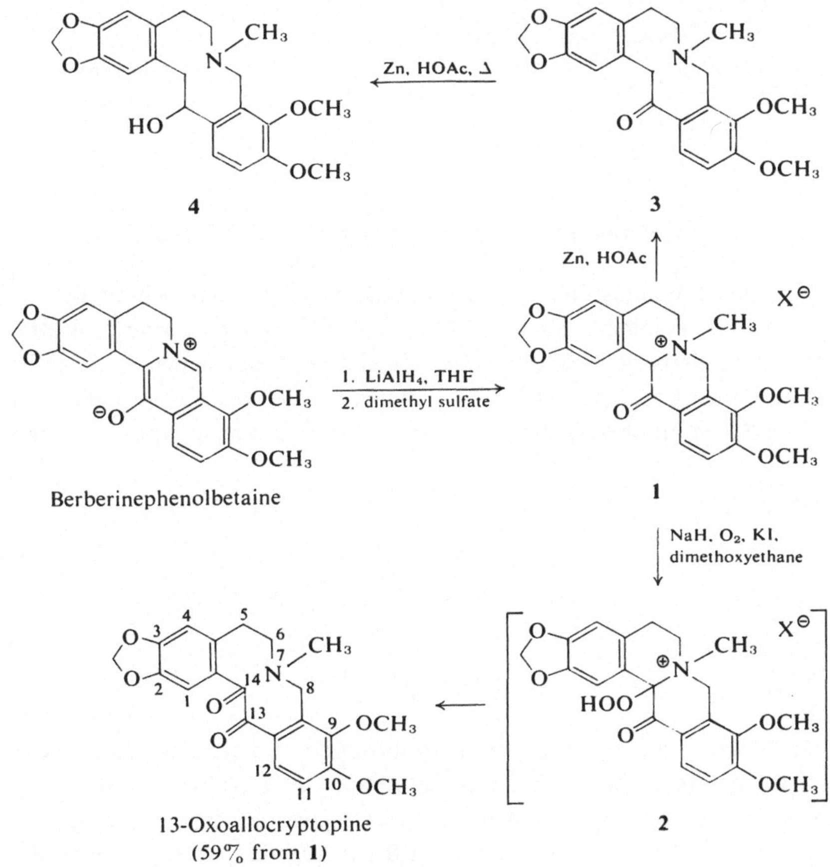

4

3

Zn, HOAc, Δ

Zn, HOAc

Berberinephenolbetaine

1. LiAlH₄, THF
2. dimethyl sulfate

1

NaH, O₂, KI, dimethoxyethane

13-Oxoallocryptopine
(59% from **1**)

2

SCHEME 23.1

conversion of a protoberberine to a protopine. It probably involves as an inter-mediate the hydroperoxide **2** which is reduced by iodide anion to the product. Alternatively, reduction of **1** with zinc in acetic acid at room temperature pro-vided the aminoketone **3**, which could be further reduced to the aminoalcohol **4** (see Scheme 23.1).[1]

A related hydroperoxide intermediate has been proposed in the photo-chemical oxidation of canadine methiodide to allocryptopine by Hanaoka and co-workers.[2]

Canadine methiodide

Allocryptopine
(16%)

23.1.2. The Conversion of an Aporhoeadane to Allocryptopine

The benzazepinoisoindole **6**, commonly referred to as Schöpf Base VI, is prepared from commercially available β-hydrastine,[3] or by zinc in acetic acid reduction of berberinephenolbetaine.[4] It can more conveniently be called an aporhoeadane derivative where the aporhoeadane nucleus is defined as in structure **5**, the numbering system corresponding to that for the rhoeadines.

5

Hofmann degradation of the methobromide of **6** gives two isomeric bases, **7** and **8**, readily separable by chromatography (IRA-400). Treatment of the *N*-oxide of the *cis* compound **8** with hot acid gives allocryptopine but only in 15% yield, whereas the trans analog of **8** has afforded allocryptopine in much higher yield (see Scheme 23.2).[5]

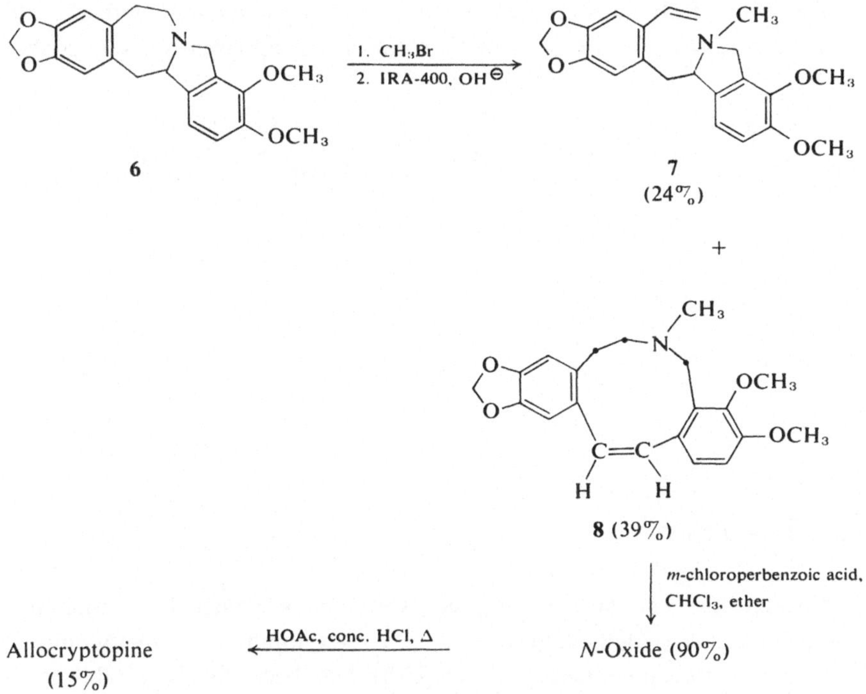

6

1. CH₃Br
2. IRA-400, OH⊖

7
(24%)

+

8 (39%)

m-chloroperbenzoic acid,
CHCl₃, ether

Allocryptopine   HOAc, conc. HCl, Δ   *N*-Oxide (90%)
(15%)

SCHEME 23.2

23.2. Reactions

Treatment of allocryptopine with cyanogen bromide in tetrahydrofuran results in cleavage of the N-7 to C-8 bond and formation of an *N*-cyano derivative.[6]

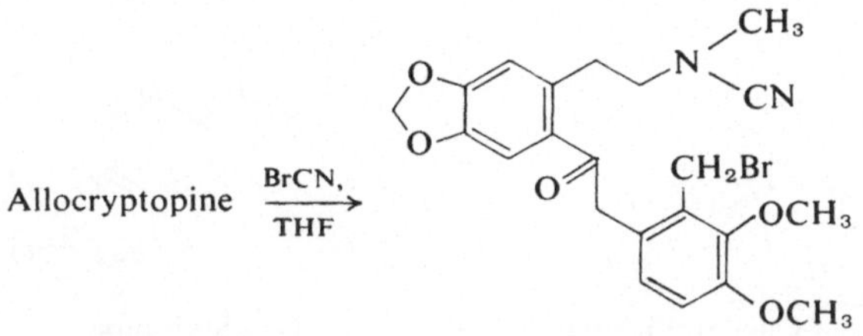

Allocryptopine   BrCN,
                        THF

The above fully substantiated result should be compared with the older report outlined below in which *N*-demethylation was apparently accomplished.[7]

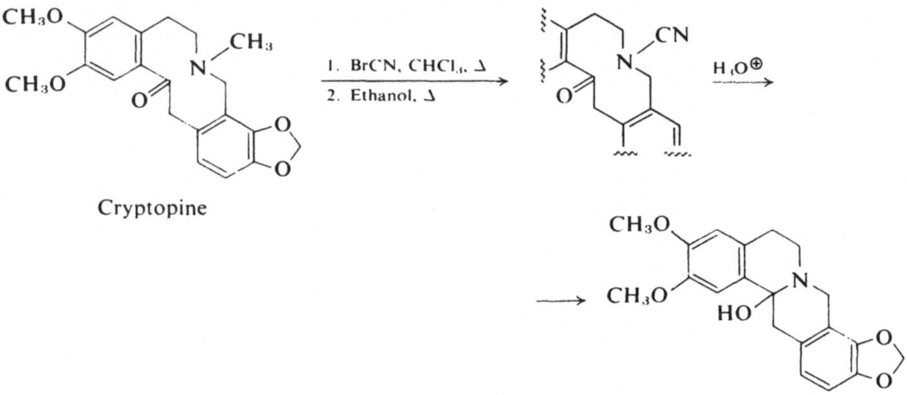

Cryptopine

23.3. Biogenesis

Through the use of labeled precursors, the sequence (+)-reticuline → (−)-scoulerine → (−)-cheilanthifoline → (−)-stylopine → (−)-stylopine *N*-metho salt → protopine (see Scheme 23.3) has been firmly established in *Chelidonium majus* L. (Papaveraceae).[8] (−)-Stylopine *N*-metho salt is also efficiently converted into protopine in *Corydalis incisa* Pers. (Fumariaceae).[9]

The indications are that the cis B/C fused (−)-stylopine *N*-metho salt rather than the trans analog is a precursor for protopine.[9a] (For a discussion of the biogenetic relationship between reticuline, scoulerine, stylopine, and the benzophenanthridines, see Sec. 21.5.)

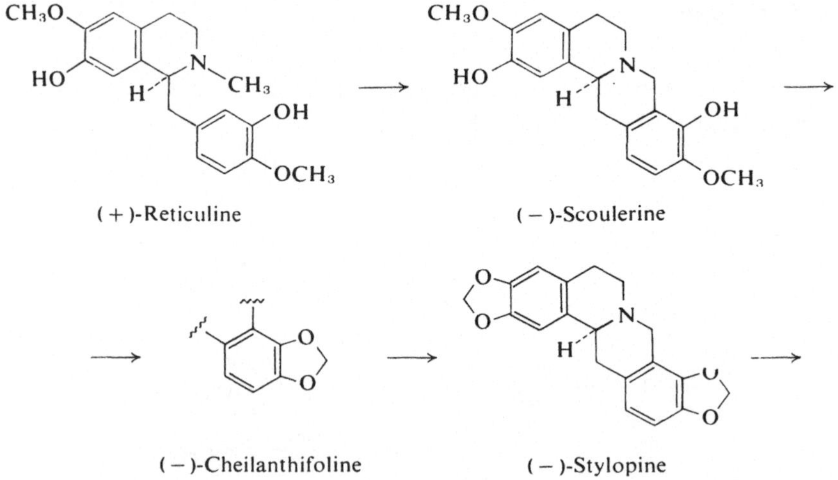

(+)-Reticuline   (−)-Scoulerine

(−)-Cheilanthifoline   (−)-Stylopine

SCHEME 23.3

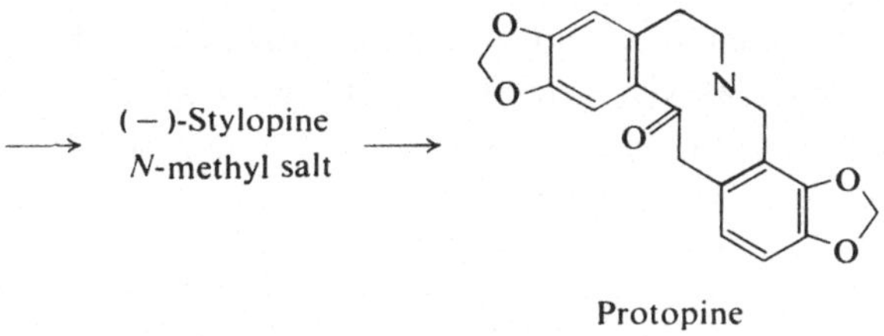

$\longrightarrow$ (−)-Stylopine $\longrightarrow$
N-methyl salt

Protopine

SCHEME 23.3 (Continued)

23.4. Pharmacology

Allocryptopine (thalictrimine) is a strong myocontractile agent in mice, and is apparently less toxic than the indole alkaloid brevicolline.[10] The pharmacology of protopines and other alkaloids of the Papaveraceae has been thoroughly covered up to 1972 in Ref. 11.

23.5. Spectral Studies

Some spectral data, including PMR and UV, that have recently become available include the three examples cited below. In PMR experiments, when the solvent is changed from deuteriochloroform to the aromatic and anisotropic benzene-d_6 or pyridine-d_5 which are capable of forming collision complexes, it is found that protons ortho to a phenolic hydroxyl show larger deshielding than protons in the meta or para positions. This phenomenon, commonly called "aromatic solvent induced shifts" (ASIS), has been used to locate the site of the phenolic function in the new alkaloid protothalipine found in *Thalictrum rugosum* Ait. (*T. glaucum* Desf.) (Ranunculaceae).[12]

The PMR spectrum of protopine itself shows a multiplet at $\delta 2.88$ and another at $\delta 2.55$, and these had been assigned previously to the C-5 and C-6 protons, respectively.[13] This assignment has now been reversed, so that the multiplet further downfield represents the C-6 methylene protons, where C-6 is bonded directly to the nitrogen atom.[14] A systematic study of the chemical shifts of the ring B protons on the carbon adjacent to nitrogen for a variety of isoquinoline alkaloids has shown that the shifts are dependent on the state of the nitrogen atom, and are in the following order of increasing δ values: Amines $\delta 2.79$–3.17, protopines $\delta 2.88$–2.95, enamines $\delta 3.03$–3.05, amides $\delta 3.50$–4.10, imines $\delta 3.82$–4.05, imides $\delta 4.04$–4.29, pyridones $\delta 4.23$–4.66, N-oxides (specifically canadine N-oxide) $\delta 4.48$, and pyridinium salts (in TFA)

$\delta 4.60$–5.12.[14] The generalization has also been drawn that the ring B methylene protons will appear as symmetrical triplets when nitrogen inversion is precluded (e.g., amides and quaternary salts), and as complex multiplets when the nitrogen atom can undergo inversion.[14]

6.62 (s)
H
5.90 (s)
H, O
H, O
CH₃ 1.82 (s)
N
6.82 (s) H
7.82 (s) O
H
H
OCH₃ 3.80 (s)
and
7.07 (d) H
OCH₃ 3.86 (s)
6.83 (d) H
J = 8.5 Hz

Not a natural product,[1] ν_{max}^{Nujol} 1685 cm^{-1} (5.93 μ);
λ_{max}^{EtOH} 237 sh and 287 nm (4.08 and 3.94)

6.57 (s) 2.50–3.00 (br. s.)
H H H
H
6.03 (s)
H, O
H, O
H
CH₃ 1.82 (s)
N
7.35 (s) H O
O
OCH₃ 3.85 (s)
and
OCH₃ 3.95 (s)
7.73 (d)
H
7.00 (d)
J = 8.5 Hz

13-Oxoallocryptopine,[1] ν_{max}^{CHCl3} 1660–1680 cm^{-1}
(5.95–6.03 μ)

2.67 (m)
H H
H
2.90 (m)
CH₃O
H
3.90 (s)
CH₃ 1.87 (s)
N
H
H
3.71 (s) or 3.75 (s)
CH₃O
O
OH 4.07 (br. s.)
3.71 (s) H
or 3.75 (s) H
OCH₃ 3.90 (s)

4 Aromatic H 6.68–7.05 (m)
Protothalipine[12] λ_{max}^{MeOH} 232 and 282 nm
(4.60 and 3.91)

The CMR spectra of five protopine alkaloids, namely protopine, allo-cryptopine, cryptopine, muramine and hunnemanine, have been determined in 5% (v/v) cyclohexane in chloroform, with the ^{13}C resonance of cyclohexane serving as the internal reference. The values for protopine and muramine are shown below. In general, a slight increase in shielding ($\approx$ 2–3 ppm) is observed for the ipso carbons upon replacement of two methoxyls by a methylenedioxy group, and substitution of a methoxyl for a hydroxyl leads to a decrease in shielding of about 4.8 ppm for the ipso carbon. In addition, the carbonyl carbon in the phenolic alkaloid hunnemanine is more shielded by about 20 ppm than the mean of the chemical shifts for the carbonyl carbons of the other alkaloids in the series. The results of this CMR study were also consistent with a transannular interaction between the carbonyl carbon and the nitrogen atom.[15]

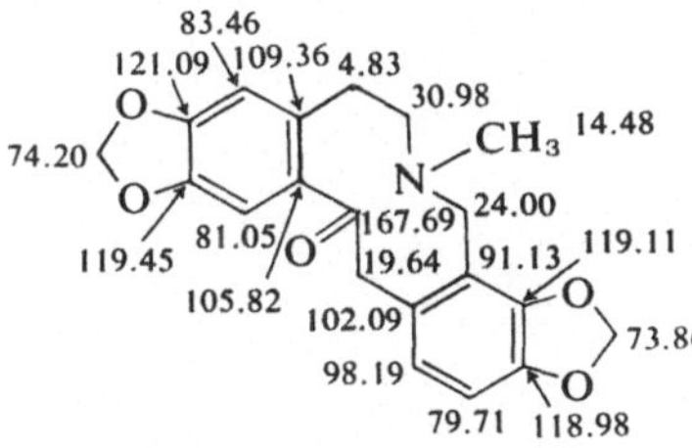

CMR values for protopine

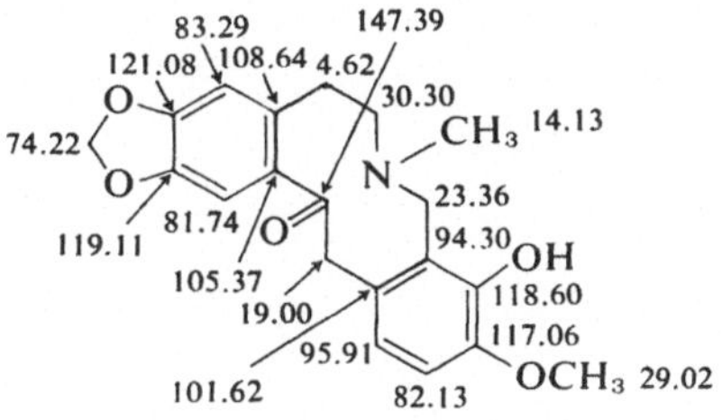

CMR values for hunnemanine

References and Notes

1. B. Nalliah, R. H. F. Manske, and R. Rodrigo, *Tetrahedron Lett.*, 1765 (1974).
2. M. Hanaoka, C. Mukai, and Y. Arata, *Heterocycles*, **4**, 1685 (1976).
3. S. Teitel, W. Klötzer, J. Borgese, and A. Brossi, *Can. J. Chem.*, **50**, 2022 (1972).
4. C. Schöpf and M. Schweickert, *Chem. Ber.*, **98**, 2566 (1965). For a total synthesis of compound **6**, see H. O. Bernhard and V. Snieckus, *Tetrahedron Lett.*, 4867 (1971). For the preparation of an analog of **6** from narceine, see J. Trojánek, Z. Koblicová, Z. Veselý, V. Suchan, and J. Holubek, *Collect. Czech. Chem. Commun.*, **40**, 681 (1975).
5. S. Teitel, J. Borgese, and A. Brossi, *Helv. Chim. Acta*, **56**, 553 (1973).
6. B. Nalliah, R. H. Manske, and R. Rodrigo, *Tetrahedron Lett.*, 2853 (1974).
7. K. W. Bentley and A. W. Murray, *J. Chem. Soc.*, **2497** (1963).
8. A. R. Battersby, J. Staunton, H. R. Wiltshire, R. J. Francis, and R. Southgate, *J. Chem. Soc. Perkin I*, 1147 (1975).
9. C. Tani and K. Tagahara, *Chem. Pharm. Bull., Tokyo*, **22**, 2457 (1974).
9a. N. Takao, K. Iwasa, M. Kamigauchi, and M. Sugiura, *Chem. Pharm. Bull., Tokyo*, **24**, 2859 (1976).
10. V. I. Popova and A. I. Leskov, *Tr. Vses. Nauch.-Issled. Inst. Lek Rast.*, **14**, 91 (1971); through *Chem. Abstr.*, **79**, 142838j (1973).

11. V. Preininger, in *The Alkaloids*, *Vol. 15*, R. H. F. Manske, ed., Academic Press, New York (1975), p. 207.
12. W.-N. Wu, J. L. Beal, G. W. Clark, and L. A. Mitscher, *Lloydia*, **39**, 65 (1976).
13. *Varian NMR Spectra Catalog*, *Vol. 1*, Varian Associates, Palo Alto (1962), Spectrum No. 339.
14. J. L. Moniot and M. Shamma, *Heterocycles*, **9**, 145 (1978).
15. T. T. Nakashima and G. E. Maciel, *Org. Magn. Reson.*, **5**, 9 (1973).

THE PHTHALIDEISOQUINOLINES

24

24.1. Introduction

The number of phthalideisoquinoline alkaloids was increased by the recent isolation of several new bases some of which do not incorporate a γ-lactone ring. These nonlactonic compounds nevertheless may be assigned to the phthalide-isoquinoline grouping on biogenetic grounds. Some of the new bases are shown below:

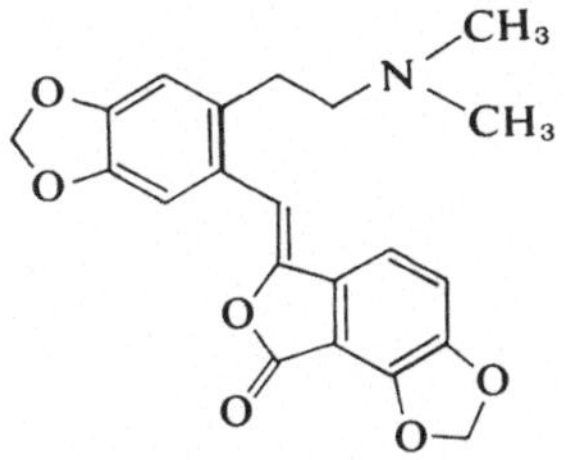

Aobamidine
Corydalis sempervirens Pers.[1]
and *C. ochotensis* Turcz.[2]

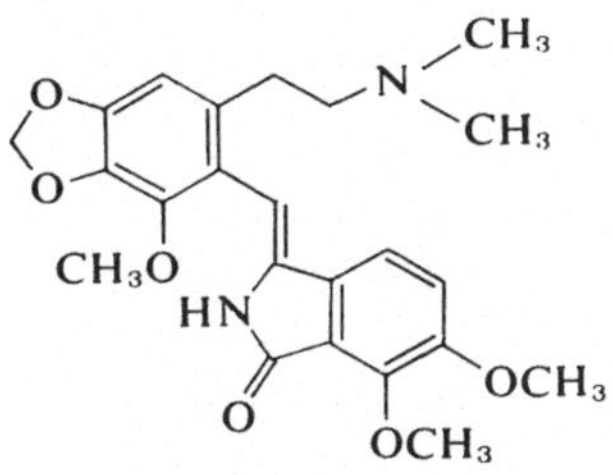

Narceine imide[3]
Papaver somniferum L.

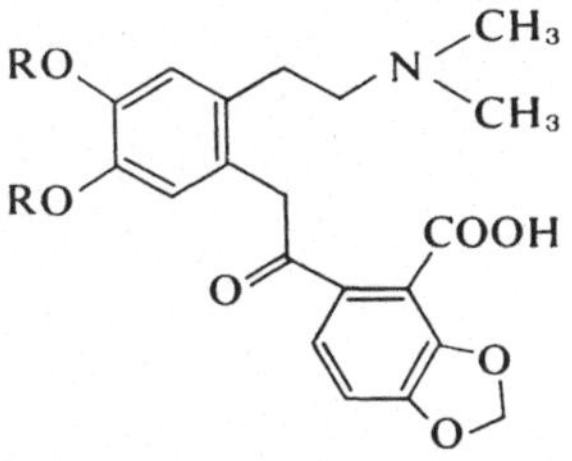

Adlumidiceine,[1] R + R = CH$_2$
Adlumiceine,[1] R = CH$_3$
C. sempervirens

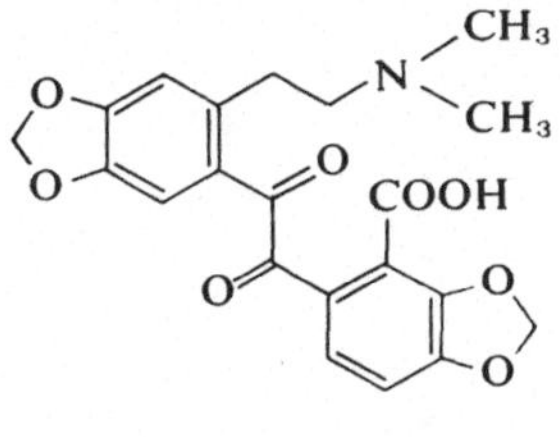

Bicucullinine[4]
C. ochroleuca Koch

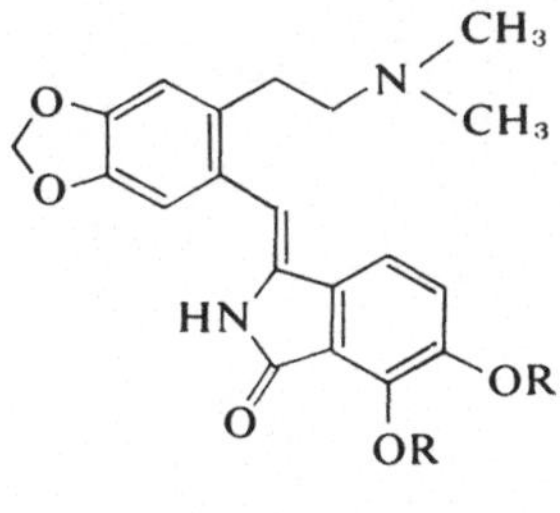

New structures[5] for
fumaridine, R = CH$_3$
and fumaramine, R + R = CH$_2$

Fumaridine and fumaramine have now been shown to be phthalideiso-quinolines as indicated above, rather than protopines, as previously believed; thus fumaridine corresponds to the known compound hydrastine imide.[5]

24.2. Synthesis

24.2.1. From a Substituted 2-Phenyl-1,3-indandione

This approach derives from work in the spirobenzylisoquinoline series in which analogs of the spirobenzylisoquinoline **1** had been prepared (see Sec. 25.2.2). The key step is the acid-catalyzed rearrangement of **1** to the isomeric phthalideisoquinolines, **2** and **3**, as shown in Scheme 24.1.[6] The minor product **3** is formed by alternative cleavage of the spirobenzylisoquinoline **1**.

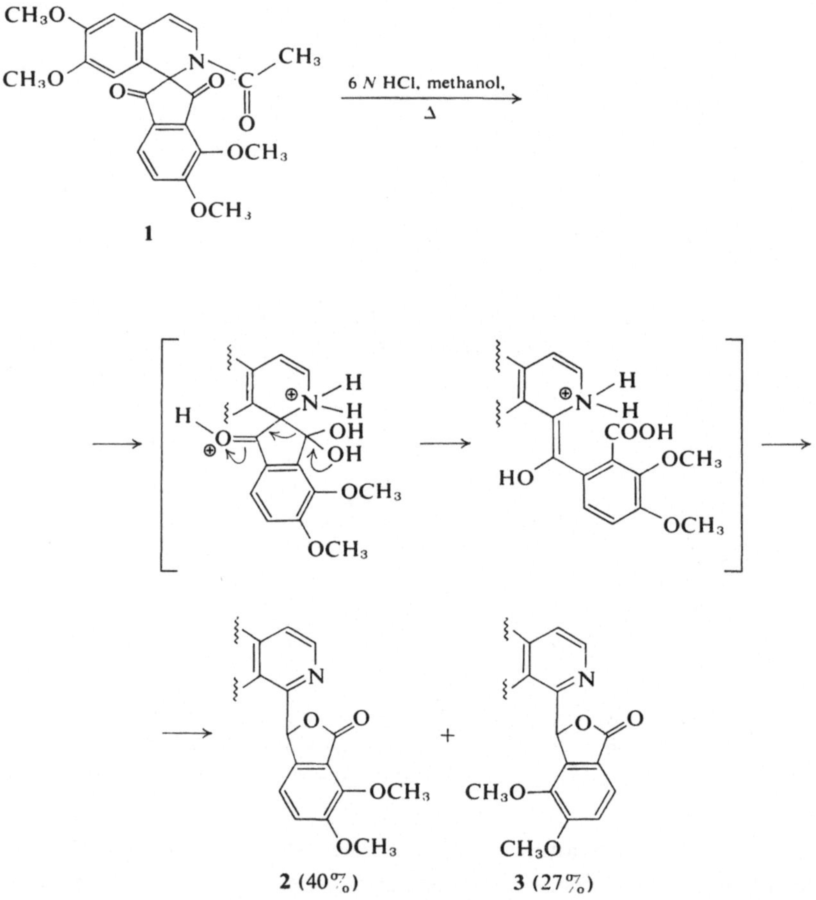

SCHEME 24.1

then:

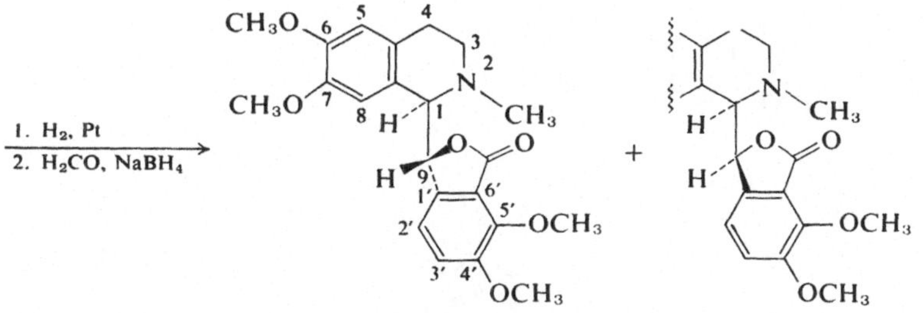

Cordrastine I (46%)   Cordrastine II (43%)

SCHEME 24.1 (Continued)

24.2.2. By Condensation of an Immonium Salt with an α-Diazotoluene

Kametani and his school have shown that condensation of the α-diazo-toluene **4**, prepared as in Scheme 24.2, with 3,4-dihydro-6,7-dimethoxyiso-quinoline methiodide yields cordrastine II. The unstable intermediate leading to cordrastine II can be represented by structures **5** or **6**.[7]

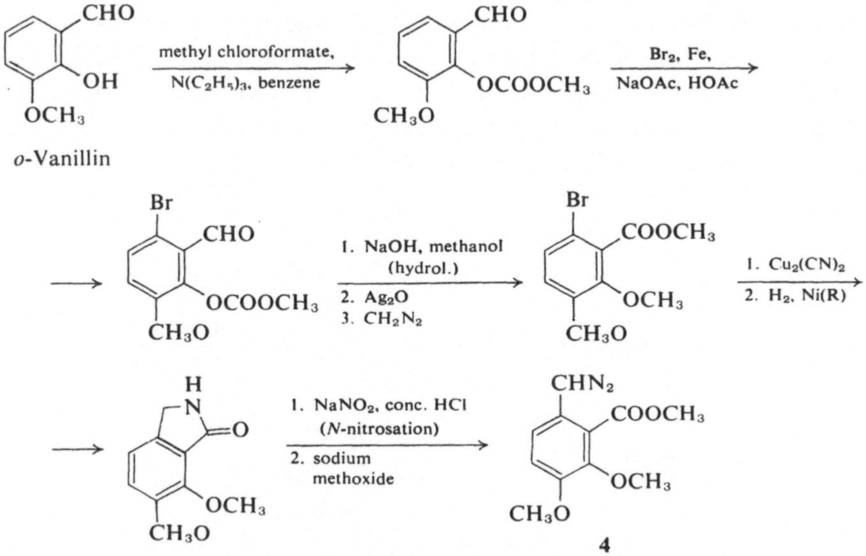

o-Vanillin

SCHEME 24.2 (Continued overleaf)

then:

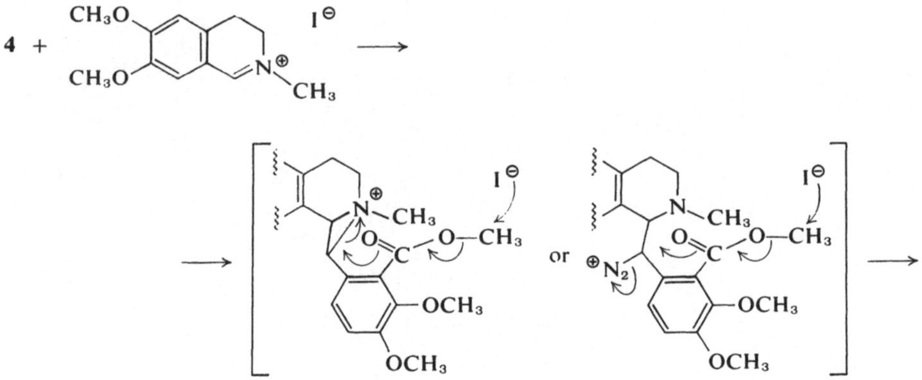

$$4 +$$

⟶ Cordrastine II (5–10% yields from **4**)

SCHEME 24.2 (*Continued*)

24.2.3. From Papaverine

A short sequence to phthalideisoquinolines starts with the known and readily available 6'-hydroxymethylpapaverine (**7**), derived in high yield from easily available papaverine. Chromium trioxide oxidation of **7** leads in 75% yield to the aromatic phthalideisoquinoline **8**, which can be reduced and *N*-methylated to phthalideisoquinolines **11** and **12**, Scheme 24.3.[8] (For the conversion of the norphthalideisoquinolines **9** and **10** to C-13 hydroxylated protoberberines, see Sec. 19.2.9.)

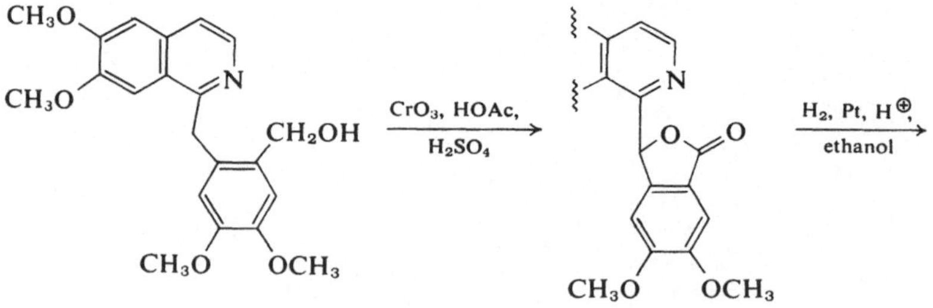

SCHEME 24.3

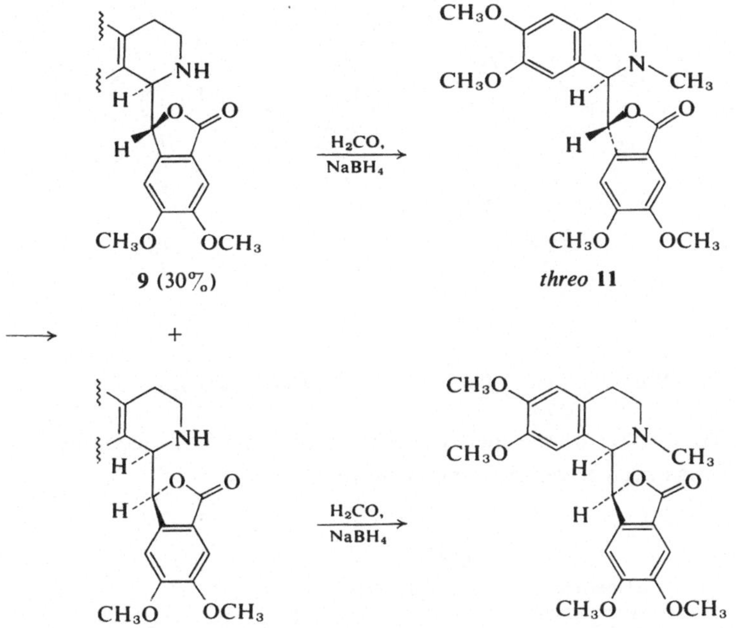

9 (30%)   threo 11

⟶   +

10 (33%)   erythro 12

SCHEME 24.3 (Continued)

24.2.4. From a Benzindenoazepine

It had previously been established that reaction of the keto immonium salt **13** with diazomethane resulted in ring expansion to the benzazepine **14**.[9] Cyclization of this benzazepine with phosphorus oxychloride produced the benzindenoazepine **15** whose mild oxidation resulted in ring contraction to the phthalideisoquinoline **16**. Compound **16** could readily be reduced to the *erythro* phthalideisoquinoline **12**[10]:

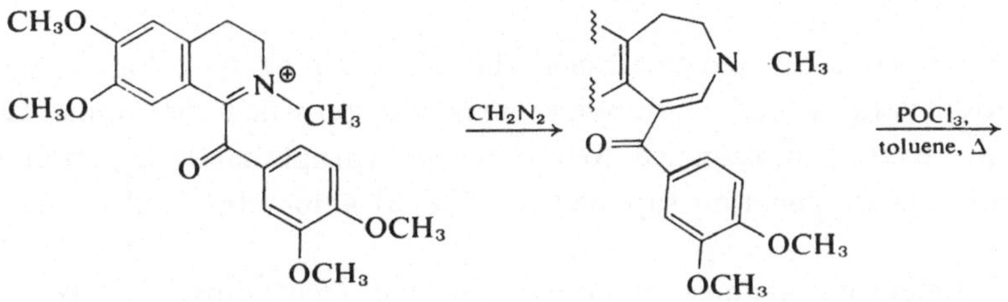

13   14

15 (90%)   **16**

24.2.5. An Improvement of the Haworth–Pinder Synthesis

A facile route to the methylenedioxy analog **17** of meconine-α-carboxylic acid (**18**) has been developed and is outlined below.[11]

Piperonal   (63%)

(71%)   (70%)

17, R + R = CH_2
18, R = CH_3

The acid **17** was converted to its acid chloride using oxalyl chloride, and further reaction with *N*-methylhomopiperonylamine provided the amide **19**. Bischler–Napieralski cyclization led to the key dehydrophthalideisoquinoline **20**, and catalytic hydrogenation supplied the desired adlumidine and bicuculline.[11]

The 6,7-dimethoxy analogs of adlumidine and bicuculline, namely adlumine and corlumine, were prepared by a parallel route.[11]

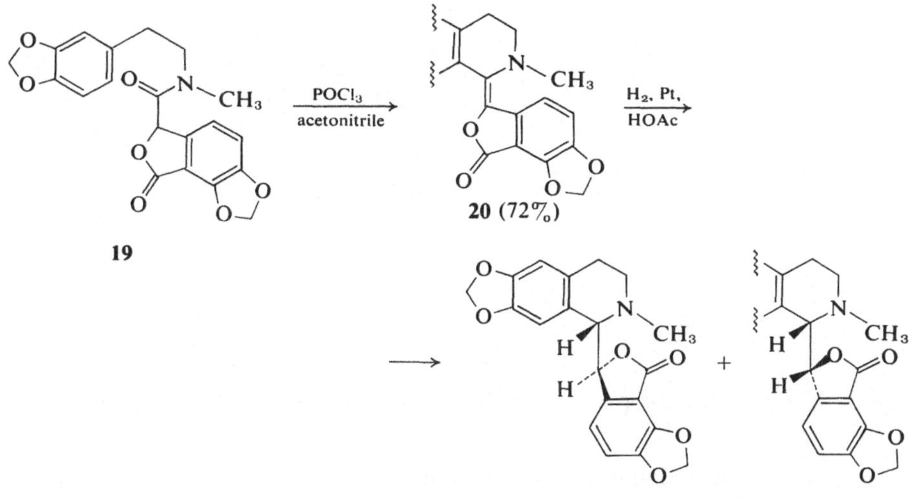

20 (72%)

19

Adlumidine (31%)   Bicuculline (34%)

24.2.6. From Protoberberinium Salts

For the derivation of (±)-α- and (±)-β-hydrastine from berberine through the intermediacy of oxybisberberine, see Sec. 19.3.5.

When dehydronorcoralydinephenolbetaine, derived from dehydronorcoralydine as shown, was briefly exposed to sunlight in the presence of rose bengal, and the product immediately reduced with borohydride, the corresponding phthalideisoquinoline **8** was obtained.[11a]

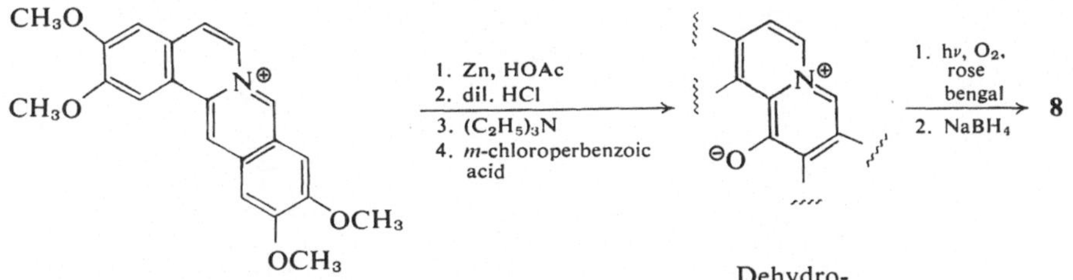

Dehydronorcoralydine

Dehydro-
norcoralydine-
phenolbetaine

A recent stereospecific conversion of berberine to β-hydrastine proceeds through the intermediacy of oxyberberine. Short photooxidation of this pyridone—readily derived from berberine—in the presence of oxygen provided a γ-lactol. N-Methylation and reduction then supplied racemic β-hydrastine, practically unadulterated with the diastereomeric α-hydrastine (see also Sec. 19.3.5).[11b]

Oxyberberine   A γ-lactol   β-Hydrastine

24.3. Chemistry of (−)-α-Narcotine and Related Compounds

A facile conversion of (−)-α-narcotine to the alkaloid nornarceine involves Hofmann elimination of *N*-benzyl-(−)-α-narcotinium bromide (**21**) followed by ethanolysis. The resulting ester **22** was debenzylated and saponified in good yield.[12]

In similar fashion, (−)-β-hydrastine (**23**) is a source of the amino acid **24** which has been taken to the aporhoeadane **25**[13]:

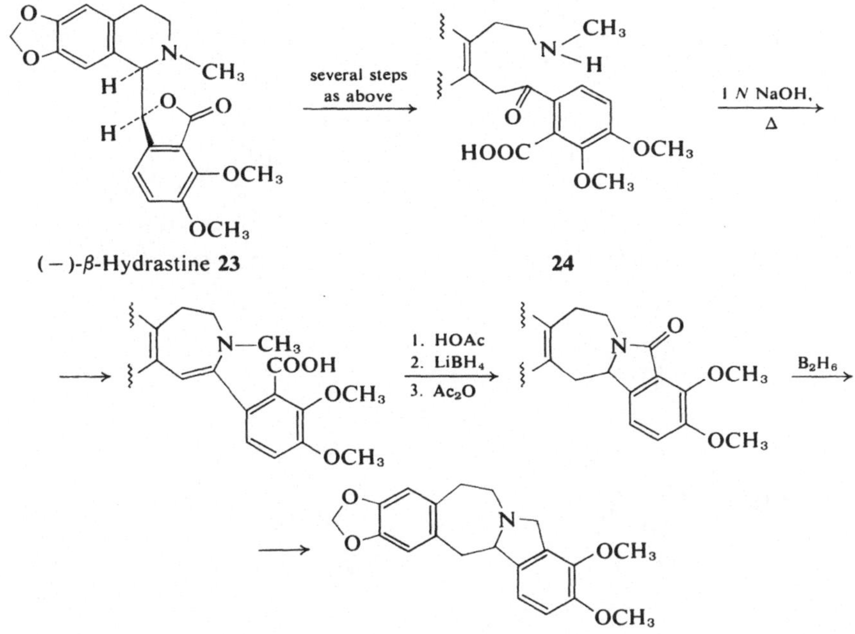

(−)-β-Hydrastine **23**

24

25

When (−)-α-narcotine N-oxide is allowed to stand in chloroform or is pyrolyzed, the product is a crystalline material which was erroneously labeled anhydro-N-oxynornarceine. This compound has now been shown to possess structure **26** through X-ray analysis of the bromo derivative **27**. (−)-β-Hydrastine N-oxide undergoes a similar rearrangement.[14]

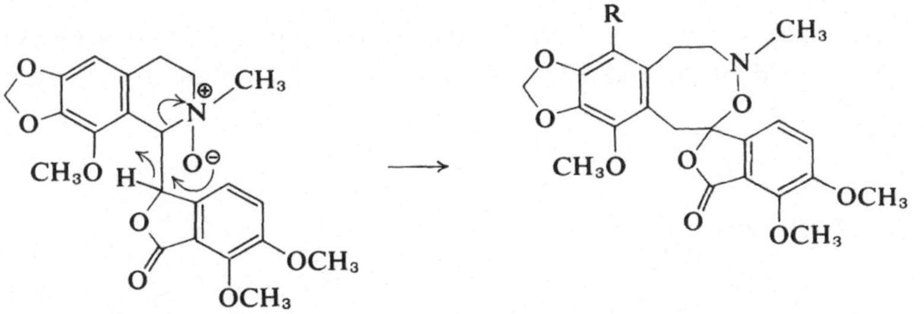

(−)-α-Narcotine N-oxide

26, Anhydro-N-oxynornarceine, R = H
27, R = Br

Thermal decomposition of narceine imide methiodide in the presence of 30% potassium hydroxide yields the Hofmann products **28** and **29**, as well as the aporhoeadane derivative **30** which can be further reduced to **31** and **32**[15]:

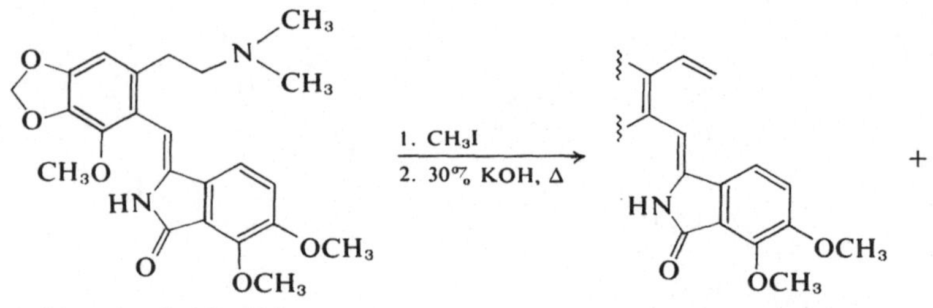

Narceine imide, Z-form

1. CH₃I
2. 30% KOH, Δ

28 Z-form (82%)

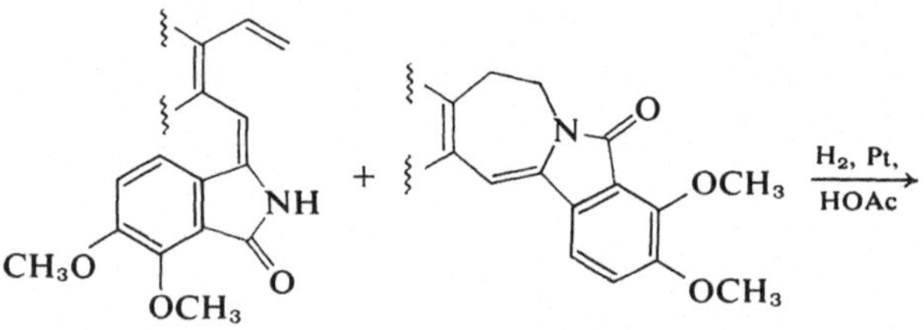

+

29 E-form (10%)

30 (< 5%)

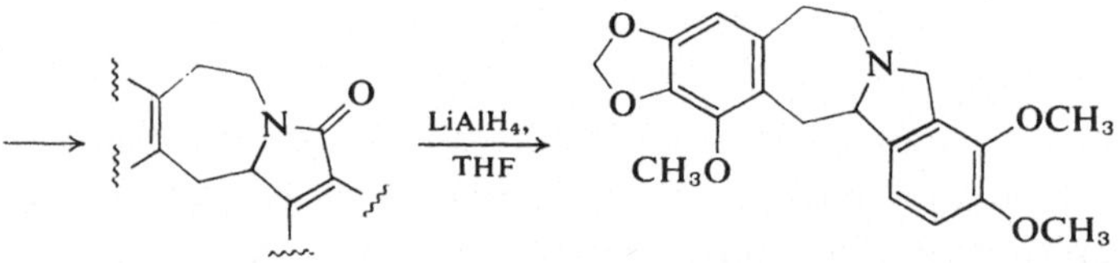

H₂, Pt, HOAc

31

LiAlH₄, THF

32

In contrast, thermal decomposition of narceine imide methohydroxide gives a 55% yield of **28** and **29**, as well as a small amount of the rearranged lactam **33**.[15]

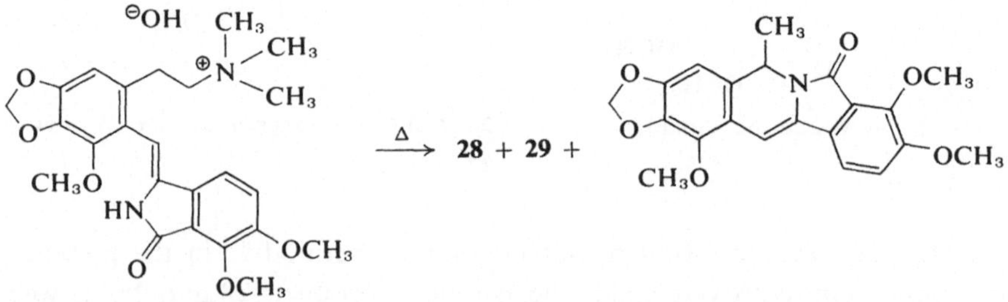

Narceine imide methohydroxide

Δ → **28 + 29** +

33

Turning back now to the aporhoeadane **32**, basic pyrolysis of its methiodide salt yielded three products in low yields, **34**, **35**, and **36**. The cis cyclic olefin **34** was then treated with perbenzoic acid to furnish an *N*-oxide which upon reaction with aqueous mineral acid gave rise to 1-methoxyallocryptopine[16]:

1-Methoxyallocryptopine

Oxidation of (−)-β-hydrastine with *N*-bromosuccinimide results in cleavage of the alkaloid at the C-1 to C-9 bond, and formation of *N*-methyl-4-bromo-6,7-methylenedioxyisoquinoline.[16a]

24.4. Interconversions Among Lactonic Phthalideisoquinolines

The methylenedioxy group in (−)-β-hydrastine (**23**) can be selectively converted to an *o*-diphenol by treatment with boron trichloride,[17] whereas *O*-demethylation without affecting the methylenedioxy function can be accomplished with pyridine hydrochloride in hot pyridine. Boron tribromide cleaves both methylenedioxy and methoxyl groups. Isomerization at C-9 is known to

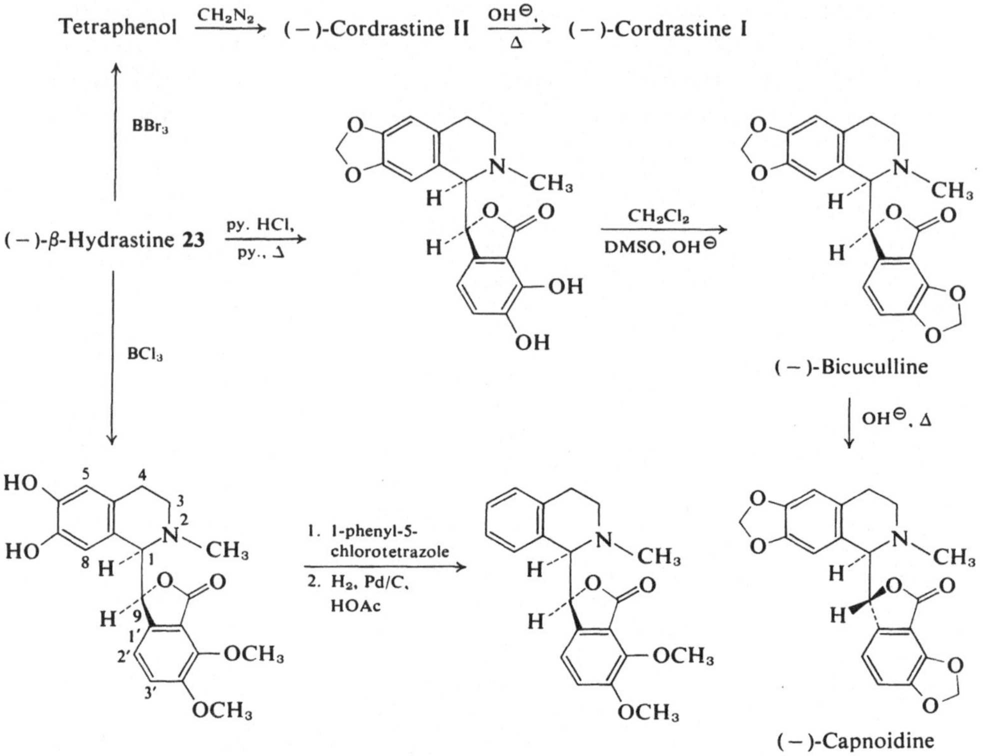

SCHEME 24.4

be possible by treatment of a nonphenolic phthalideisoquinoline with hot hydroxylic base. The transformations shown in Scheme 24.4 have thus been carried out by Teitel, O'Brien, and Brossi.[18] The structure and stereochemistry of (−)-bicuculline was further confirmed by X-ray analysis of the hydrobromide salt.[18] This work also presented for the first time the correct stereo structures for the cordrastines I and II (see Sec. 24.2.1).

24.5. Pharmacology

The discovery by Curtis and colleagues that (+)-bicuculline antagonizes both the depressant effects of microiontophoretically applied γ-aminobutyric acid (GABA), and certain strychnine-resistant central inhibitions, has provided support for the view that GABA might be a central transmitter.[19] The next development was the finding that (+)-bicuculline methochloride is a very effective GABA antagonist. Since this salt is much more soluble in water than the alkaloid-free base, its use, particularly in microiontophoresis experiments, is more convenient and gives more reproducible results.[20]

(+)-Bicuculline, R + R = CH₂
(+)-Corlumine, R = CH₃

Tritoqualin

Significantly, (−)-bicuculline methochloride is ineffective as a GABA antagonist; only the (+)-form of bicuculline methochloride is active.[21] In addition, (+)-corlumine also is an effective GABA antagonist, and it possesses the same absolute configuration as (+)-bicuculline.[20]

Tritoqualin is a synthetically derived drug which is a histidine decarboxylase inhibitor, and may thus find use in the treatment of allergies.[22]

24.6. PMR and CMR Spectra of Phthalideisoquinolines; Conformational Analysis

The CD and ORD curves of phthalideisoquinolines have been recorded and interpreted.[23,24]

The fact that pairs of diastereoisomeric norphthalideisoquinolines and phthalideisoquinolines were readily available from papaverine has allowed a comparison of favored conformations. It had formerly been shown that in the tetrahydrobenzylisoquinoline series ring C lies close to the nitrogen atom if the nitrogen is secondary, but that following N-methylation ring C is found in the proximity of ring A, and away from the relatively bulky N-methyl group. This same steric factor, i.e., N-methylation, has been found to apply equally to the phthalideisoquinolines, so that N-methylation forces rings C and D away from the nitrogen, and into proximity with ring A.[8]

The *erythro*-nor base **10** exists mostly in conformation **10a**. The 2'-H and 3'-CH₃O signals are relatively upfield at $\delta 5.84$ and $\delta 3.62$, respectively, due to ring A shielding. In the *erythro* base **12**, on the other hand, the 8-H signal appears upfield at $\delta 6.20$ because of shielding by ring D, and conformation **12a** prevails.[8]

erythro-nor **10**   **10a**

erythro **12**   **12a**

The chemical shift of H-8 can also be used as a probe in determining conformation in the *threo*-nor and *threo* series. This signal is relatively upfield, at δ6.32, in the case of the *threo* base **11** because of shielding by the lactone carbonyl in the favored conformation **11a**. It is further downfield at δ6.52 in the *threo*-nor base **9** since no shielding is involved in this instance, and conformation **9a** is paramount.[8]

threo-nor **9**   **9a**

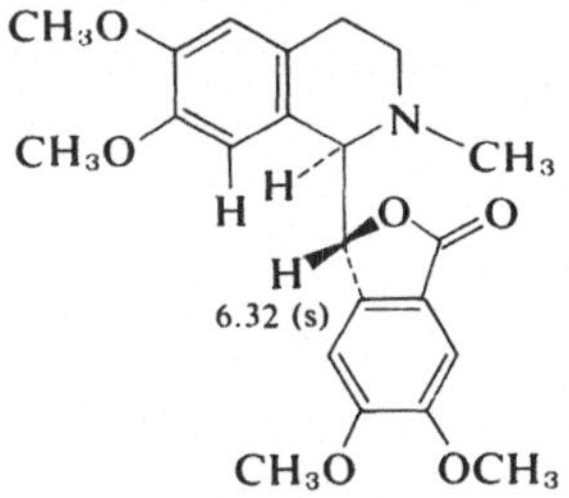

CH₃O—

CH₃O—

H H'

H

6.32 (s)

N—CH₃

O O

CH₃O OCH₃

threo 11

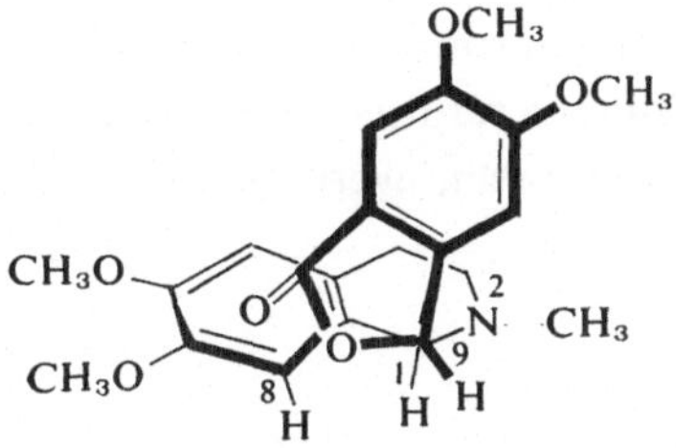

OCH₃

OCH₃

CH₃O—

CH₃O—

O

N CH₃

2

9

8

H H H

11a

An interesting case is that of the erythro alkaloid (−)-α-narcotine which has the same substitution and stereochemistry as (−)-β-hydrastine except for an additional methoxyl substituent at C-8. This substituent, however, forces the molecule into a **10a** type conformation.[25] Two opposing steric factors, namely N-methylation and substitution at C-8 can, therefore, influence the conformation of phthalideisoquinolines. N-Methylation tends to move rings C and D towards ring A, whereas substitution at C-8 in ring A forces rings C and D towards the nitrogen atom.[8,26]

The CMR spectral assignments shown below have been made for β-hydrastine, α-hydrastine, corlumine, and adlumine. Chemical shifts at C-3 and C-4 were found to be diagnostic of the relative stereochemistry at C-1 and C-9.[27]

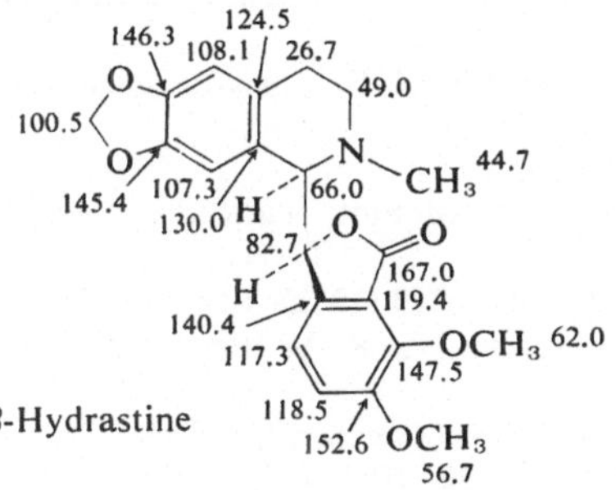

(−)-β-Hydrastine

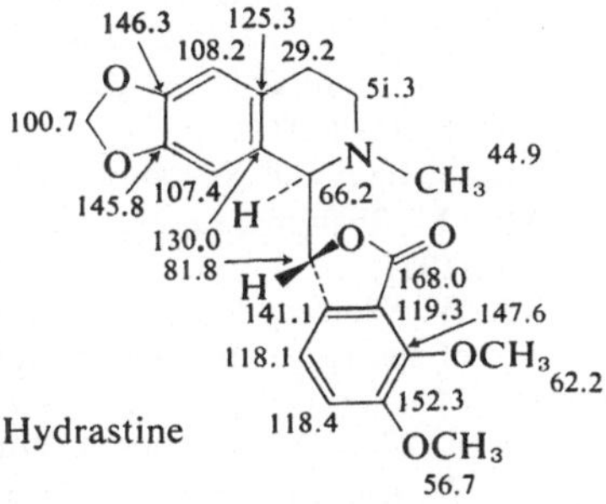

146.3 124.5
108.1 26.7
100.5 49.0
145.4 107.3 66.0 N CH₃ 44.7
130.0 H
82.7
H 167.0
140.4 119.4
117.3 OCH₃ 62.0
147.5
118.5 OCH₃
152.6
56.7

(−)-α-Hydrastine

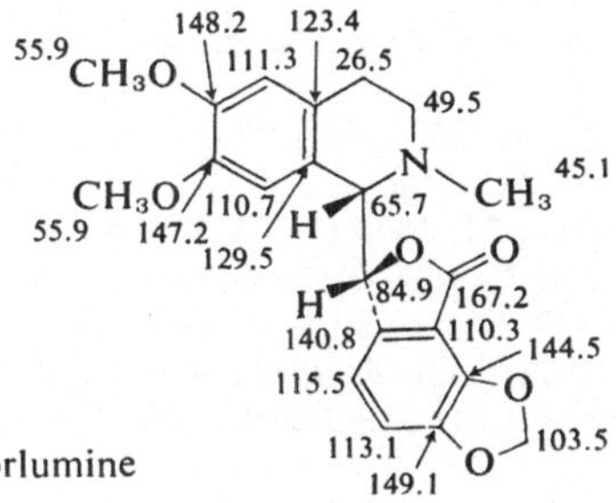

146.3 125.3
108.2 29.2
100.7 51.3
145.8 107.4 66.2 N CH₃ 44.9
130.0 H
81.8
H 168.0
141.1 119.3 147.6
118.1 OCH₃
152.3 62.2
118.4 OCH₃
56.7

(+)-Corlumine

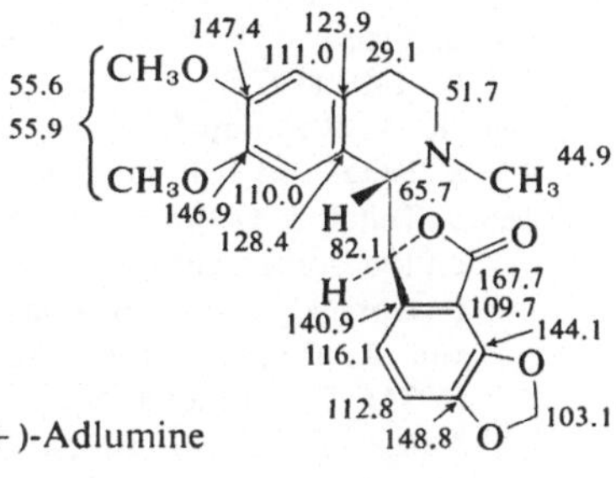

148.2 123.4
55.9 111.3 26.5
CH₃O
49.5
CH₃O 110.7 65.7 N CH₃ 45.1
55.9 147.2 H
129.5
H 84.9 167.2
140.8 110.3 144.5
115.5 O
113.1 O 103.5
149.1

(+)-Adlumine

147.4 123.9
111.0 29.1
55.6 CH₃O
51.7
55.9
CH₃O 110.0 65.7 N CH₃ 44.9
146.9 H
128.4 82.1
H 167.7
140.9 109.7
116.1 144.1
112.8 O 103.1
148.8 O

Irradiation of tetrahydrofuran solutions of $(-)$-α-narcotine and of $(-)$-β-hydrastine resulted in epimerization at C-1 and C-9, so that racemic mixtures of diastereomers were obtained in each case.[28]

References and Notes

1. V. Preininger, V. Šimánek, O. Gašić, F. Šantavý, and L. Dolejš, *Phytochemistry*, **12**, 2513 (1973).
2. T. Kametani, M. Takemura, M. Ihara, and K. Fukumoto, *Heterocycles*, **4**, 723 (1976).
3. H. Hodková, Z. Veselý, Z. Koblicová, J. Holubek, and J. Trojánek, *Lloydia*, **35**, 61 (1972).
4. R. G. A. Rodrigo, R. H. F. Manske, H. L. Holland, and D. B. MacLean, *Can. J. Chem.*, **54**, 471 (1976).
5. M. Shamma and J. L. Moniot, *Chem. Commun.*, 89 (1975).
6. V. Smula, N. E. Cundasawmy, H. L. Holland, and D. B. MacLean, *Can. J. Chem.*, **51**, 3287 (1973).
7. T. Kametani, T. Honda, H. Inoue, and K. Fukumoto, *Heterocycles*, **3**, 1091 (1975); and *J. Chem. Soc. Perkin I*, 1221 (1976).
8. M. Shamma and V. St. Georgiev, *Tetrahedron Lett.*, 2339 (1974); and *Tetrahedron*, **32**, 211 (1976).
9. T. Kametani, S. Hirata, F. Satoh, and K. Fukumoto, *J. Chem. Soc. Perkin I*, 2509 (1974).
10. T. Kametani, S. Hirata, M. Ihara, and K. Fukumoto, *Heterocycles*, **3**, 405 (1975). See also T. Kametani, T. Ohsawa, S. Hirata, M. S. Premila, M. Ihara, and K. Fukumoto, *Chem. Pharm. Bull., Tokyo*, **25**, 321 (1977).
11. B. C. Nalliah, D. B. MacLean, R. G. A. Rodrigo, and R. H. F. Manske, *Can. J. Chem.*, **55**, 922 (1977).
11a. J. Imai and Y. Kondo, *Heterocycles*, **6**, 959 (1977).
11b. M. Shamma, D. M. Hindenlang, T.-T. Wu, and J. L. Moniot, *Tetrahedron Lett.*, 4285 (1977).
12. W. Klötzer, S. Teitel, and A. Brossi, *Monatsch.*, **103**, 1210 (1972). For an older study of the Hofmann degradation of phthalideisoquinolines, see K. W. Bentley and A. W. Murray, *J. Chem. Soc.*, 2491 (1963).
13. S. Teitel, W. Klötzer, J. Borgese, and A. Brossi, *Can. J. Chem.*, **50**, 2022 (1972).
14. W. Klötzer and W. E. Oberhansli, *Helv. Chim. Acta*, **56**, 2107 (1973).
15. J. Trojánek, Z. Koblicová, Z. Veselý, V. Suchan, and J. Holubek, *Collect. Czech. Chem. Commun.*, **40**, 681 (1975).
16. Z. Veselý, J. Holubek, H. Kopecká, and J. Trojánek, *Collect. Czech. Chem. Commun.*, **40**, 1403 (1975).
16a. K. V. Rao and L. S. Kapicak, *J. Heterocycl. Chem.*, **13**, 1073 (1976).
17. S. Teitel and J. P. O'Brien, *J. Org. Chem.*, **41**, 1657 (1976).
18. S. Teitel, J. O'Brien, and A. Brossi, *J. Org. Chem.*, **37**, 1879 (1972). For other studies on the isomerization of phthalideisoquinolines at C-1 and C-9, see G. Gaál, P. Kerekes, P. Gorecki, and R. Bognár, *Pharmazie*, **26**, 431 (1971); as well as G. Gaál, P. Kerekes, and R. Bognár, *Magy. Kem. Foly.*, **77**, 533 (1971); and *J. Prakt. Chem.*, **313**, 935 (1971). For another X-ray study of bicuculline, see R. D. Gilardi, *Nature (London)*, *New Biol.*, **233**, 87 (1971).

19. D. R. Curtis, A. W. Duggan, D. Felix, and G. A. R. Johnston, *Nature*, **226**, 1222 (1970).
20. G. A. R. Johnston, P. M. Beart, D. R. Curtis, C. J. A. Game, R. M. McCulloch, and R. M. Maclachlan, *Nature (London)*, *New Biol.*, **240**, 219 (1972). See also S. Pong and L. Graham, *Brain Res.*, **42**, 486 (1972).
21. J. F. Collins and R. G. Hill, *Nature*, **249**, 845 (1974).
22. For lead references, see C. Hörig, H. Koch, E. Mayr, and M. Selchan, *Pharmazie*, **26**, 509 (1971).
23. G. Snatzke, G. Wollenberg, J. Hrbek, Jr., F. Šantavý, K. Bláha, W. Klyne, and R. J. Swan, *Tetrahedron*, **25**, 5059 (1969).
24. T. Kikuchi, T. Nishinaga, M. Inagaki, M. Niwa, and K. Kuriyama, *Chem. Pharm. Bull., Tokyo*, **23**, 416 (1975).
25. S. Safe and R. Y. Moir, *Can. J. Chem.*, **42**, 160 (1964).
26. For the PMR spectrum of bicuculline *N*-methyl cation and a discussion of its conformation, see P. R. Andrews and G. A. R. Johnston, *Nature (London)*, *New Biol.*, **243**, 29 (1973); through *Chem. Abstr.*, **79**, 53641x (1973).
27. D. W. Hughes, H. L. Holland, and D. B. MacLean, *Can. J. Chem.*, **54**, 2252 (1976).
28. T. Kametani, H. Inoue, T. Honda, T. Sugahara, and K. Fukumoto, *J. Chem. Soc. Perkin I*, 374 (1977).

THE SPIROBENZYLISOQUINOLINES

25.1. Introduction

It is a characteristic of all spirobenzylisoquinolines known,[1] including those recently reported and indicated below, to possess a methylenedioxy group in the bottom ring.

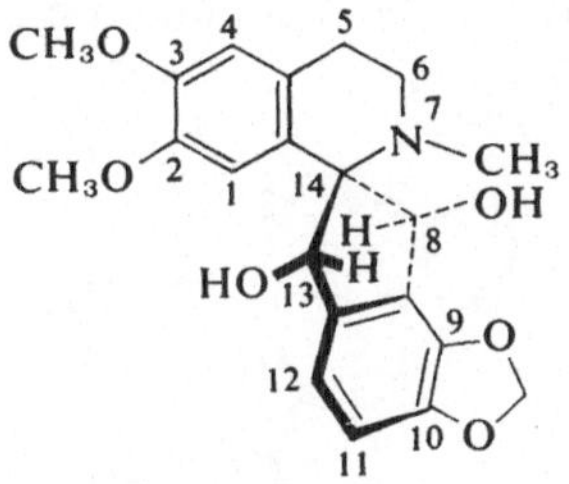

(+)-Yenhusomine[2]
Corydalis ochotensis Turcz.[2]

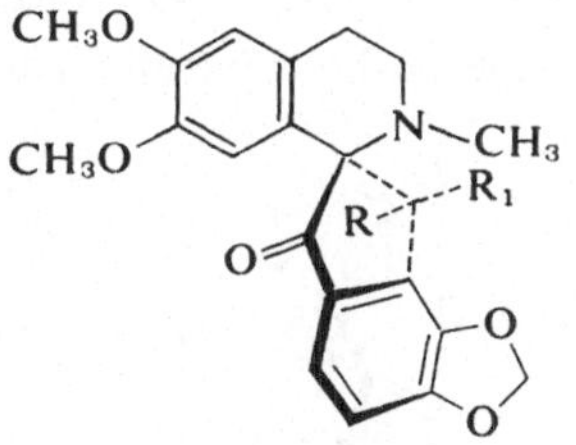

Yenhusomidine,[2]
R = H, R₁ = OH
C. ochotensis.
Raddeanone,[3]
R = OH, R₁ = H
C. ochotensis, var. raddeana.

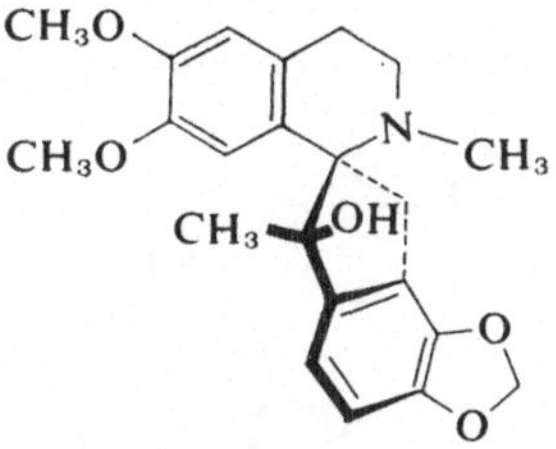

Raddeanamine[3]
C. ochotensis var. raddeana.

25.2. Synthesis

25.2.1. Modified 1,2-Indandione and 1,2,3-Indantrione Approaches

The original indandione approach has been modified by introduction of a bromine substituent into the 1,2-indandione, thus providing a synthesis of spirobenzylisoquinolines bearing two oxygen functions on ring C such as 1[4]:

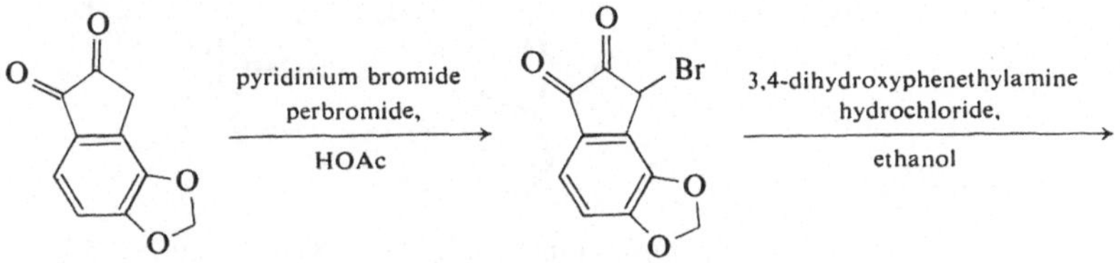

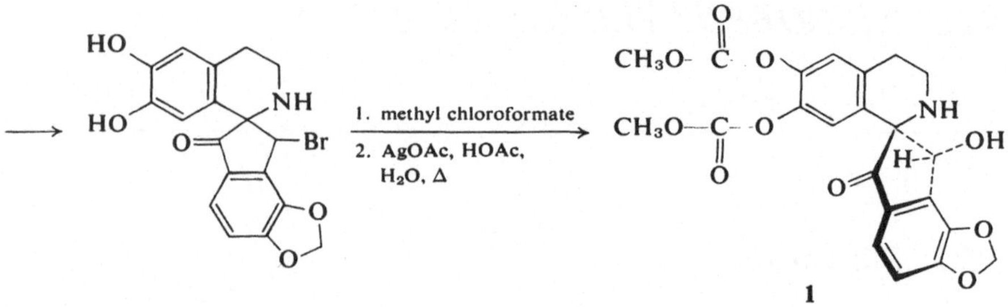

1

Attempted Eschweiler–Clarke *N*-methylation of the related norspiro-benzylisoquinoline **2**, however, did not lead to *N*-methylation due to formation of the oxazolidine **3**. But **3** could be reduced to its *N*-methyl derivative, cory-daine, using sodium cyanoborohydride. Yenhusomidine, the 2,3-dimethoxy analog of corydaine, was prepared by a parallel route.[5]

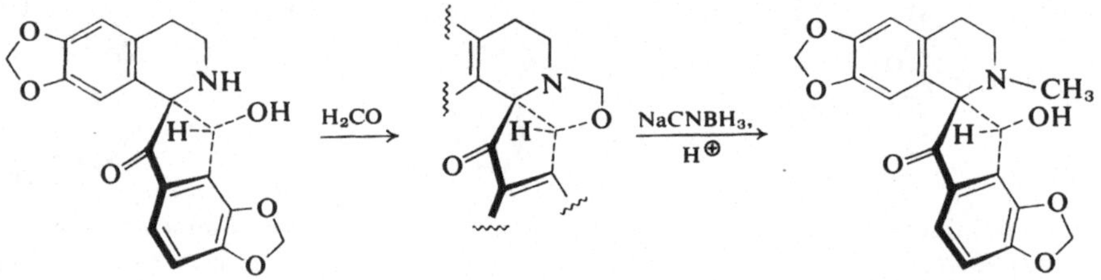

2   **3**   Corydaine

Yenhusomidine and its corresponding trans diol, yenhusomine, had actually been prepared earlier in good yield by Irie's team through the more orthodox sequence depicted below[6]:

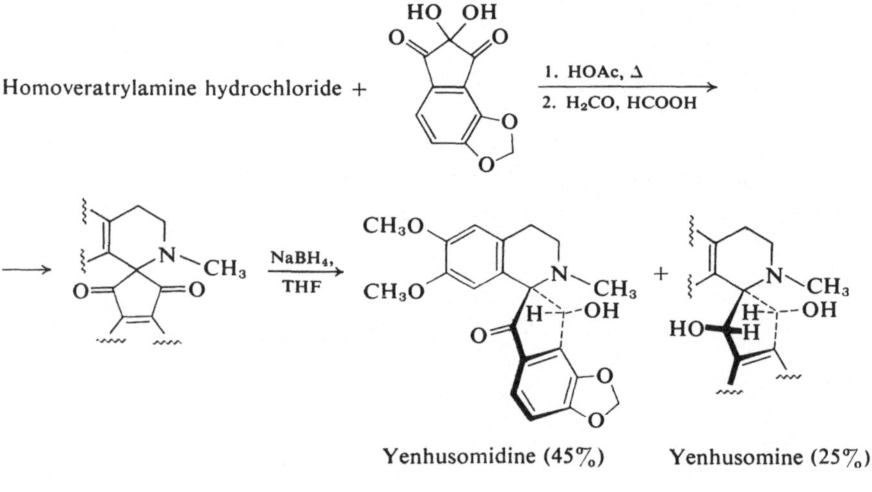

Homoveratrylamine hydrochloride +

1. HOAc, Δ
2. H₂CO, HCOOH

Yenhusomidine (45%)   Yenhusomine (25%)

25.2.2. From Substituted 2-Phenyl-1,3-indandiones

Cundasawmy and MacLean have taken advantage of the fact that substituted 2-phenyl-1,3-indandiones such as **4** are available by the route described below to achieve a new synthesis of ochrobirine.[7]

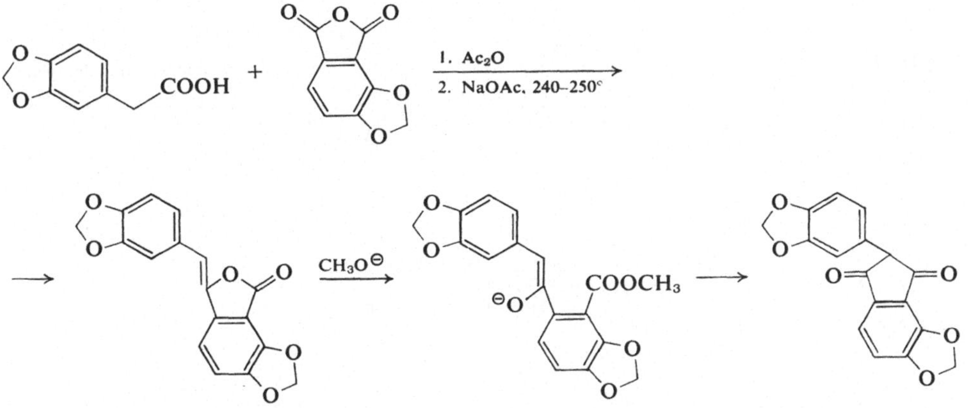

(60–65%)   **4**

Construction of the required spiro system was performed through a modified Pomeranz–Fritsch sequence (see Scheme 25.1).[7] The nitrogen function had to be acetylated before cyclization with formation of ring B could be carried out. The sodium borohydride reduction to furnish ochrobirine proceeded with a high degree of specificity.

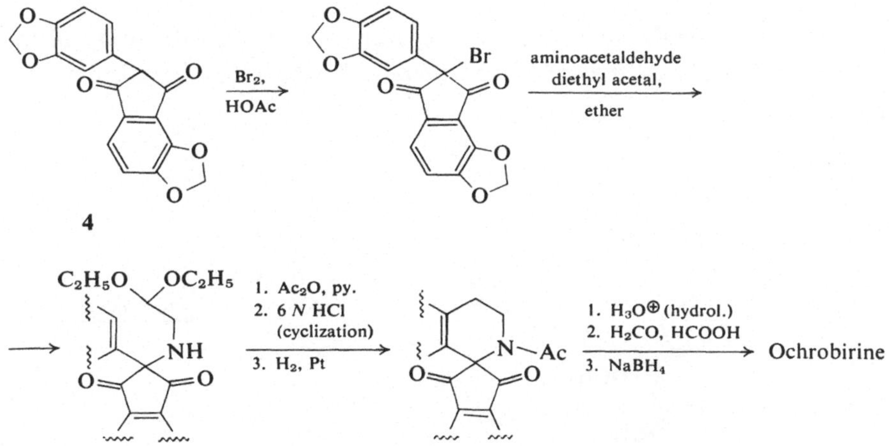

SCHEME 25.1

25.2.3. By Conversion of a 1-Indanone to a 1,3-Indandione

Manske, Rodrigo, and co-workers have shown that the 2-phenyl-1-indanone **5**, obtained by the sequence shown below, can be cyclized to the known ketonic spirobenzylisoquinoline **6** in a three-step sequence in 28% overall yield from **5**.[8]

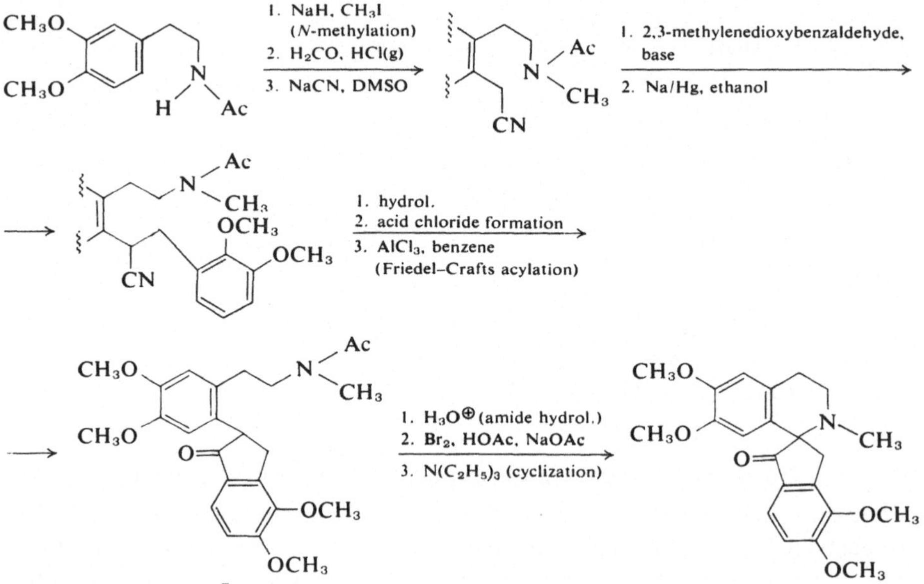

More interestingly, however, basic hydrolysis of the amide group in the keto amide **5** leads to the indenobenzazepine **7** which can be taken to the 1,3-indandione **9**[8]:

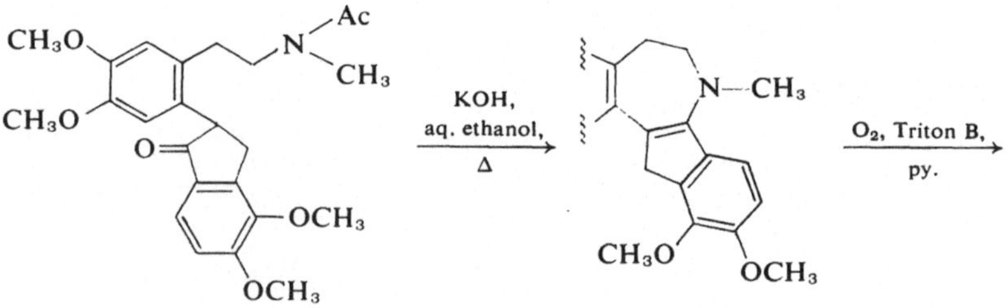

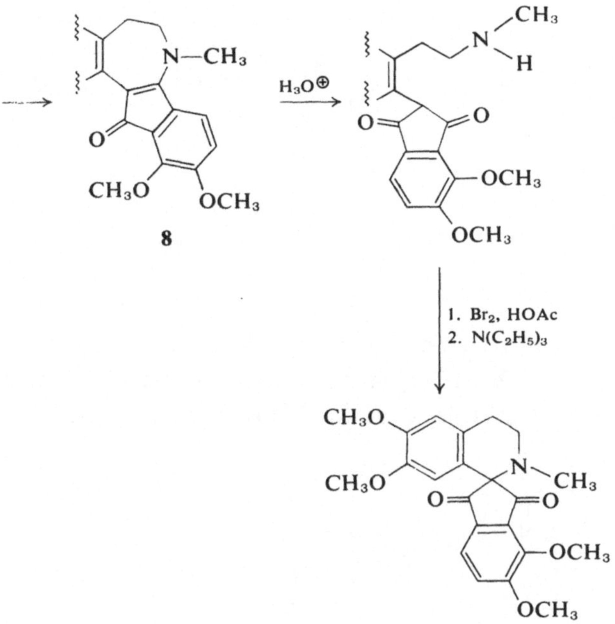
N—CH3
CH3O
OCH3
H3O⊕
CH3
N
H
O
OCH3
1. Br2, HOAc
2. N(C2H5)3
CH3O
CH3O
N
CH3
O
O
OCH3
OCH3

8

(67% from 8)

9

25.2.4. By Thermolysis of Benzocyclobutenes

The use of quinone methides in the synthesis of spirobenzylisoquinolines[9] has been extended by Kametani and his colleagues who have utilized substituted benzocyclobutenes as precursors for *o*-quinodimethides. They converted the known benzocyclobutene **10** to the amide **11** which upon treatment under Bischler–Napieralski conditions furnished the known spirobenzylisoquinoline **13** probably via the *o*-quinodimethide **12**[10]:

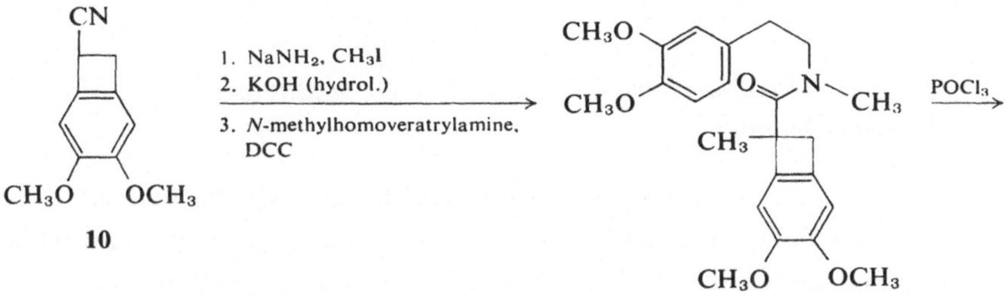
CN
CH3O
OCH3
1. NaNH2, CH3I
2. KOH (hydrol.)
3. N-methylhomoveratrylamine, DCC
CH3O
CH3O
O
N
CH3
CH3
CH3O
OCH3
POCl3

10

11

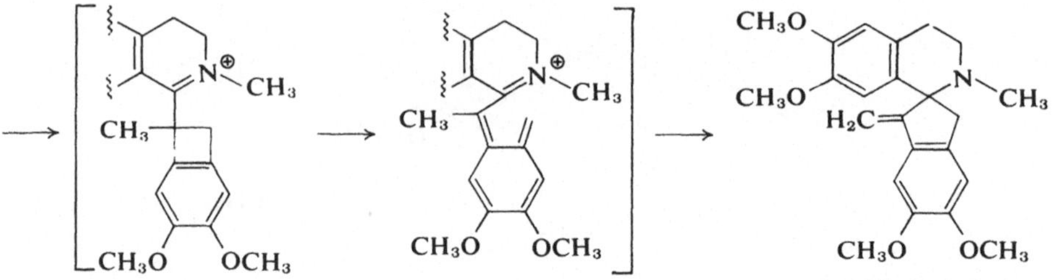

12   13 (14% from 3)

Ring C hydroxylated spirobenzylisoquinolines can also be obtained by this route. When the imine **14** was allowed to stand at room temperature in chloroform in the presence of air for 48 hr, the product was the spiroketone **15**.[11]

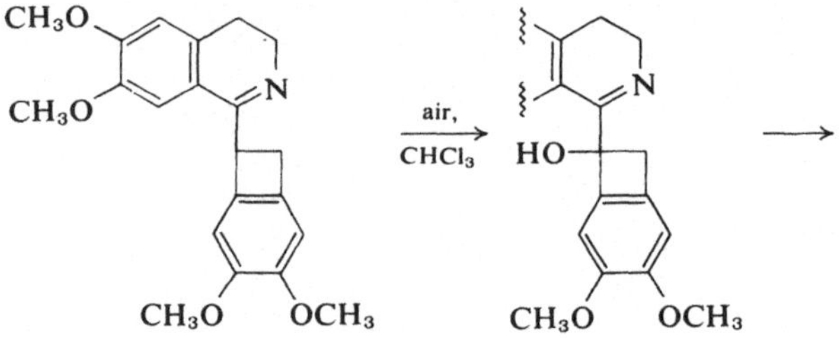

14

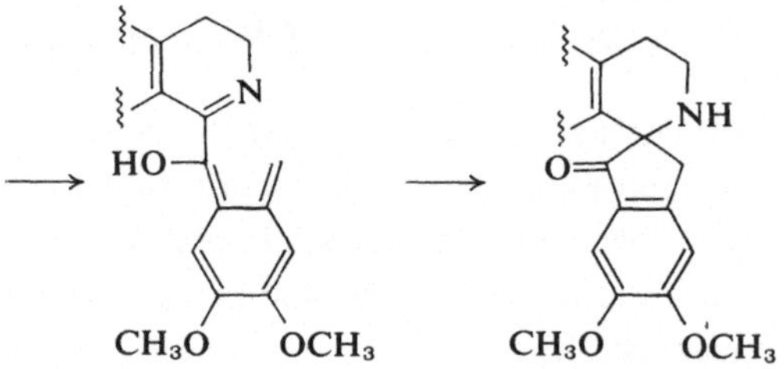

15 (64%)

25.2.5. Photolytic Protoberberine → Spirobenzylisoquinoline Rearrangements

Photolysis of the 13-ketotetrahydroprotoberberine salt **16**, derived from berberinephenolbetaine, generated in basic medium the spirobenzylisoquinoline **17** in 45% yield (see Scheme 25.2). Here again, an o-quinodimethide is a probable intermediate.[12]

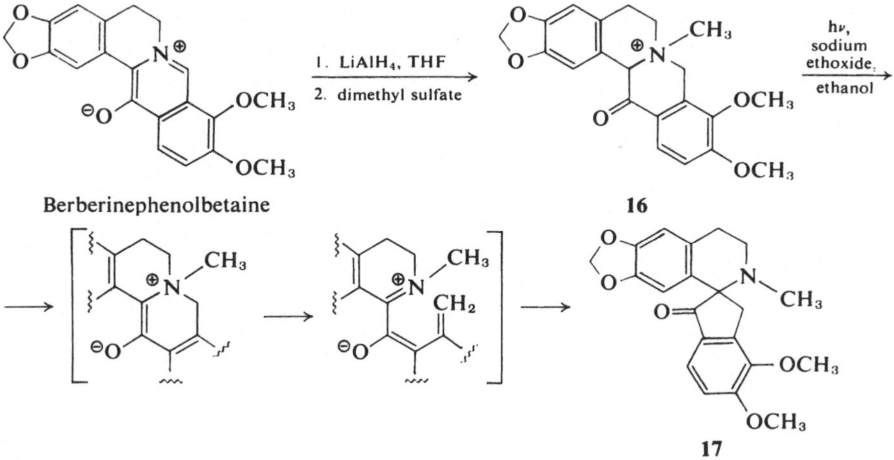

Berberinephenolbetaine

16

17

SCHEME 25.2

As the result of a separate study, it was established that oxidative photolysis of 8-methoxyberberinephenolbetaine in methanol furnishes as the main product a spiro monoketal formulated as **17a** which can be hydrolyzed to the spirodione **17b**.[12a]

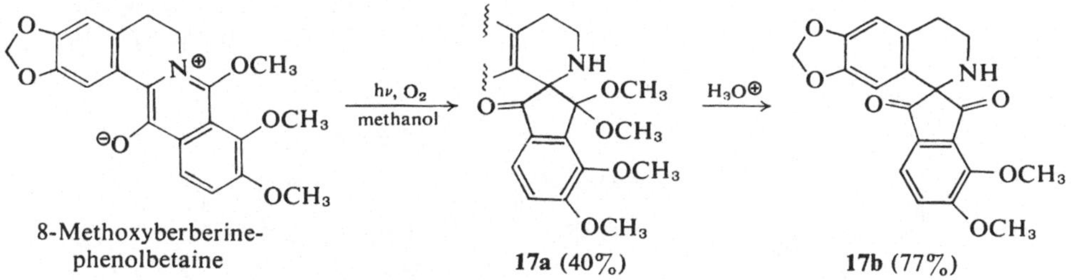

8-Methoxyberberine-
phenolbetaine

17a (40%)

17b (77%)

25.2.6. From Phthalideisoquinolines

In a preliminary paper by a Canadian team, it was reported that treatment of the known dehydrocordrastine with diisobutylaluminum hydride (Dibal) leads to equimolar amounts of the diastereomeric spirobenzylisoquinolines **18** and **19** in 76% yield. The reaction proceeds by reductive opening of the oxygen ring, and reclosure of the resulting enolate.[13]

In a subsequent and more complete study, Nalliah, MacLean, Rodrigo, and Manske have shown that diisobutylaluminum hydride reduction of the dehydrophthalideisoquinoline **20** gives corydaine and sibiricine. A novel and significant finding was that sibiricine can be isomerized into corydaine by simple treatment with alkali via a retroaldol–aldol mechanism. Corydaine is the more stable isomer due to intramolecular hydrogen bonding with the basic nitrogen. Sodium borohydride reduction of this isomer provides solely the trans diol ochrobirine.[14]

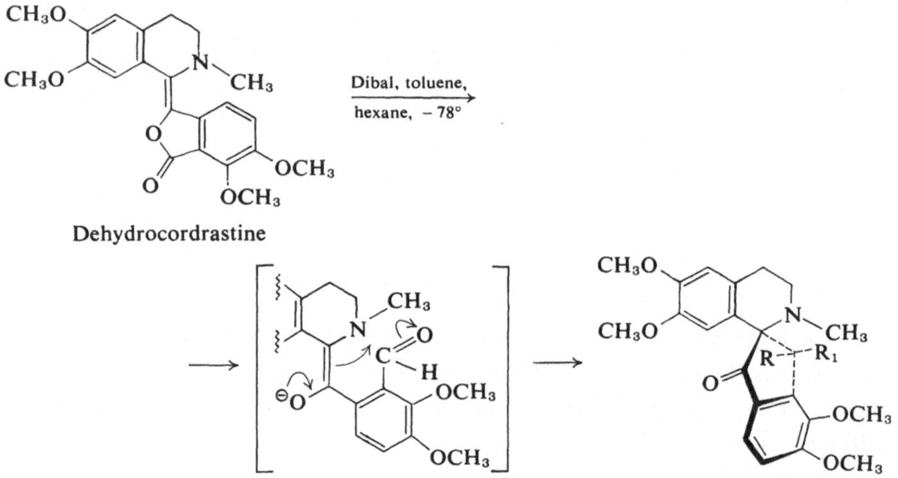

Dehydrocordrastine

18, R = H and R_1 = OH
19, R = OH and R_1 = H

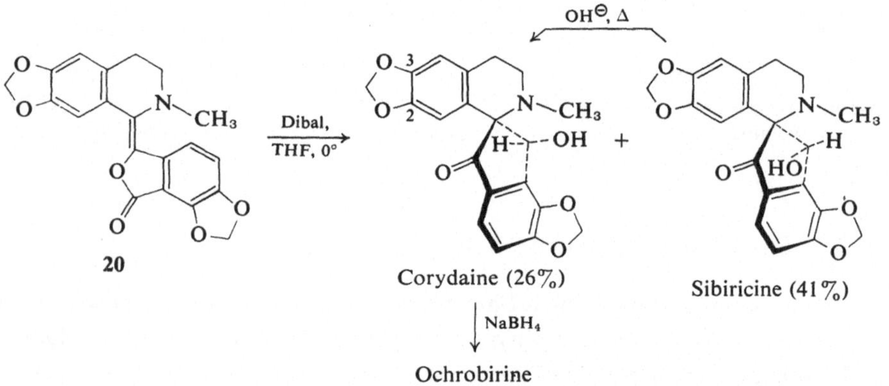

20

Corydaine (26%)

Sibiricine (41%)

Ochrobirine

The 2,3-dimethoxy analogs of corydaine, sibiricine, and ochrobirine, namely yenhusomidine, raddeanone, and yenhusomine, were synthesized by a route similar to that described above.[14]

Indandiones analogous to those discussed in Sec. 25.2.3 above may be obtained in high yield from phthalideisoquinolines. The key step is the base-catalyzed rearrangement of the enelactone derived from a phthalide to give an indan-1,3-dione. The overall sequence is shown below[14a]:

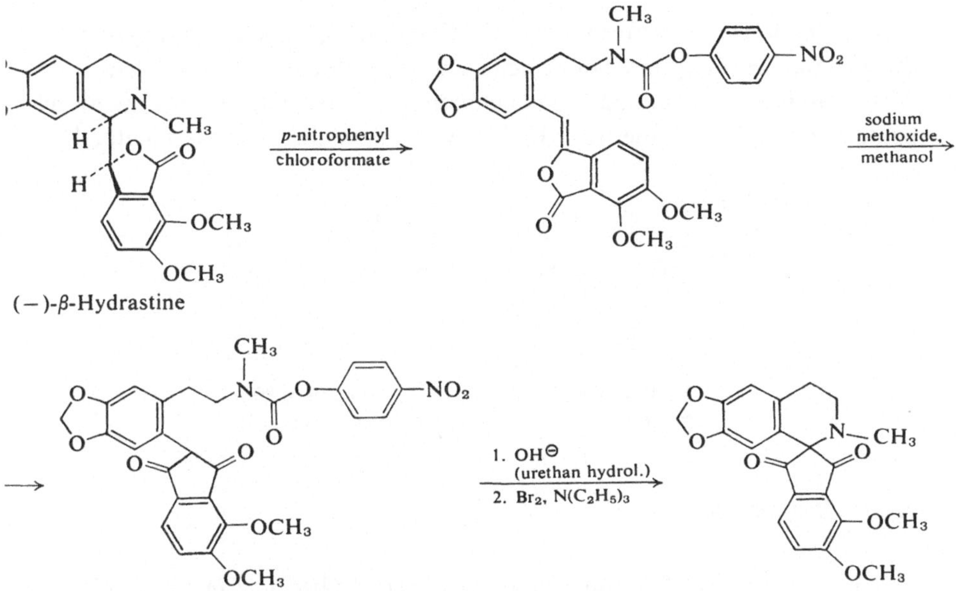

(−)-β-Hydrastine

25.2.7. From Tetrahydroprotoberberinium Salts

Three Japanese groups have studied the base-catalyzed rearrangement of tetrahydroprotoberberine *N*-metho salts.[15–17] In the presence of a strong base such as butyllithium or dimsyl anion the methochloride of canadine is converted to the corresponding spirobenzylisoquinoline free base. The rearrangement is stereospecific and the spirobenzylisoquinoline possessing the sign of rotation opposite to that of the starting methochloride salt is obtained. (See also Secs. 26.2.2 and 26.3.)

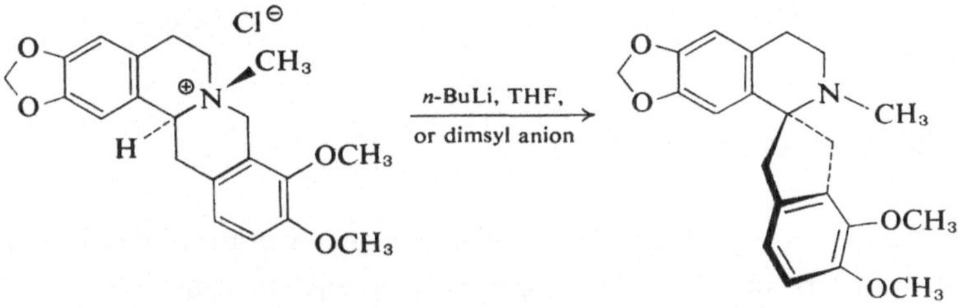

(−)-Canadine methochloride   (+)

25.3. Reductive Rearrangement of Spirobenzylisoquinolines to Dibenzocyclopentazepines

A reductive transformation of diketonic spirobenzylisoquinolines to dibenzocyclopentazepines can be achieved using zinc in hot acetic acid. Under these conditions, **21** and **22** were each converted to mixtures of dibenzocyclopentazepines in which the trans B/C fused isomers always predominated.[18]

21, R = CH₃
22, R = H

(overall ≈ 80%, cis/trans 1 : 3)

25.4. Biogenesis

The biogenesis of ochotensimine has been studied in *Corydalis ochotensis* (Fumariaceae), and in conjunction with that of the tetrahydroprotoberberine alkaloid corydaline in *C. solida*. Both compounds are derived from two molecules of tyrosine and the *S*-methyl group of methionine. The *S*-methyl group of methionine supplies, among others, the C-13 methyl group of corydaline and the exocyclic methylene group of ochotensimine.[19] (See also Sec. 19.4.)

Corydaline

Ochotensimine

25.5. Mass Spectroscopy

The mass spectral breakdown of spirobenzylisoquinolines is appreciably affected by the position of the alkoxy substituents in ring D.[20] If one or two ketonic functions are present in ring C, decarbonylation readily occurs.[20,21]

References and Notes

1. For a review on the spirobenzylisoquinolines, see S. McLean and J. Whelan, *MTP Int. Rev. of Sci.*, *Alkaloids, Org. Chem.*, *Series One*, K. Wiesner, ed., Vol. 9, Butterworths, London (1973), p. 161.

2. S.-T. Lu, T.-L. Su, T. Kametani, and M. Ihara, *Heterocycles*, 3, 301 (1975); and S.-T. Lu, T.-L. Su, T. Kametani, A. Ujiie, M. Ihara, and K. Fukumoto, *J. Chem. Soc. Perkin I*, 63 (1976).

3. T. Kametani, M. Takemura, M. Ihara, and K. Fukumoto, *Heterocycles*, 4, 723 (1976); and *J. Chem. Soc. Perkin I*, 390 (1977).

4. S. McLean and J. Whelan, *Can. J. Chem.*, 51, 2457 (1973).

5. S. McLean and D. Dime, *Can. J. Chem.*, 55, 924 (1977).

6. H. Irie, A. Kitagawa, A. Kuno, J. Tanaka, and N. Yokotani, *Heterocycles*, 4, 1083 (1976).

7. N. E. Cundasawmy and D. B. MacLean, *Can. J. Chem.*, 50, 3028 (1972).

8. S. O. de Silva, K. Orito, R. H. Manske, and R. Rodrigo, *Tetrahedron Lett.*, 3243 (1974).

9. M. Shamma and J. F. Nugent, *Tetrahedron*, 29, 1265 (1973).

10. T. Kametani, T. Takahashi, and K. Ogasawara, *J. Chem. Soc. Perkin I*, 1464 (1973). For an improved procedure, see T. Kametani, Y. Hirai, H. Nemoto, and K. Fukumoto, *J. Heterocycl. Chem.*, 12, 185 (1975).

11. T. Kametani, H. Takeda, Y. Hirai, F. Satoh, and K. Fukumoto, *J. Chem. Soc. Perkin I*, 2141 (1974).

12. B. Nalliah, R. H. F. Manske, R. Rodrigo, and D. B. MacLean, *Tetrahedron Lett.*, 2795 (1973).

12a. M. Hanaoka and C. Mukai, *Heterocycles*, 6, 1981 (1977).

13. H. L. Holland, D. B. MacLean, R. G. A. Rodrigo, and R. H. F. Manske, *Tetrahedron Lett.*, 4323 (1975).

14. B. C. Nalliah, D. B. MacLean, R. G. A. Rodrigo, and R. H. F. Manske, *Can. J. Chem.*, in press.

14a. H. L. Holland, M. Curcumelli-Rodostamo, and D. B. MacLean, *Can. J. Chem.*, 54, 1472 (1976).

15. J. Imai, Y. Kondo, and T. Takemoto, *Heterocycles*, 3, 467 (1975); and *Tetrahedron*, 32, 1973 (1976).

16. S. Kano, T. Yokomatsu, E. Komiyama, S. Tokita, Y. Takahagi, and S. Shibuya, *Chem. Pharm. Bull.*, *Tokyo*, 23, 1171 (1975).

17. T. Kametani, S.-P. Huang, A. Ujiie, M. Ihara, and K. Fukumoto, *Heterocycles*, 4, 1223 (1976); and T. Kametani, A. Ujiie, S.-P. Huang, M. Ihara, and K. Fukumoto, *J. Chem. Soc. Perkin I*, 394 (1977).

18. T. Kametani, S. Hirata, S. Hibino, H. Nemoto, M. Ihara, and K. Fukumoto, *Heterocycles*, 3, 151 (1975); and T. Kametani, S. Hirata, H. Nemoto, M. Ihara, S. Hibino, and K. Fukumoto, *J. Chem. Soc. Perkin I*, 2028 (1975). For the pyrolytic rearrangement of spirobenzylisoquinoline esters to dibenzocyclopentazepines, see N. M. Mollov and G. I. Yakimov, *Dokl. Bolg. Akad. Nauk*, 24, 1325 (1971).

19. H. L. Holland, M. Castillo, D. B. MacLean, and I. D. Spenser, *Can. J. Chem.*, 52, 2818 (1974).

20. A. Kato, K. Akagi, H. Irie, and S. Uyeo, *J. Pharm. Soc. Japan*, 95, 1058 (1975).

21. T. Kametani, S. Hibino, S. Shibuya, and S. Takano, *J. Heterocycl. Chem.*, 9, 47 (1972).

THE RHOEADINES

26.1. Introduction

Rhoeadine and papaverrubine (norrhoeadine) alkaloids which were previously believed to occur only within the genus *Papaver* have now been found to be present in two other Papaveraceae genera, namely *Bocconia* and *Meconopsis*. In *Bocconia frutescens* L. rhoeadine alkaloids accompany a variety of protopine, protoberberine, and benzophenanthridine bases.[1]

26.2. Synthesis

26.2.1. From Phthalideisoquinolines

In a variation of the known conversion of phthalideisoquinolines into rhoeadines,[3,4] the δ-lactone **1** derived from β-hydrastine was converted to its *N*-oxide. The secondary amine **2** was obtained from the *N*-oxide by reaction with trifluoroacetic anhydride, in a modified Polonovski reaction. The papaverrubine **3** was then generated from the amine **2** as shown below[5]:

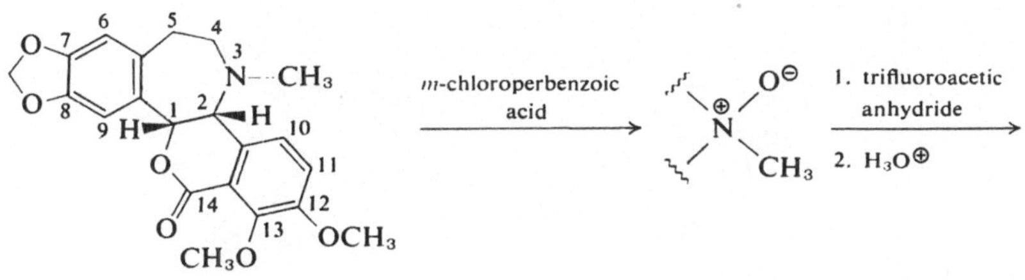

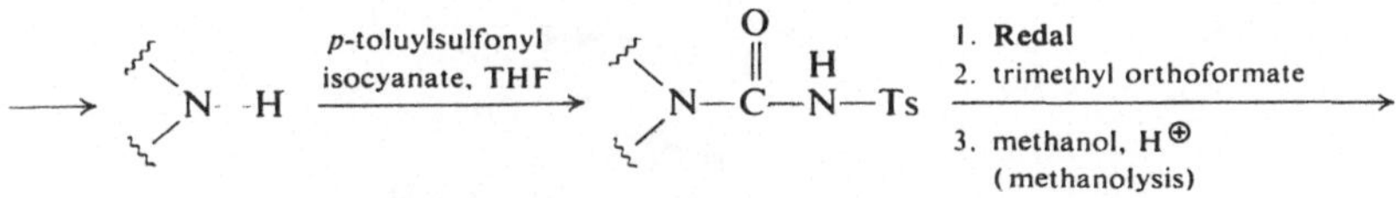

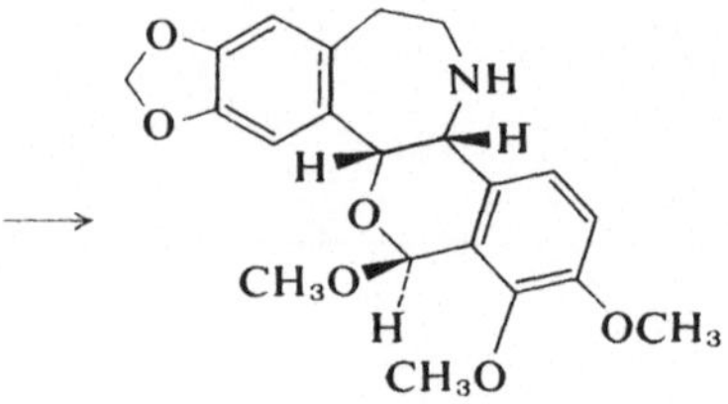

3

26.2.2. From Spirobenzylisoquinolines and Dibenzocyclopentazepines

Full details are now available from Irie's laboratory on the known conversion of appropriately substituted spirobenzylisoquinolines into rhoeagenine-diol[6]:

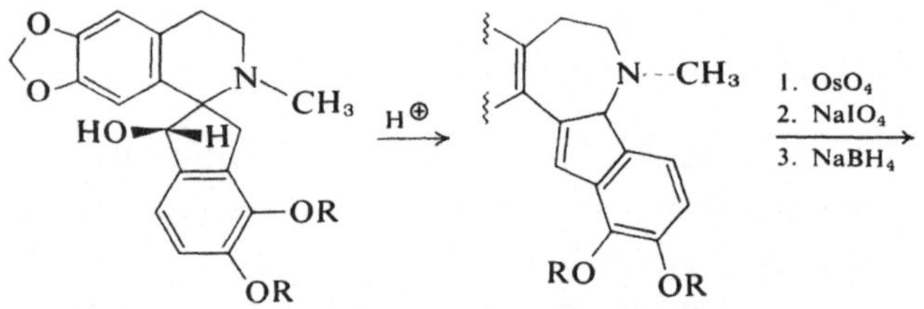

Two *rac.* series:
$R = CH_3$ and $R + R = CH_2$

A dibenzocyclopentazepine

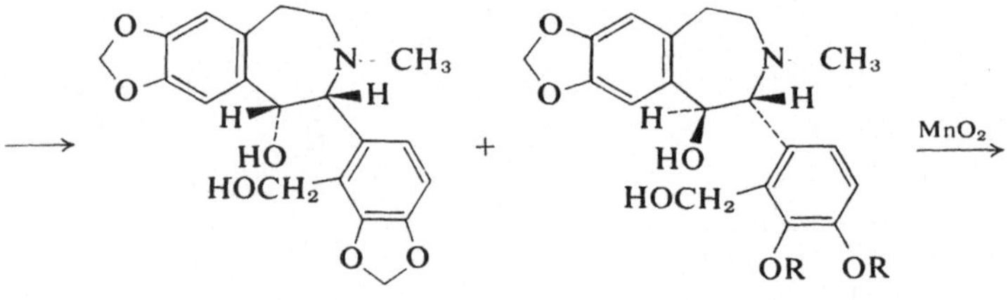

Rhoeageninediol

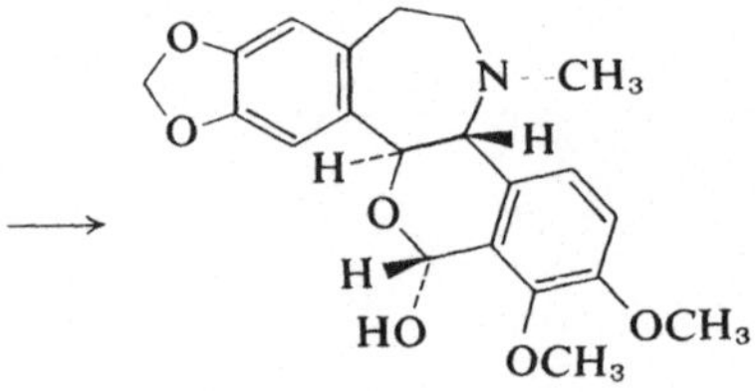

The dibenzocyclopentazepine **4** which had previously been obtained by Irie, was synthesized by Orito, Manske, and Rodrigo by a new route in which the yield on each step was around 70%[7]:

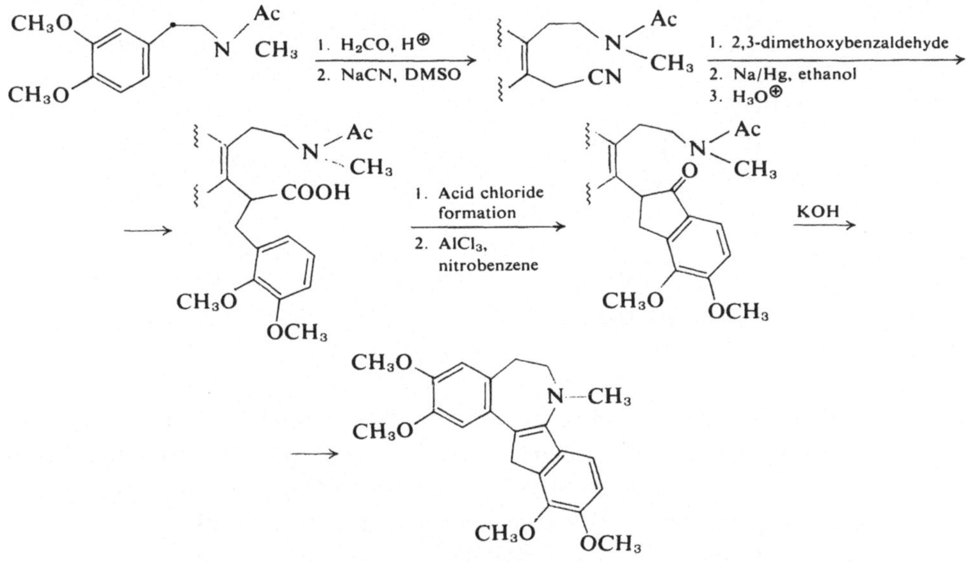

Compound **4** was subsequently converted to rhoeadine bases. The key step in this sequence was the photosensitized oxidation of the enaminoketone **5** to form the ketonic γ-lactone **6**. Analogs of this γ-lactone had previously been converted into rhoeadines by the Brossi–Klötzer–Teitel team, so that a similar sequence was followed in the present case to obtain *cis*-alpinine and *cis*-alpinigenine[7]:

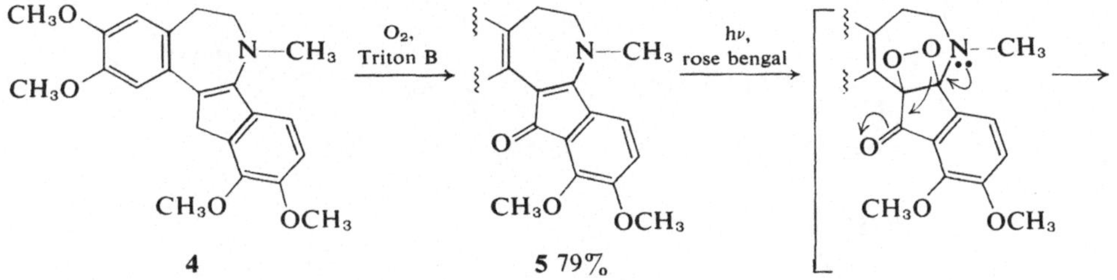

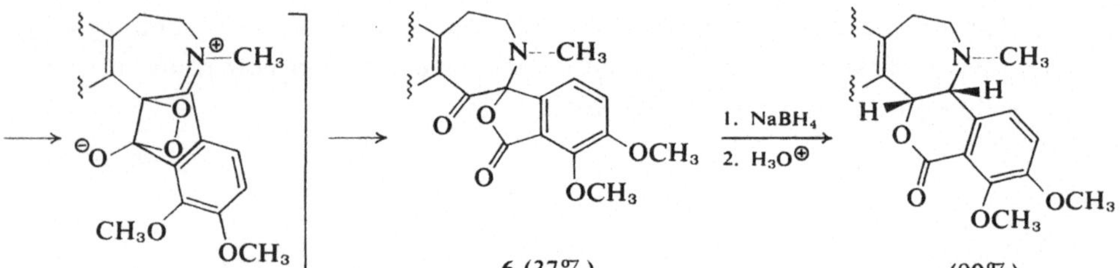

6 (37%)　　　　　　　　　　　　　　(90%)

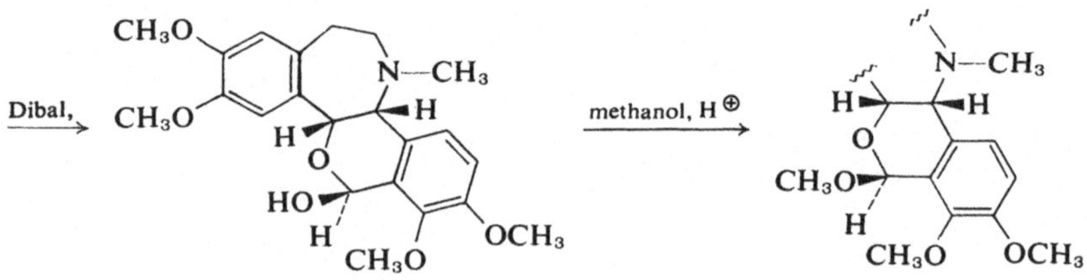

cis-Alpinigenine　　　　　　　　　　*cis*-Alpinine

Parallel to the conversion of canadine to allocryptopine and its cyanogen bromide cleavage product (discussed in Secs. 23.1 and 23.2), tetrahydropalmatine was converted to the *N*-metho salt **7** which was hydrogenolyzed to the keto amine **8**. Subsequent cyanogen bromide cleavage, succeeded by alkaline ethanolysis furnished the dibenzocyclopentazepine **4**[8]:

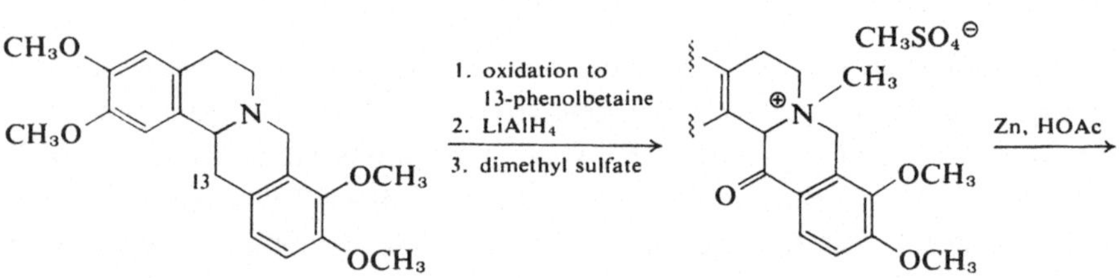

Tetrahydropalmatine　　　　　　　　　　　　　　　　7

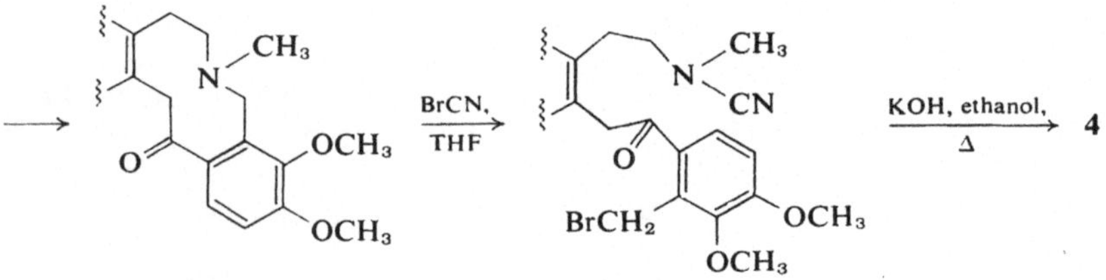

8 (57%)　　　　　　　　　9

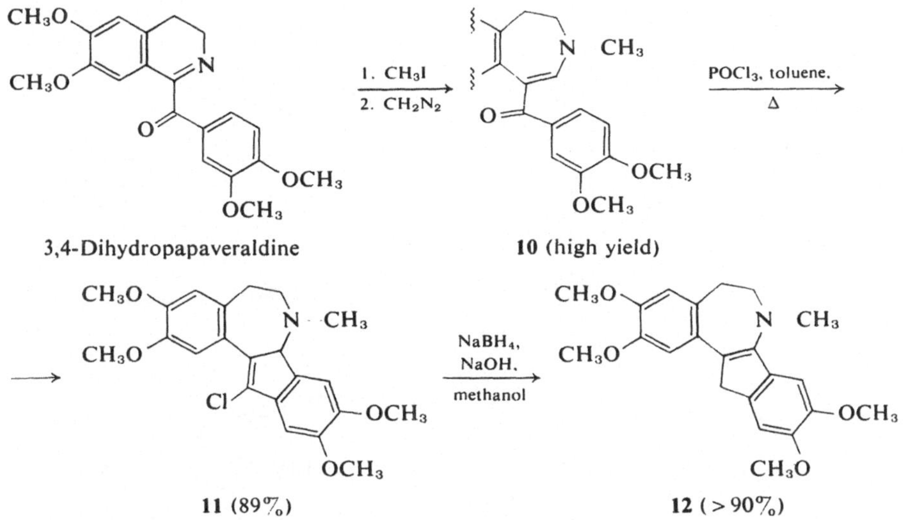

3,4-Dihydropapaveraldine

10 (high yield)

11 (89%)

12 (> 90%)

SCHEME 26.1

Kametani and co-workers have demonstrated that 3,4-dihydropapaveral-dine can be an alternative source for dibenzocyclopentazepines. Reaction of the methiodide of this iminoketone with diazomethane supplied the benzazepine 10 which upon treatment with phosphorus oxychloride gave rise to the di-benzocyclopentazepine 11. Sodium borohydride reduction then supplied the nonchlorinated dibenzocyclopentazepine 12 (see Scheme 26.1).[9]

Additional chemistry of the dibenzocyclopentazepine 11 has recently been described by Kametani's group. Permanganate oxidation of this amine in the presence of piperidine gave the diketospirobenzylisoquinoline 12a as well as the lactone 12b. A probable mechanism for these transformations is[9a]:

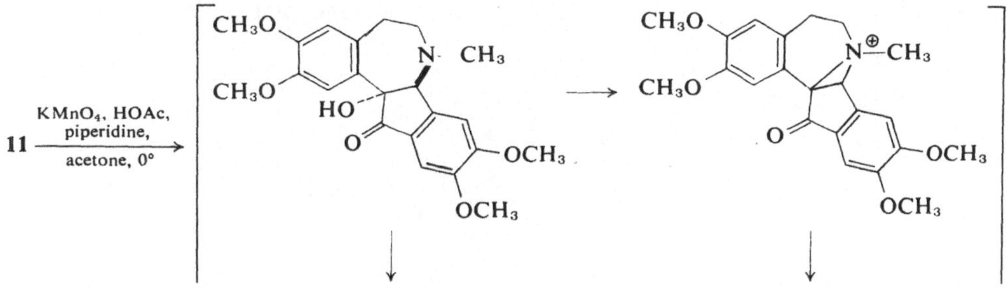

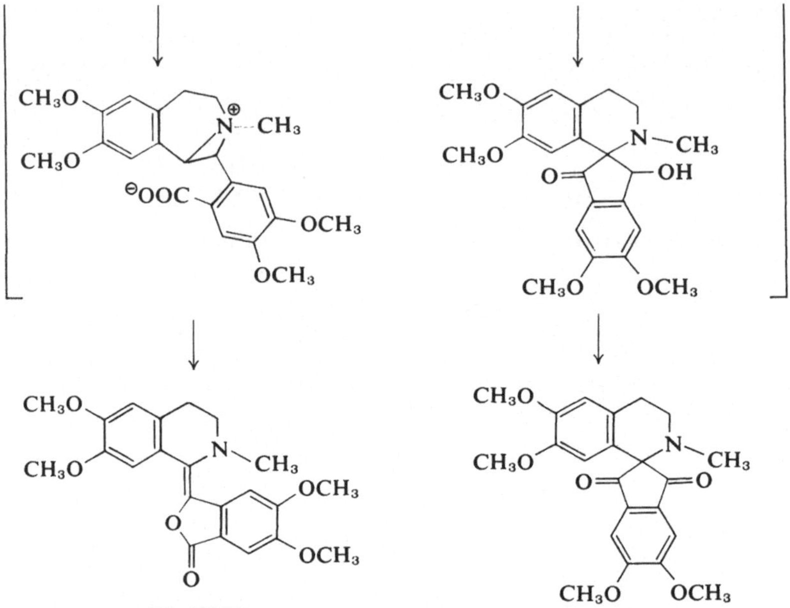

12b (22%)

12a (3%)

26.2.3. By Ring Expansion of a Pseudobase

A novel route to the rhoeadines starts with the readily available immonium salt **13** which was subjected to Schotten–Baumann conditions; the aldehyde **15** was formed in 97% yield via the intermediate *N*-benzoylated pseudobase **14**. Intramolecular aldol condensation of the aldehyde was induced by a strong base, thus generating the lactonic rhoeadine amide **16** in 56% yield. This short series of transformations represents a new benzazepine synthesis.

Reduction with Redal (bismethoxyethoxyaluminum hydride) at 4° of the lactone **16** supplied the hemiacetal **17** which was *O*-methylated and subjected again to Redal reduction, this time at 20°. The product was the secondary amine **18** which was *N*-methylated to the trans-fused rhoeadine analog **19**[10]:

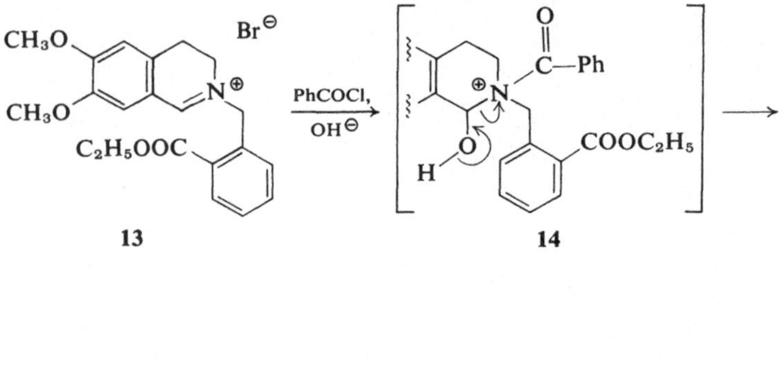

13   14

26.2.4. From a Tetrahydroprotoberberine

The known nonnaturally occurring tetrahydroprotoberberine **19a** was N-methylated and then subjected to Hofmann elimination. The resulting *trans*-azocine was converted to the corresponding glycol which was cleaved quantitatively with periodic acid to a dialdehyde. Subsequent photolysis afforded directly *cis*-alpinigenine [10a]:

(51% from **19a**)

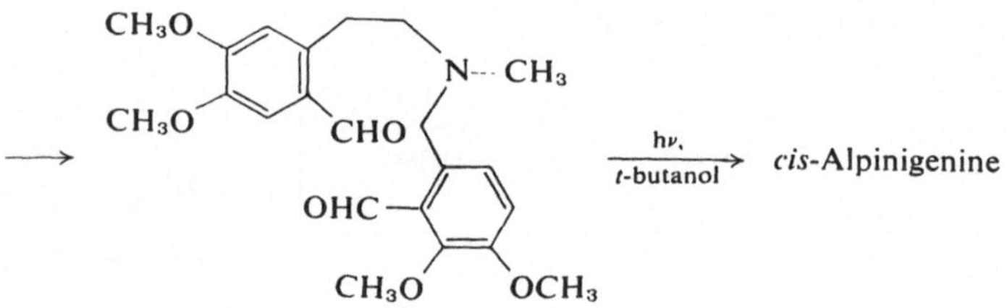

$$\longrightarrow \qquad \xrightarrow[\text{t-butanol}]{h\nu,} \qquad \text{cis-Alpinigenine}$$

26.3. Benzazepine Preparations of Potential Use in Rhoeadine Synthesis[11]

The isomeric cis and trans benzazepinols **21** and **22** have been synthesized in good yields from the known keto ester **20**[12]:

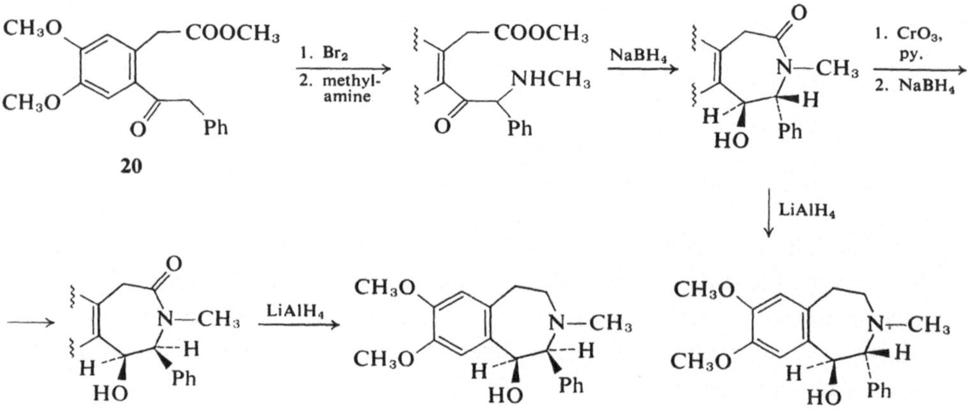

Reduction with zinc in propionic acid of the ketoimmonium or the keto-ammonium salts **23** and **24** lead to benzazepine **25**.[13]

On the other hand, if **23** is treated first with diazomethane and then with phosphorus oxychloride, the enolic dibenzocyclopentazepine **26** is isolated.[9,14] (Compare this result with that obtained using toluene as solvent in Scheme 26.1.)

A substituted benzazepine was formed by Friedel–Crafts cyclization of the toluenesulfonamide **27**. The resulting ketone was reduced to the alcohol, and detosylation was achieved using sodium in liquid ammonia. The amino ketone **29** was also obtained by oxidation of **28** after protection of the amino function.[15]

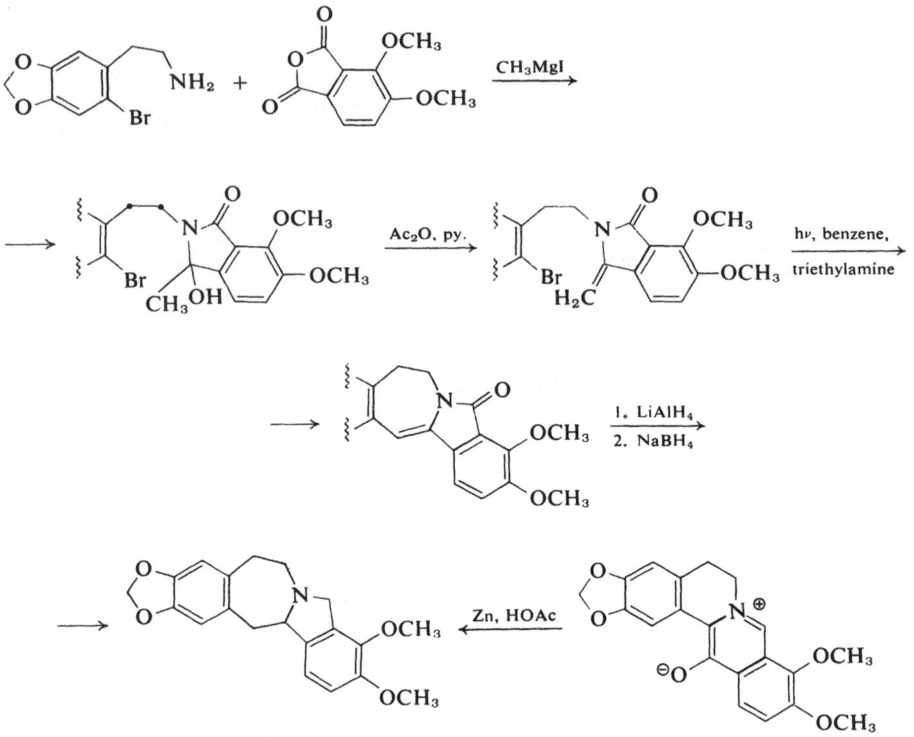

30

Berberinephenolbetaine

SCHEME 26.2

A photochemical synthesis of benzazepines has allowed an alternate preparation of the aporhoeadane derivative **30**, also known as Schöpf Base VI.[16] Compound **30** had originally been formed by reduction of berberinephenolbetaine with zinc in acetic acid (see Scheme 26.2).[17] (For the chemistry of the aporhoeadane **30**, see Sec. 23.1.2.)

Finally, diketospirobenzylisoquinolines can be rearranged in one step to dibenzocyclopentazepines by means of zinc in hot acetic acid.[18] (See Sec. 25.3.)

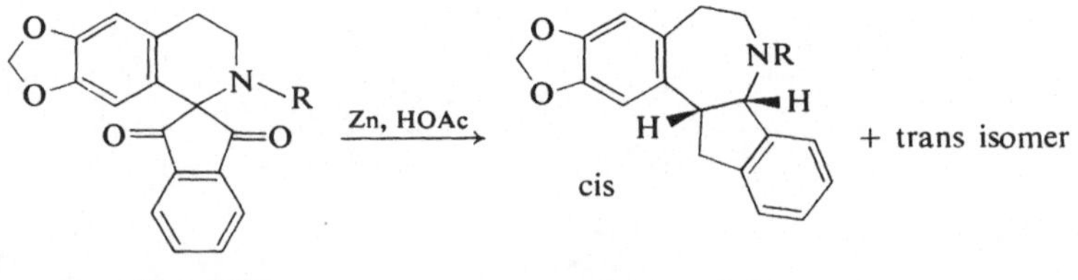

R = H or CH_3

cis/trans 1 : 3

26.4. Reactions

Hofmann degradation of rhoeagenine methiodide yields a complex mixture from which compounds **31–34** have been isolated and characterized.[19]

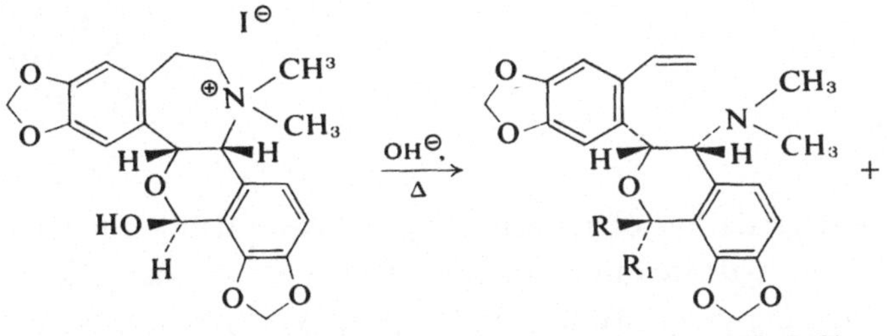

(+)-Rhoeagenine methiodide

31, R = OH, R₁ = H (21%)
32, R = H, R₁ = OH (1%)

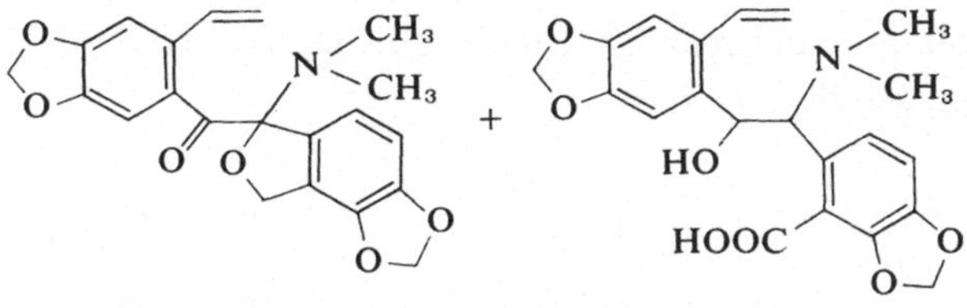

33 (12%)   **34** (9%)

On the other hand, Hofmann degradation of the methiodide salt of the closely related (+)-rhoeadine yields two products, **35** and **36**.[20]

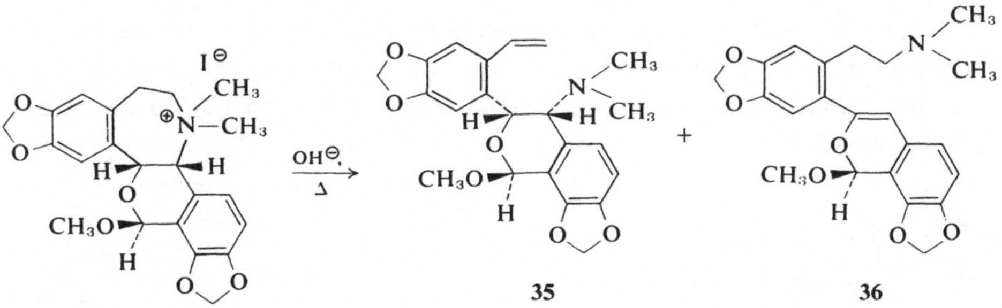

(+)-Rhoeadine methiodide

35
Main product

36

Product **35** can be reduced to its dihydro derivative, which upon hydrolysis and further reduction yields the aminodiol **37**.[20]

37

Emde degradation of rhoeadine type bases usually affords two products resulting from hydrogenolysis of rings B and D. In the typical case of (+)-isorhoeadine, the main product **38** is optically active because the integrity of the C-1 asymmetric center has been preserved, whereas the minor product **39** is achiral. These Emde studies have also reconfirmed the fact that acid catalyzed isomerization of trans-B/D fused rhoeadines occurs at C-1 rather than at C-2.[21] (For the interrelation between rhoeadine and peshawarine through a similar Emde reduction, see Sec. 20.2.)

(+)-Isorhoeadine   (+)-**38**   **39**

Lithium aluminum hydride reduction of rhoeagenine is known to yield rhoeageninediol. Catalytic hydrogenolysis of rhoeageninediol diacetate leads to the expected substituted benzazepine **40** as well as to the aporhoeadane derivative **41**.[20]

Rhoeageninediol diacetate   **40**   **41**

26.5. Stereochemistry and Absolute Configuration

The stereochemical assignments for the rhoeadine alkaloids have been confirmed by X-ray analysis of (+)-rhoeagenine methiodide which led as expected to the accepted configuration.[22] A detailed discussion of the CD curves of rhoeadine alkaloids and their degradation products has appeared.[20]

26.6. Biogenesis

Several new suggestions concerning the biogenesis of the rhoeadines have appeared. Using *Papaver bracteatum* Lindl. plants, it has been demonstrated that [N-^{14}CH$_3$, 8-^{14}C]-($\pm$)-tetrahydropalmatine methiodide is incorporated with high efficiency into alpinigenine. A mechanism was therefore proposed which involves C-14 hydroxylation of the quaternary tetrahydroprotoberberine salt, followed by cleavage of the bond between N-7 and C-14.[23]

Tetrahydropalmatine methiodide

Alpinigenine

A second suggestion, which is not based on any work with labeled precursors, is that a salt such as tetrahydropalmatine methiodide could undergo oxidation at C-13 with subsequent fission of the N-7 to C-14 bond. Further cleavage of the N-7 to C-8 linkage is required before the rhoeadine skeleton can be formed[8]:

Tetrahydropalmatine methiodide → ... → ... →

→ ... → ... → Alpinigenine

A third and most significant study involves tracer experiments with *Papaver rhoeas* L. Both 13S- and 13R-tritium labeled scoulerines were incorporated into rhoeadine, but the rhoeadine isolated from plants fed with 13S-labeled scoulerine had lost 79% of the tritium present in the precursor, whereas the 13R-labeled scoulerine afforded rhoeadine which had retained 74% of the original tritium. A stereospecific loss of the pro-S hydrogen must, therefore, have occurred from the C-13 of scoulerine at some specific stage of its biotransformation into rhoeadine. It seems likely that scoulerine is converted first into stylopine which is oxidized to **42**. *N*-Methylation and rearrangement then provide the correct skeleton from which rhoeadine can arise[24]:

(−)-Scoulerine   (−)-Stylopine

42

(+)-Rhoeadine

Similarly, in a recent investigation, [13-T$_2$]-protopine and [N-^{13}CH$_3$]-cis-tetrahydrocoptisine methochloride were administered to *Papaver rhoeas* L., and radioactive rhoeadine was isolated. The following hypothetical biogenetic route was thus proposed.[25]

cis-Tetrahydrocoptisine   Protopine
N-metho salt

Rhoeadine

There is an interesting relationship between protoberberines, phthalide-isoquinolines, and rhoeadines. Phthalideisoquinolines, with the glaring exception of the ring A trioxygenated (−)-α-narcotine, have their analogs or counterparts in terms of substitution pattern among the rhoeadines. It can, therefore, be postulated that, when the protoberbinoid precursor reaches the stage in the oxidative process where ring expansion to a rhoeadine may occur, a substituent at C-1 in ring A interferes sterically with that process. With rhoeadine formation thus precluded, the precursor may oxidize to the phthalideisoquinoline (−)-α-narcotine, or alternatively may undergo a second Mannich condensation to generate a retroprotoberberine. Indeed, all of the retroprotoberberine alkaloids found to date, just like the phthalideisoquinoline (−)-α-narcotine, are trioxygenated in ring A.

26.7. CMR Spectroscopy

The CMR chemical shifts for (+)-rhoeadine have been determined.[25]

CMR chemical shifts for
(+)-rhoeadine.

References and Notes

1. J. Slavík and L. Slavíkova, *Collect. Czech. Chem. Commun.*, **40**, 3206 (1975).
2. For a discussion of synthetic approaches to the rhoeadines, see T. Kametani and K. Fukumoto, *Heterocycles*, **3**, 931 (1975).
3. W. Klötzer, S. Teitel, J. Blount, and A. Brossi, *Monatsch.*, **103**, 435 (1972).
4. W. Klötzer, S. Teitel, and A. Brossi, *Helv. Chim. Acta*, **55**, 2228 (1972).
4a. For a detailed discussion of pseudobase formation from isoquinoline salts, see V. Šimánek and V. Preininger, *Heterocycles*, **6**, 475 (1977).
5. R. Hohlbrugger and W. Klötzer, *Chem. Ber.*, **107**, 3457 (1974).
6. H. Irie, S. Tani, and H. Yamane, *J. Chem. Soc. Perkin I*, 2986 (1972).
7. K. Orito, R. H. Manske, and R. Rodrigo, *J. Am. Chem. Soc.*, **96**, 1944 (1974).
8. B. Nalliah, R. H. Manske, and R. Rodrigo, *Tetrahedron Lett.*, 2853 (1974).
9. T. Kametani, S. Hirata, H. Nemoto, M. Ihara, S. Hibino, and K. Fukumoto, *J. Chem. Soc. Perkin I*, 2028 (1975); T. Kametani, M. S. Premila, S. Hirata, H. Seto, H. Nemoto, and K. Fukumoto, *Can. J. Chem.*, **53**, 3824 (1975). See also Ref. 14 below.
9a. T. Kametani, T. Ohsawa, S. Hirata, M. S. Premila, M. Ihara, and K. Fukumoto, *Chem. Pharm. Bull.*, *Tokyo*, **25**, 321 (1977).
10. M. Shamma and L. Töke, *Tetrahedron*, **31**, 1991 (1975).
10a. S. Prabhakar, A. M. Lobo, and I. M. C. Oliveira, *Chem. Commun.*, 419 (1977).
11. For a review on the preparation of benzazepines, see S. Kasparek, *Adv. Heterocycl. Chem.*, **17**, 45 (1974). See also B. Pecherer, R. C. Sunbury, and A. Brossi, *J. Heterocycl. Chem.*, **9**, 609 (1972).
12. Y. Inubushi, T. Harayama, and K. Takeshima, *Chem. Pharm. Bull.*, *Tokyo*, **20**, 689 (1972).
13. C. Reby, J. Likforman, and J. Gardent, *C. R. Acad. Sci. Paris, Serie C*, **269**, 45 (1969); and C. Reby and J. Gardent, *Bull. Soc. Chim. France*, 1574 (1972).
14. T. Kametani, S. Hirata, F. Satoh, and K. Fukumoto, *J. Chem. Soc. Perkin I*, 2509 (1974).

15. M. Lennon, A. McLean, G. R. Proctor, and I. W. Sinclair, *J. Chem. Soc. Perkin I*, 622 (1975). See also G. Hazebroucq and J. Gardent, *C. R. Acad. Sci. Paris, Serie C*, **257**, 923 (1963).
16. H. O. Bernhard and V. Snieckus, *Tetrahedron Lett.*, 4867 (1971). For an alternate synthesis of the Schöpf Base VI, see S. Teitel, W. Klötzer, J. Borgese, and A. Brossi, *Can. J. Chem.*, **50**, 2022 (1972).
17. C. Schöpf and M. Schweickert, *Chem. Ber.*, **98**, 2566 (1965).
18. T. Kametani, S. Hirata, S. Hibino, H. Nemoto, M. Ihara, and K. Fukumoto, *Heterocycles*, **3**, 151 (1975).
19. M. D. Rozwadowska and J. W. ApSimon, *Tetrahedron*, **28**, 4125 (1972).
20. J. Hrbek, Jr., L. Hruban, V. Šimánek, F. Šantavý, and G. Snatzke, *Collect. Czech. Chem. Commun.*, **38**, 2799 (1973).
21. V. Šimánek, L. Hruban, V. Preininger, A. Neméckova, and A. Klásek, *Collect. Czech. Chem. Commun.*, **40**, 705 (1975); V. Šimánek, A. Klásek, L. Hruban, V. Preininger, and F. Šantavý, *Tetrahedron Lett.*, 2171 (1974); H. Rönsch, *Tetrahedron Lett.*, 4431 (1972); and V. Šimánek, A. Klásek, and F. Šantavý, *Tetrahedron Lett.*, 1779 (1973).
22. C. S. Huber, *Acta Crystallogr.* (*Copenhagen*), **B26**, 373 (1970); and **B28**, 982 (1972).
23. H. Rönsch, *Eur. J. Biochem.*, **28**, 123 (1972); and *Pharmazie*, **29**, 71 (1974).
24. A. R. Battersby and J. Staunton, *Tetrahedron*, **30**, 1707 (1974).
25. C. Tani and K. Tagahara, *J. Pharm. Soc. Japan*, **97**, 93 (1977).

EMETINE AND RELATED BASES

27.1. Introduction

No new alkaloids in this series have been isolated. However, the structure of ankorine has been revised[1] and that of alangiside has been established with certainty.[2]

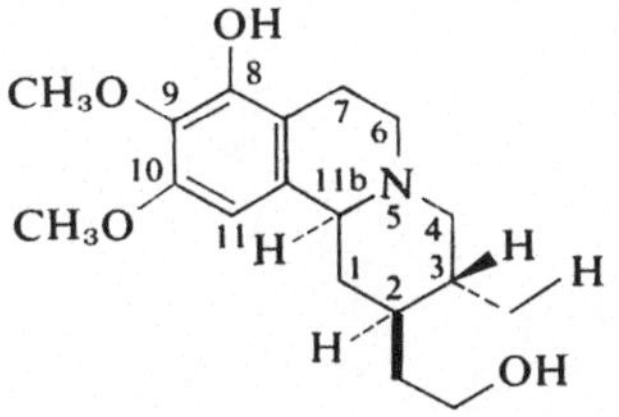

(−)-Ankorine (revised structure)
Alangium lamarckii Thw. (Alangiaceae)

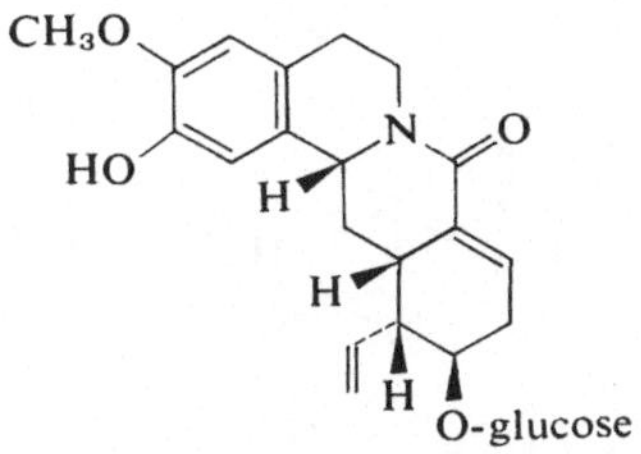

(−)-Alangiside
A. lamarckii

27.2. Structural Elucidation and Synthesis[2a]

27.2.1. Ankorine

Szántay's group has synthesized all four possible racemates of the originally proposed general structure **1** for ankorine by the route indicated in Scheme 27.1.[1]

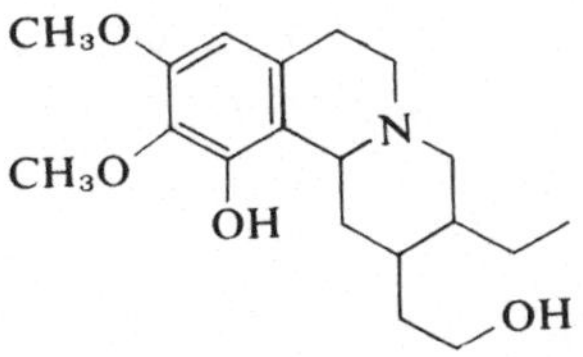

1

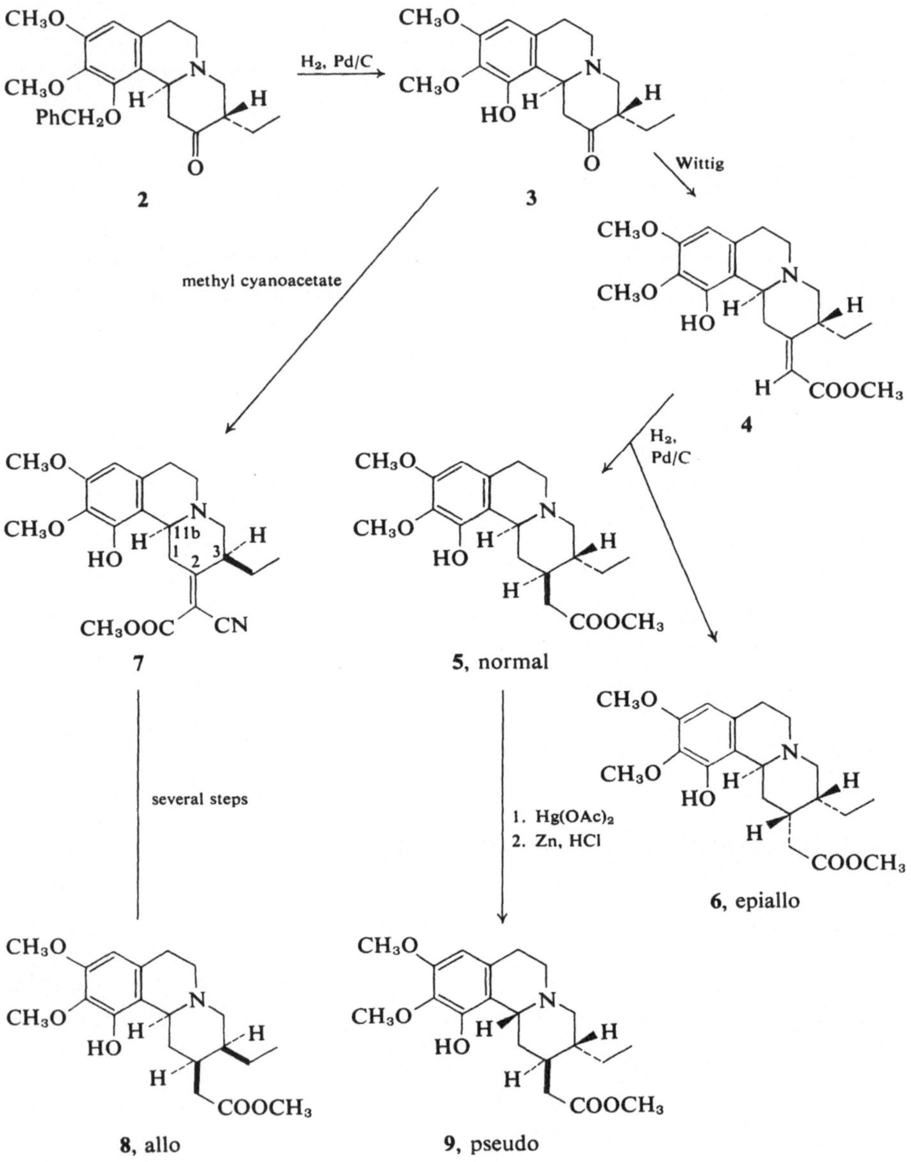

H₂, Pd/C
Wittig
methyl cyanoacetate
H₂, Pd/C
several steps
1. Hg(OAc)₂
2. Zn, HCl
2
3
4
7
5, normal
6, epiallo
8, allo
9, pseudo

SCHEME 27.1

The tricyclic ketone **2** was debenzylated to the phenol **3** which upon Wittig reaction afforded the unsaturated ester **4**. Catalytic reduction of **4** gave rise to a separable mixture of the normal and epiallo esters, **5** and **6**. Alternative reaction of the phenol **3** with methyl cyanoacetate led to the unsaturated ester **7**, epimerized at C-3. Borohydride reduction of **7**, followed by hydrolysis and decarboxylation, supplied an intermediate nitrile which on treatment with methanolic hydrogen chloride generated the allo ester **8**. In analogy with the yohimbine series, the normal ester **5** was transformed by mercuric acetate oxidation and zinc in hydrochloric acid reduction to the pseudo ester **9**. The respective alcohols derived by lithium aluminum hydride reduction of the esters **5, 6, 8**, and **9**, all proved to be nonidentical to natural ankorine. A revised structure bearing an 8-hydroxyl rather than an 11-hydroxyl group was therefore proposed, and the relative stereochemistry was assigned as the normal type based on comparative mass spectral studies in this and the yohimbine series.[3] Similarly, the structure of alangicine[4] was proposed to correspond to 8-hydroxypsychotrine, and the related base alangimarckine[5] should also reflect this structural change.

(+)-Alangicine, R =

(−)-Alangimarckine, R =

The correctness of the revised structure and relative stereochemistry[4] of ankorine was confirmed by a total synthesis[6] closely paralleling the emetine synthesis by Preobrazhenskii.[7]

A partial synthesis from the known optically active ester **10** derived from commercial (+)-cinchonine has established the absolute configuration of (−)-ankorine (**13a**).[8] The key step in this sequence is the thermal isomerization of the cis acid **11** to the trans acid **12**, proceeding in 90% yield to furnish a 37:63 mixture of **11** and **12**:

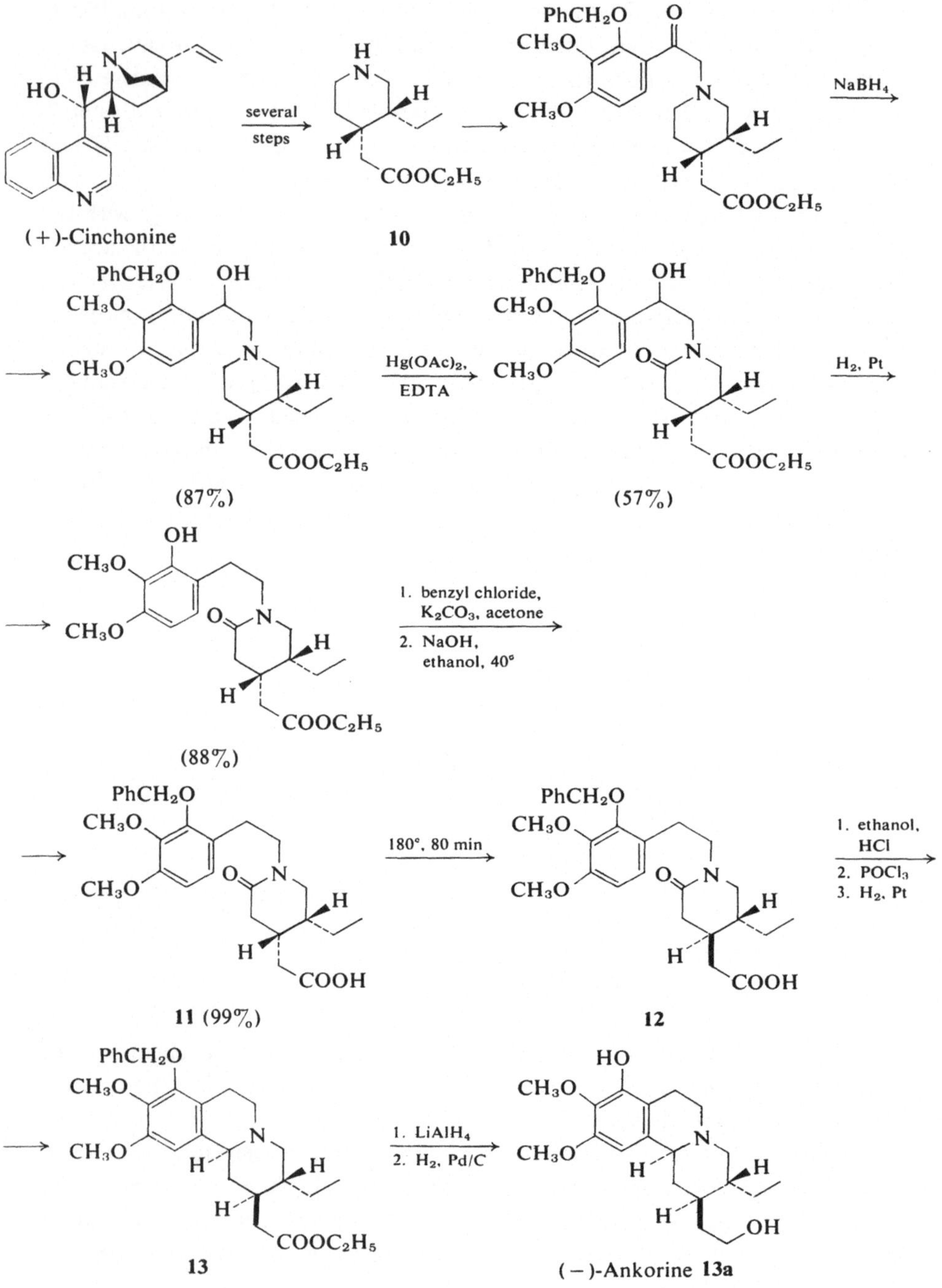
(+)-Cinchonine
several steps
10
NaBH$_4$
(87%)
Hg(OAc)$_2$, EDTA
(57%)
H$_2$, Pt
(88%)
1. benzyl chloride, K$_2$CO$_3$, acetone
2. NaOH, ethanol, 40°
11 (99%)
180°, 80 min
12
1. ethanol, HCl
2. POCl$_3$
3. H$_2$, Pt
13
1. LiAlH$_4$
2. H$_2$, Pd/C
(−)-Ankorine 13a

The intermediate ester **13** has also been used in the synthesis of the related base alangicine.[9]

By a parallel series of conversions, with the crucial step again being the thermal isomerization from the cis to the trans stereochemistry at C-2 and C-3, the known precursor ester (−)-**14**, has been prepared.[10] This ester had previously been shown to lead to optically active *O*-methylpsychotrine, emetine, psychotrine, protoemetine, and tubulosine.

14

27.2.2. Alangiside

Condensation of 3-hydroxy-4-methoxyphenethylamine with secologanin followed by lactam formation (catalyzed by sodium carbonate) generated mainly the alangiside analog **15**. The optical rotation and PMR spectrum of **15** were different from those of (−)-alangiside (**16**), but diazomethane *O*-methylation of **15** or **16** gave *O*-methylalangiside (**17**) thus locating the phenolic hydroxyl of alangiside at C-10[2]:

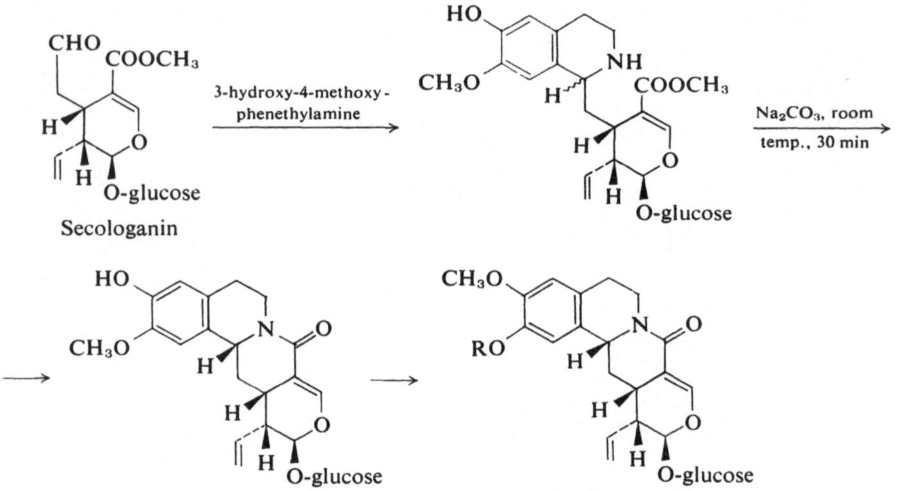

Secologanin

15

16, (−)-Alangiside, R = H
17, (−)-O-Methylalangiside.

27.2.3. Emetine

Takano's group has recently reported on a high-yield synthesis of emetine starting with the tetrahydroprotoberberine **18**.[11]*N*-Chorosuccinimide treatment of **19** afforded the thermodynamically more stable isomer **20** which underwent a stereospecific catalytic reduction to the ketone **21**. The thioketal **22** was prepared through the intermediacy of the pyrrolidine enamine, and was then cleaved in high yield. Subsequent esterification with diazomethane and desulfurization supplied the known protoemetine precursor **23** which was converted to emetine by established methods:

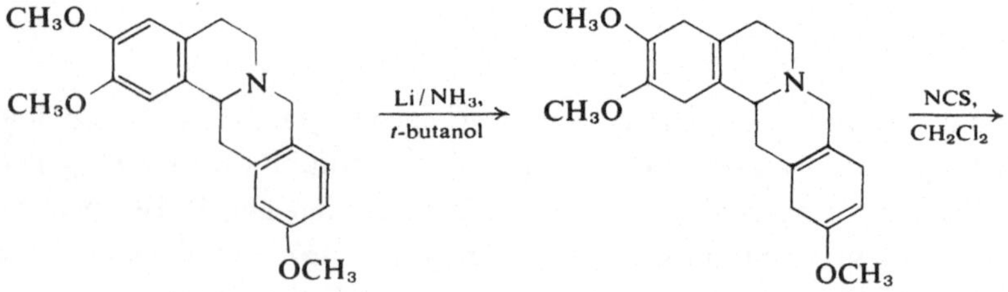

18

19 (76%)

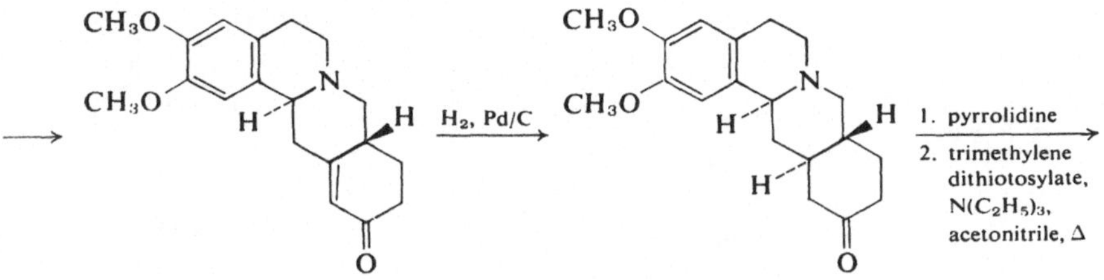

20 (63%)   21 (89%)

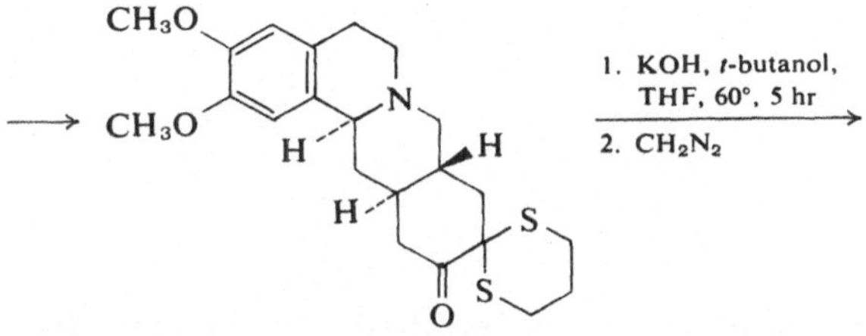

22 (65%)

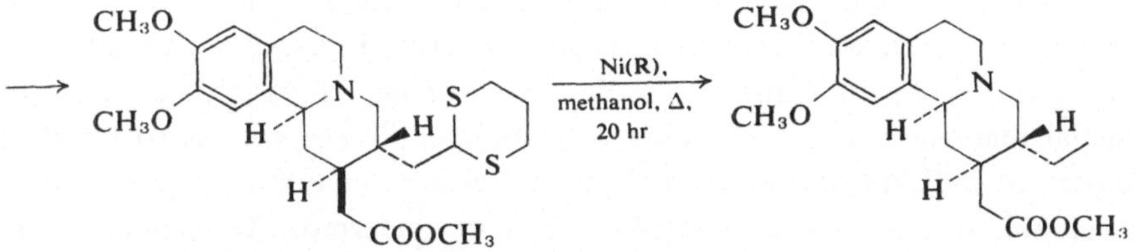

(94%)   23 (92%)

A synthesis of didehydroemetine has been accomplished starting from homoveratrylamine and using the previously known tricyclic ketone **24**. The stereochemistry of the didehydroemetine was not completely defined[11a]:

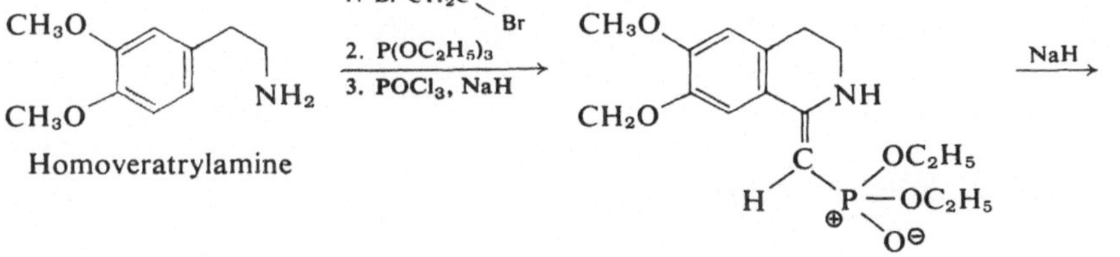

Homoveratrylamine

A phosphonate ester

A didehydroemetine

27.3. Pharmacology

Emetine is a strong inhibitor of protein synthesis, functioning through prevention of the translation of amino–acyl transfer RNA to ribosomal peptide. It also inhibits *in vitro* aerobic glycolysis in cardiac muscle.

In Phase II clinical trials, emetine as an antineoplastic agent was shown to exert moderate and reversible toxicity, consisting mainly of muscular weakness, local pain, and transient cardiac arrhythmias. However, no complete or partial regression of solid tumors was observed.[12] The studies so far do not appear to indicate that emetine treatment results in tumor response with an acceptable degree of drug related toxicity in epidermoid bronchogenic carcinoma.[13]

2-Dehydroemetine was found to elicit good response in cutaneous leshmaniasis whether administered in the form of its hydrochloride or other salts, and apparently led to prompt healing with very moderate scarring, even in cases which responded poorly to other medications.[14] 2-Dehydroemetine is excreted unchanged and in appreciable amounts in the urine of rats administered this antiamoebic drug by the intravenous route.[15] N-Chloroacetylemetine and N-dichloroacetylemetine, prepared by simple acylation of emetine, have shown about the same level of antiamoebic activity as emetine.[16]

27.4. CMR Spectroscopy

The following CMR chemical shift assignments have been made for emetine[17]:

References and Notes

1. C. Szántay, E. Szentirmay, and L. Szabó, *Tetrahedron Lett.*, 3725 (1974).
2. A. Shoeb, K. Raj, R. S. Kapil, and S. P. Popli, *J. Chem. Soc. Perkin I*, 1245 (1975).
2a. For discussions of methods of synthesis of benz[a]quinolizines, which correspond to rings A, B, and C of emetine, see F. D. Popp and R. F. Watts, *Heterocycles*, 6, 1189 (1977); T. Kametani, H. Terasawa, M. Ihara, and K. Fukumoto, *Heterocycles*, 6, 37 (1977); T. Kametani, H. Terasawa, and M. Ihara, *J. Chem. Soc. Perkin I*, 2547 (1976); and R. F. Watts and F. D. Popp, *Heterocycles*, 6, 47 (1977).
3. M. Bárczai-Beke, G. Dörnyei, J. Tamás, and C. Szántay, *Chem. Ber.*, 105, 3244 (1972).
4. S. C. Pakrashi and E. Ali, *Tetrahedron Lett.*, 2143 (1967).
5. A. R. Battersby, R. S. Kapil, D. S. Bhakuni, S. P. Popli, J. R. Merchant, and S. S. Salgar, *Tetrahedron Lett.*, 4965 (1966).
6. T. Fujii, S. Yoshifuji, and K. Yamada, *Tetrahedron Lett.*, 1527 (1975).
7. R. P. Evstigneeva, R. S. Livshits, L. I. Zakharin, M. S. Bainova, and N. A. Preobrazhenskii, *Dokl. Akad. Nauk SSSR*, 75, 539 (1950); R. P. Evstigneeva and N. A. Preobrazhenskii, *Tetrahedron*, 4, 223 (1958); and L. I. Zakharin and N. A. Preobrazhenskii, *Zh. Obshch. Khim.*, 22, 1890 (1952), and 23, 153 (1953).
8. S. Yoshifuji and T. Fujii, *Tetrahedron Lett.*, 1965 (1975).
9. T. Fujii, K. Yamada, S. Yoshifuji, S. C. Pakrashi, and E. Ali, *Tetrahedron Lett.*, 2553 (1976). For a related synthesis of alangimarckine, see T. Fujii, S. Yoshifuji, and H. Kogen, *Tetrahedron Lett.*, 3477 (1977).
10. T. Fujii and S. Yoshifuji, *Tetrahedron Lett.*, 731 (1975).
11. S. Takano, M. Sasaki, H. Kanno, K. Shishido, and K. Ogasawara, in press.
11a. A. Buzas, J. P. Finet, J. P. Jacquet, and G. Lavielle, *Tetrahedron Lett.*, 2433 (1976).
12. S. Siddiqui, D. Firat, and S. Olshin, *Cancer Chemother. Rept.*, Part 1, 57, 423 (1973).
13. R. C. Kane, M. H. Cohen, L. E. Broder, M. I. Bull, P. J. Creaven, and B. E. Fossieck, *Cancer Chemother. Rept.*, Part 1, 59, 1171 (1975).
14. *Progress in Drug Research*, E. Jucker, ed., Vol. 18, Birkhauser Verlag, Basel and Stuttgart (1974), p. 325.
15. R. K. Johnson, W. T. Wynn, and W. R. Johndorf, *Biochem. J.*, 125, 26P (1971).
16. C. Viel, P. Rumpf, and L. Lamy, *Chim. Therap.*, 228 (1966).
17. M. C. Koch, M. M. Plat, N. Preaux, H. E. Gottlieb, E. W. Hagaman, F. M. Schell, and E. Wenkert, *J. Org. Chem.*, 40, 2836 (1975).

THE PHENETHYLISOQUINOLINES

28

28.1. Synthesis and Chemistry

To date, only one naturally occurring phenethylisoquinoline alkaloid is known, namely autumnaline, isolated from *Colchicum cornigerum* (Schweinf.) Tackh. et Drar. (Liliaceae).[1] However, some interesting chemistry on synthetic bases of this type has been discussed.[2–9]

The phenethylisoquinoline homopetaline which possesses the unusual 7,8-dioxygenation pattern in the isoquinoline moeity, has been synthesized along classical lines, utilizing a bromine blocking group to direct the Bischler–Napieralski cyclization.[2]

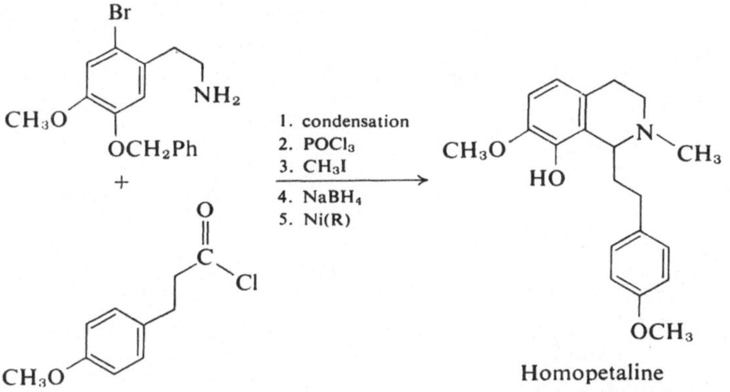
Br
NH2
CH3O
OCH2Ph
+
O
C
Cl
CH3O
1. condensation
2. POCl3
3. CH3I
4. NaBH4
5. Ni(R)
CH3O
HO
N
CH3
OCH3

Homopetaline
(not a natural product)

Enzymic oxidation of homoorientaline with potato peelings, followed by acetylation generated the triacetate **1**, whose structure was confirmed by an alternate synthesis.[3]

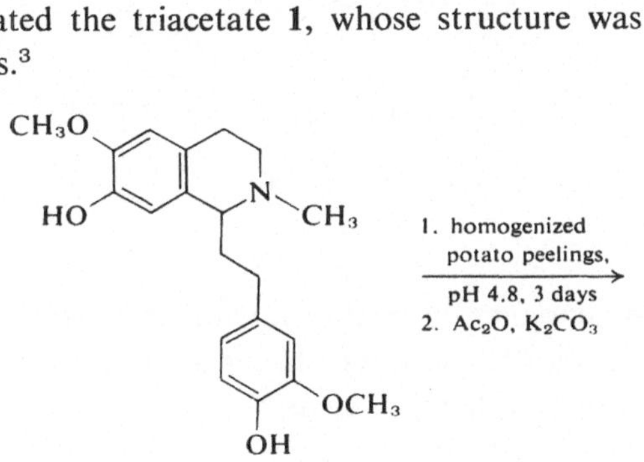
CH3O
HO
N
CH3
OCH3
OH
1. homogenized
potato peelings,
pH 4.8, 3 days
2. Ac2O, K2CO3

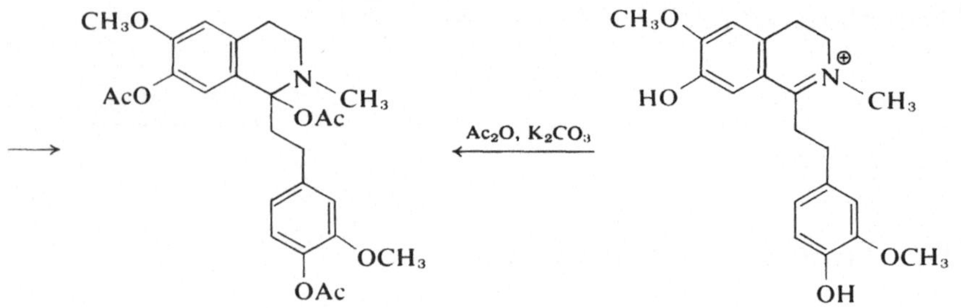

1

The dimsyl sodium induced benzyne reaction of 1'-bromo substituted phenethylisoquinolines of type **2** resulted in the formation of 2-benzazecines (**3**).[4] Ring opening, presumably by quinone methide formation, of the initially formed quaternary salt was followed by anion capture:

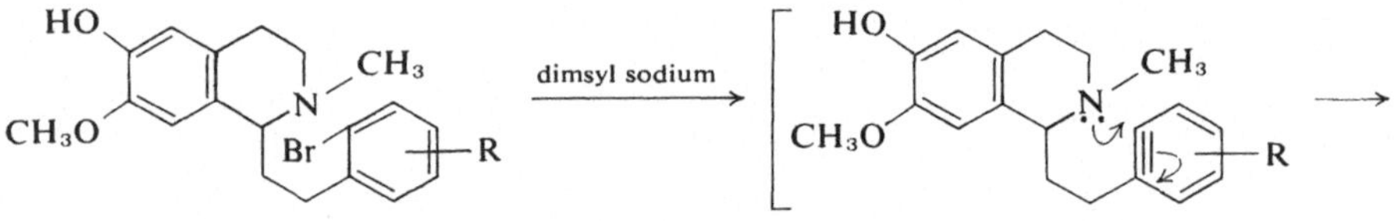

2   Benzyne intermediate

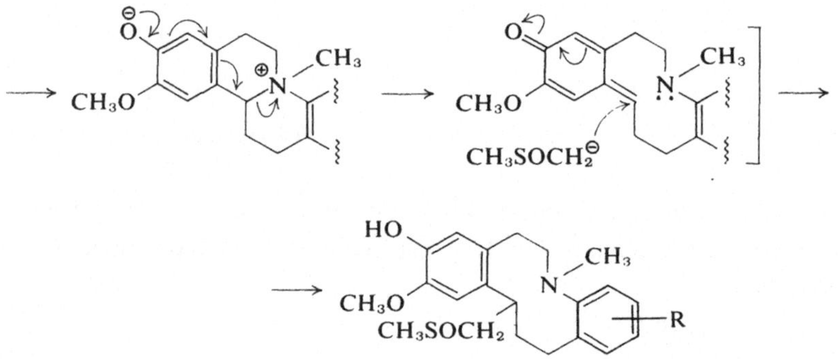

3

1-Phenethylisoquinolines on treatment with lead tetraacetate are converted to the corresponding *p*-quinol acetates which can undergo smooth halogenation at C-8 upon treatment with halogen acids.[5]

X = Cl, Br, or I

Heating nitrated phenethylisoquinolines **4** and **5** with triethylphosphite resulted in reductive cyclization to the benzimidazoisoquinoline **6** and the benzacridine **7**, respectively. Nitrene intermediates have been proposed for these transformations (see Scheme 28.1).[6]

A benzimidazoisoquinoline, **6**

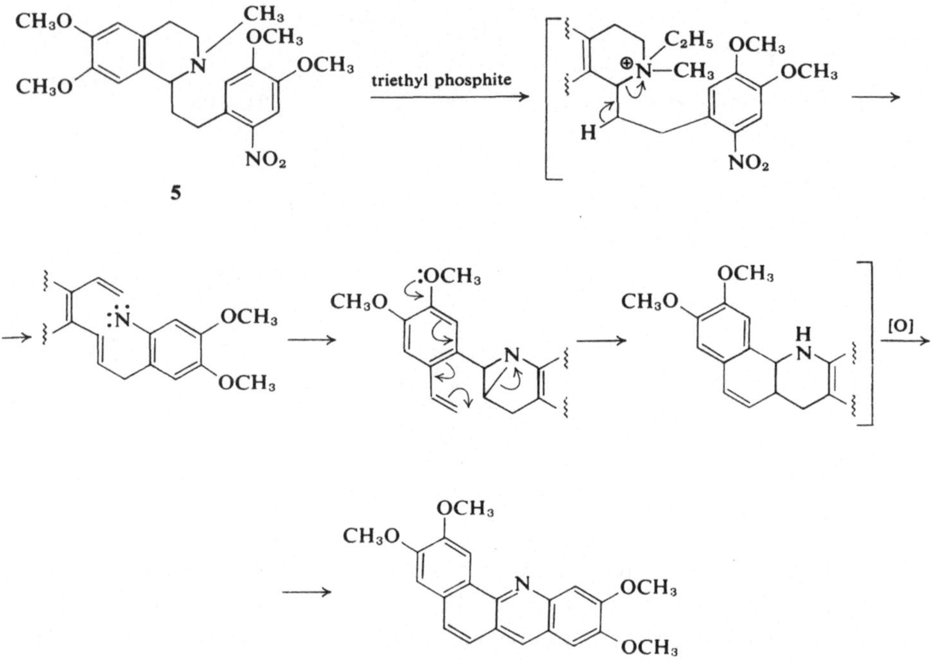

5

A benzacridine, **7**

SCHEME 28.1

The electrooxidation of 1-phenethylisoquinolines has failed so far to lead to homomorphinandienone alkaloids of the androcymbine type.[7]

The vanadium oxytrihalide oxidation of phenethylisoquinolines to neo-homoproaporphines and homoproaporphines is discussed in Sec. 29.2.2.

28.2. Pharmacology

The pharmacological properties of the phenethylisoquinolines are reviewed in Ref. 8. The most interesting of the synthetic phenylisoquinolines is metho-pholine (Versidyne, 4'-chloro-2-methyl-6,7-dimethoxytetrahydroisoquinoline), whose analgesic properties were investigated at the Hoffman–La Roche laboratories.

28.3. Yolantinine, a New Dimeric Phenethylisoquinoline Alkaloid

As mentioned above, the only known naturally occurring monomeric phenethylisoquinoline alkaloid is autumnaline.[1] There is also one fully characterized, naturally occurring bisphenylisoquinoline, namely, (−)-melanthioidine.[1] Very recently, the new alkaloid yolantinine (jolantinine) was obtained from *Merendera yolantae* E. Czerniak (Liliaceae), and was assigned the structure shown below.[9]

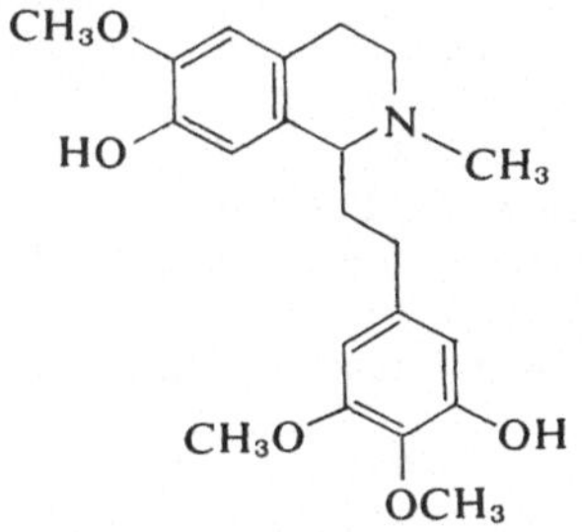

Autumnaline

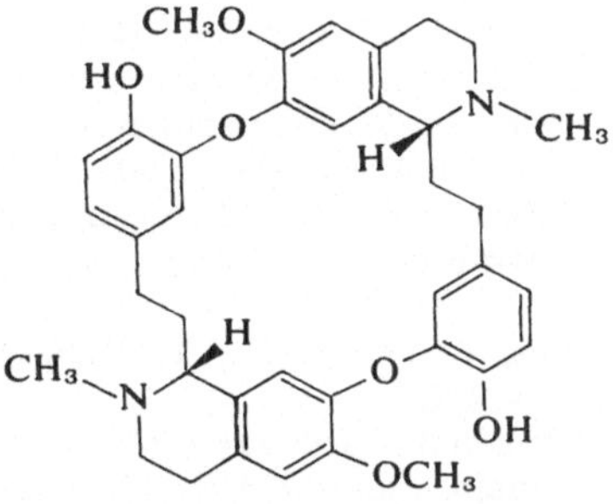

(−)-Melanthioidine

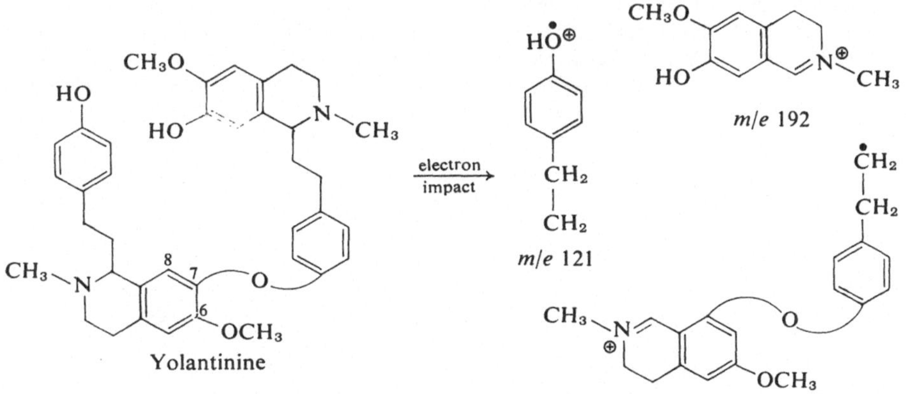

Yolantinine

m/e 192

m/e 121

m/e 295 + H• = *m/e* 296

The PMR spectrum of yolantinine exhibits 12 aromatic protons (δ6.38–6.75), two *O*-methyl singlets (δ3.72 and 3.74), and two *N*-methyl singlets (δ2.22 and 2.33). The most significant feature of the CMR spectrum is the presence of 12 substituted aromatic carbons, and 12 unsubstituted aromatic carbons. The mass spectrum shows fragmentation peaks at *m/e* 296, 192, and 121, which were assumed to possess the structures indicated.[9]

The structural assignment for yolantinine is not totally convincing since (a) The structure appears to violate the simple rules for biogenetic biaryl ether formation, and (b) the *m/e* 296 ion substituted at C-6 and C-8 cannot arise from the expression ascribed to yolantinine, which incorporates a 6,7-disubstituted tetrahydroisoquinoline moiety.

References and Note

1. For recent reviews of the phenethylisoquinolines, see T. Kametani and H. Koizumi, in *The Alkaloids*, *Vol. 14*, R. H. F. Manske, ed., Academic Press, New York (1973), p. 265; and T. Kametani and K. Fukumoto in *MTP Int. Rev. Sci., Org. Chem., Ser. One*, Vol. 9, K. Wiesner, ed., Butterworths, London (1973), pp. 181–234.
2. T. Kametani, K. Fukumoto, and M. Fujihara, *J. Org. Chem.*, **36**, 1293 (1971).
3. T. Kametani, M. Mizushima, S. Takano, and K. Fukumoto, *Tetrahedron*, **29**, 2031 (1973).
4. S. Kano, T. Ogawa, T. Yokomatsu, Y. Takahagi, E. Komiyama, and S. Shibuya, *Heterocycles*, **3**, 129 (1975).
5. H. Hara, O. Hoshino, and B. Umezawa, *Heterocycles*, **3**, 123 (1975).
6. T. Kametani, Y. Fujimoto, and M. Mizushima, *Heterocycles*, **3**, 619 (1975); and *Chem. Pharm. Bull., Tokyo*, **23**, 2025 (1975).
7. A. Najafi and M. Sainsbury, *Heterocycles*, **6**, 459 (1977).
8. A. Brossi, H. Bessendorf, L. A. Pirk, and A. Reiner, Jr., in *Medicinal Chemistry*, G. de Stevens, ed., Vol. 5, Academic Press, New York (1965), p. 281.
9. A. M. Usmanov, M. K. Yusupov, and K. A. Aslanov, *Khim. Prir. Soedin.*, 422 (1977).

THE HOMOAPORPHINES AND HOMOPROAPORPHINES

29.1. Introduction

Several homoaporphines and reduced homoproaporphines have recently been isolated from members of the Liliaceae, mostly by the Russian school. The assigned structures are reproduced below. However, in the case of the reduced proaporphines, the data available are fragmentary, and some of the stereochemical assignments still remain to be made or to be confirmed.

Bechuanine[2] and merenderine[3] are identical in all respects, and correspond to the S-(+)-enantiomer of the known homoaporphine (−)-floramultine. The reduced homoproaporphine (+)-yolantamine[1] (jolantamine, jolanthamine) is diastereomeric with (+)-bulbocodine.[7] (+)-Kesselringine[16,16a] and (+)-luteine[16] are diastereomers possessing identical stereochemistry at C-6a.

Comparative mass spectral studies of yolantamine, bulbocodine, kesselringine, and regelamine have been carried out.[3a]

Homoaporphines

(+)-O-Methylkreysigine[4]

(+)-Szovitsamine[5]

(+)-Bechuanine[2]
((+)-Merenderine)[3,6]

Szovitsinine[6a]

Szovitsine[6a]

(Continued)

Homoproporphines

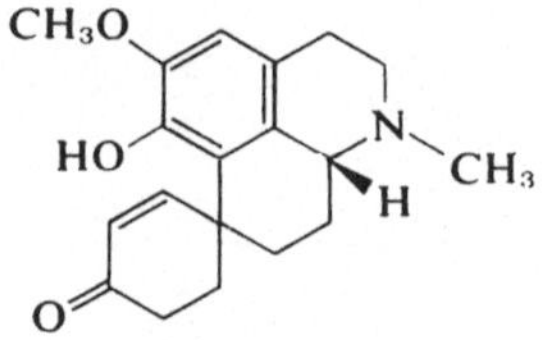

(+)-Bulbocodine[7]
(+)-Yolantamine[1]

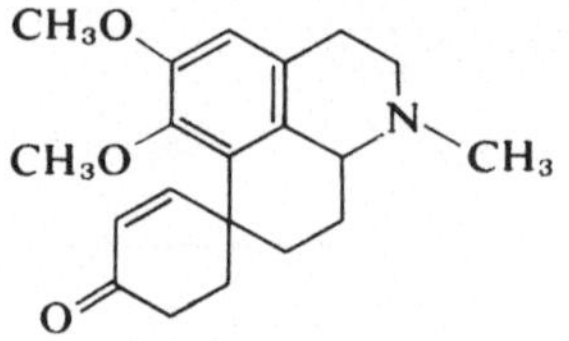

Crociflorinone[8]

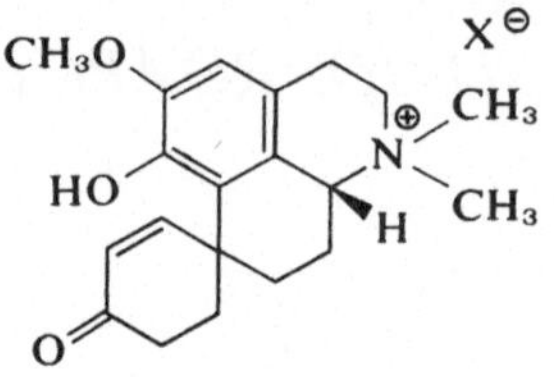

(+)-Jolantine[9]
(iolantine, jolantine)

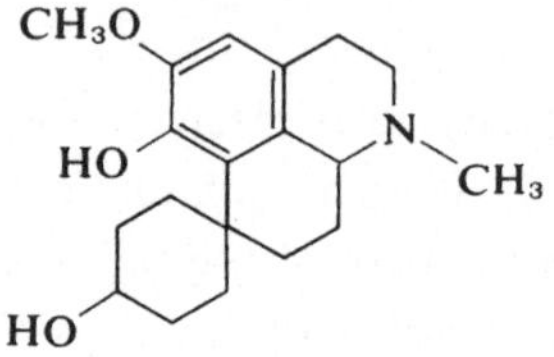

(−)-Trigamine[10]

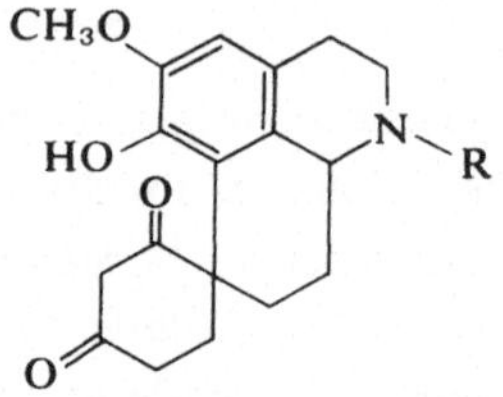

Yolantimine,[11] R = H
Luteinone,[12] R = CH₃

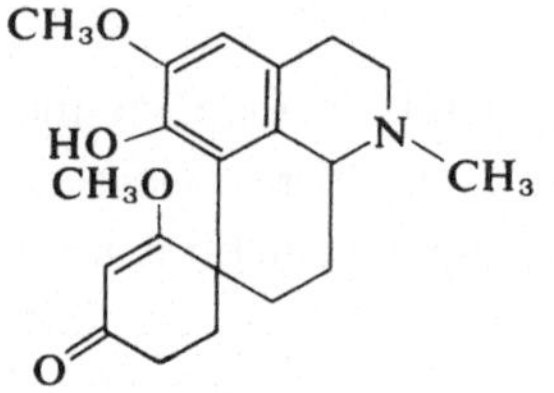

Luteidine[12]

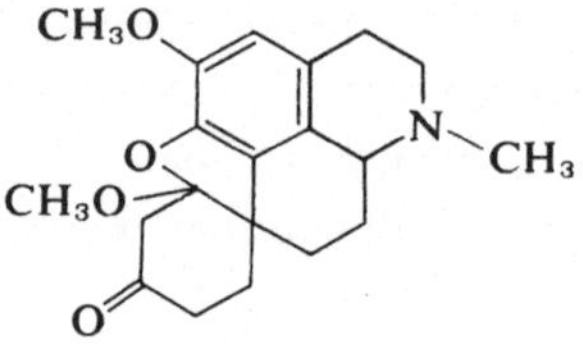

Luteicinone[13]

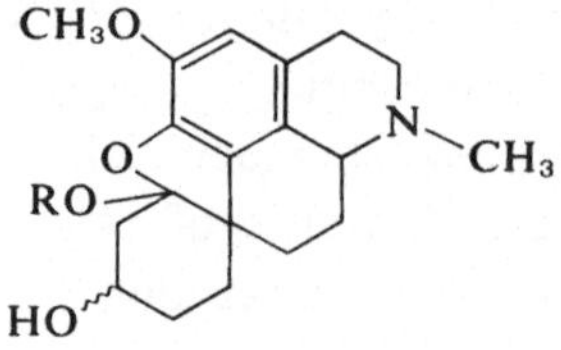

Luteicine,[13] R = CH₃
Norluteicine,[13] R = H

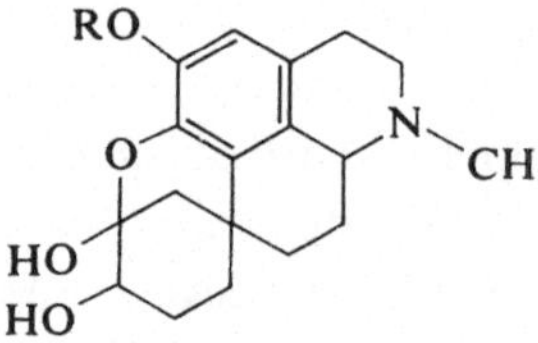

(−)-Kesselridine,[14] R = H
(+)-Regelamine,[15] R = CH₃

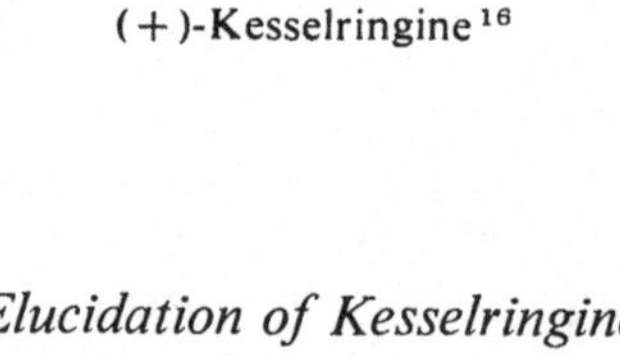

(+)-Kesselringine[16]

29.1.1. The Structural Elucidation of Kesselringine

A detailed study of the chemistry and spectroscopy of (+)-kesselringine has appeared.[16a] This alkaloid, $C_{19}H_{25}NO_4$ [λ_{max}^{EtOH} 231 and 293 nm (3.85 and

3.44)], obtained from *Colchicum kesselringi* RGL, is a monophenolic base which incorporates one *N*-methyl and one aliphatic methoxyl. The main reactions of the alkaloid are summarized below.[16a]

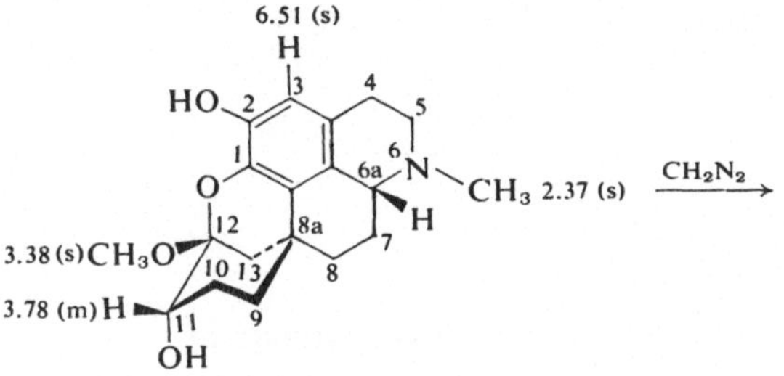

6.51 (s)

3.38 (s)CH$_3$O

3.78 (m)H

(+)-Kesselringine

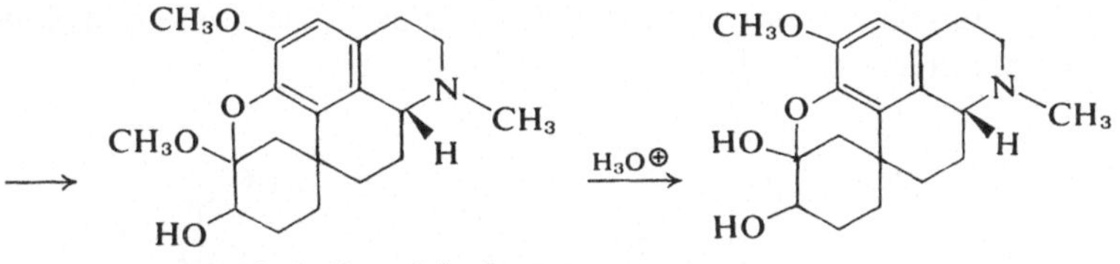

(+)-*O*-Methylkesselringine

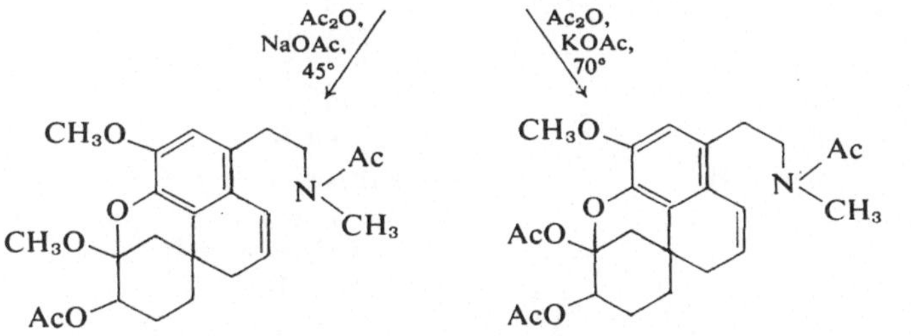

The mass spectrum of kesselringine shows the following ions, further emphasizing the fact that one is dealing with a homoproaporphine base.

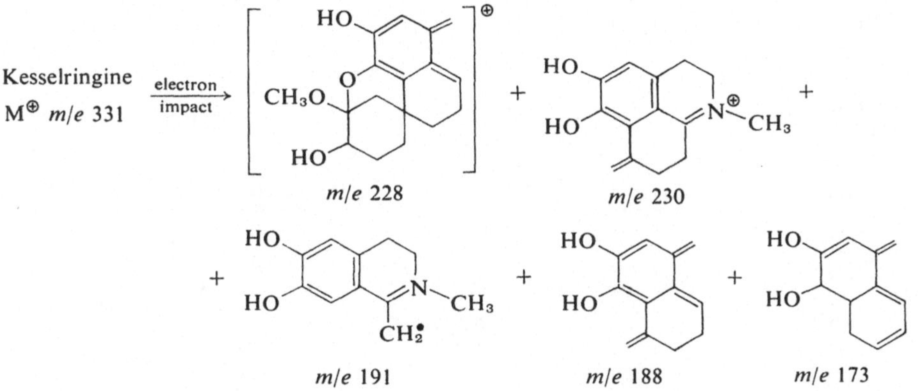

Kesselringine M$^\oplus$ *m/e* 331 $\xrightarrow[\text{impact}]{\text{electron}}$

m/e 228

m/e 230

m/e 191

m/e 188

m/e 173

The C-11 hydroxyl was tentatively assigned as being axial. Kesselringine shows negative CD Cotton effects at 250 and 300 nm, and a positive effect at 220 nm. By analogy with the reduced proaporphines, the absolute configuration of kesselringine was assumed to be as shown.[16a]

29.2. Synthesis

29.2.1. Direct Cyclization of a Tetrahydrophenethylisoquinoline

Base-catalyzed photolyses of 2'-bromo-7-hydroxytetrahydrophenethyl-isoquinolines have resulted in the synthesis of the homoaporphines kreysigine[17] and alkaloid CC-24.[18]

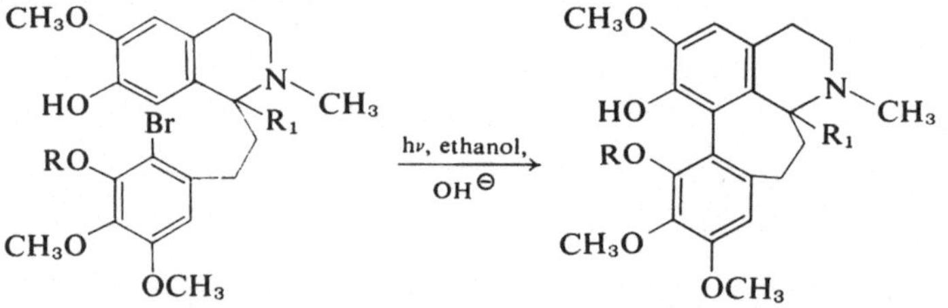

R = CH₃, R₁ = β-H
R = H, R₁ = H

(−)-Kreysigine, R = CH₃,
R₁ = β-H
(±)-CC-24, R = H,
R₁ = H (5%)

Similarly, photolysis of the diazonium salt **1**, led to the homoaporphine **2**, also obtained from acid-catalyzed dienone–phenol rearrangement of kreysigi-none.[19]

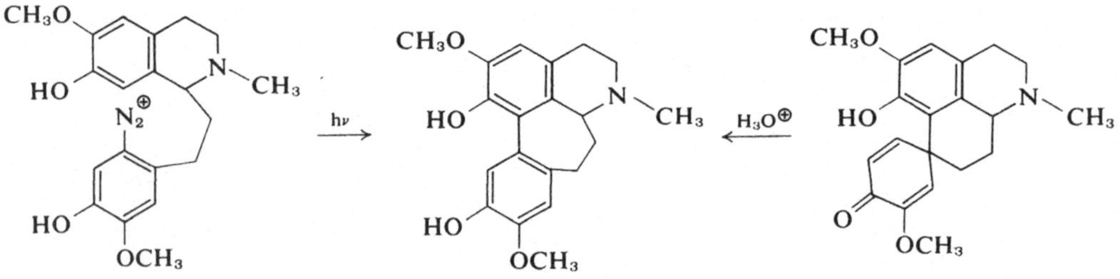

1   2   Kreysiginone

In analogy with the aporphine series (see Sec. 10.2.6), the *p*-quinol acetate **4**, generated from lead tetraacetate oxidation of **3**, was cyclized in acetic anhydride and sulfuric acid to the homoaporphine kreysigine acetate.[20]

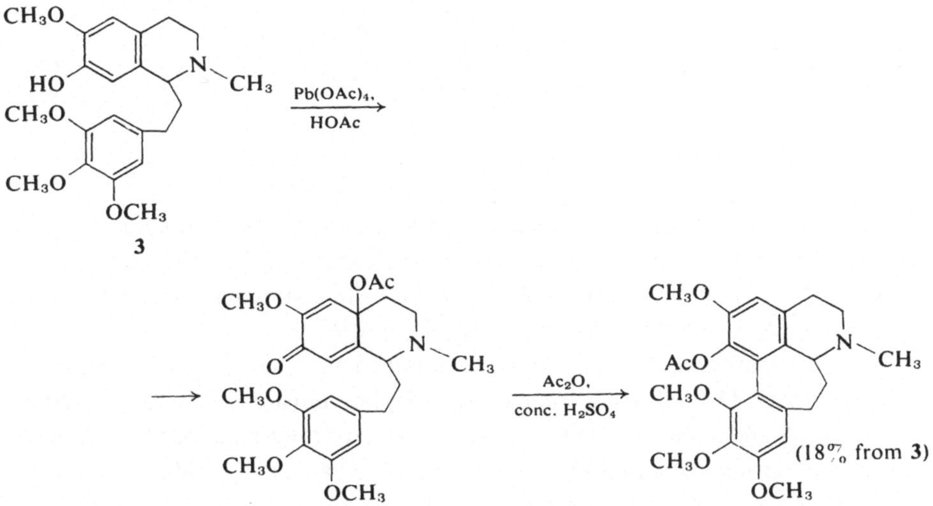

3

4

Kreysigine acetate

$(18\% \text{ from } 3)$

29.2.2. Through the Intermediacy of Dienones

The first synthesis of a homoaporphine by dienone–phenol rearrangement of a homoneoproaporphine is that of Marino and Samanen. The phenethyltetrahydroisoquinoline trifluoroacetamide **5** was oxidized with vanadium oxytrichloride to supply the dienone **6.** This dienone upon acid-catalyzed rearrangement furnished the homoaporphine trifluoroacetamide **7** in high yield.[21]

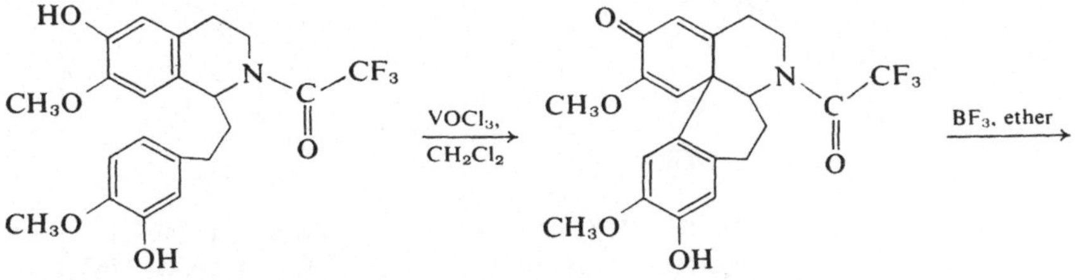

5

6 (65%)

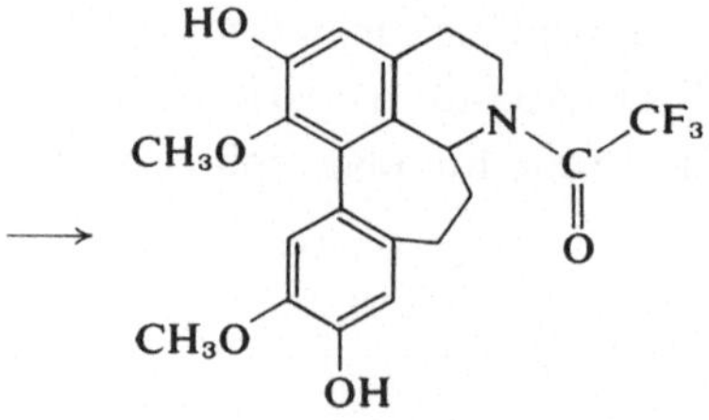

7 (65%)

In a related study, Kupchan and his group utilized vanadium oxytri-fluoride in trifluoroacetic acid to oxidize monophenolic phenethyltetrahydro-isoquinoline trifluoroacetamides **8** and **9**, and the analogous tertiary bases **14** and **15** (see Schemes 29.1 and 29.2, respectively).[22]

Homoaporphines **12** and **13** were obtained together with homoproaporphines **10** and **11**, indicating some direct coupling during the oxidation. When substrates **8** and **9** were treated with the oxidant for longer periods (30–50 min), the homoaporphines **12** and **13** were produced in 77 and 54% yields, respectively.

Interestingly, when the tertiary bases **14** and **15** were similarly oxidized, the ratio of products was the reverse of that for the aforementioned amides.

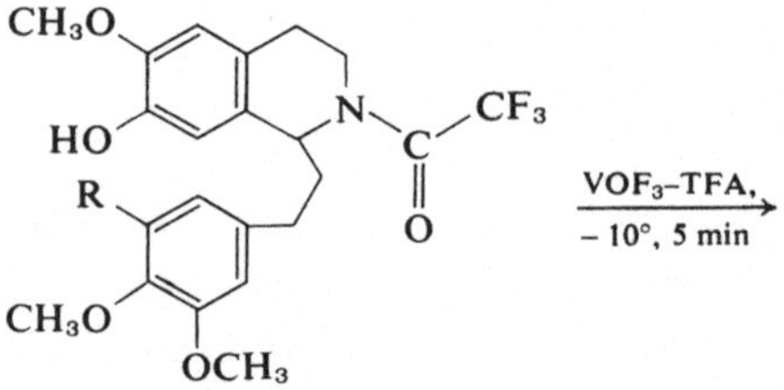

8, R = H
9, R = OCH₃

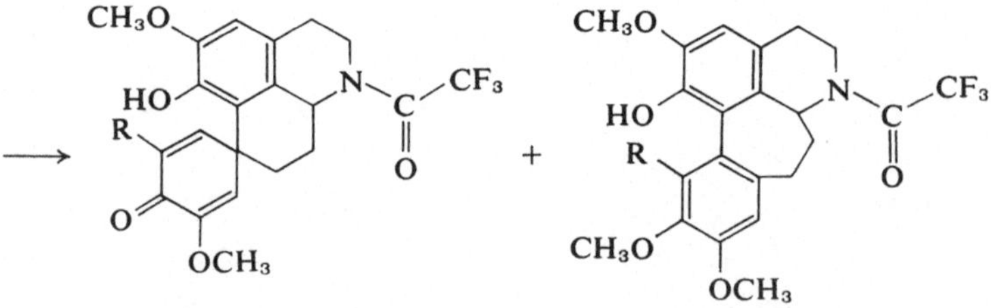

10, R = H (18%)
11, R = OCH₃ (4%)

12, R = H (40%)
13, R = OCH₃ (46%)

SCHEME 29.1

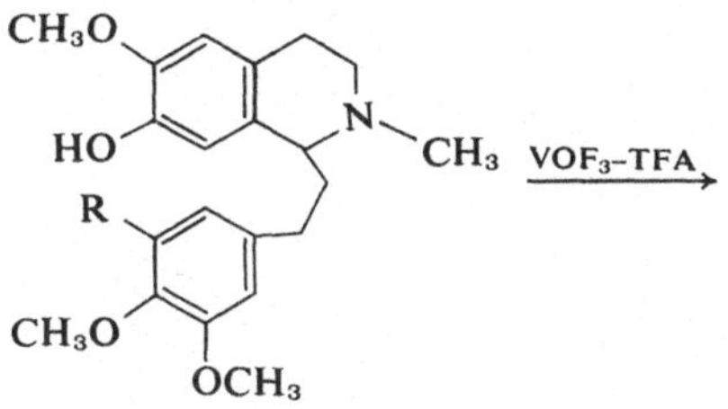

14, R = H
15, R = OCH$_3$

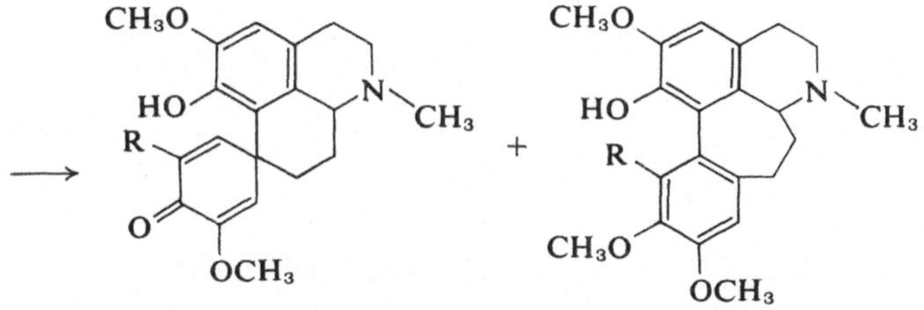

R = H (42%)   R = H (14%)
R = OCH$_3$ (54%)   R = OCH$_3$ (16%)

SCHEME 29.2

All homoproaporphines so obtained were smoothly transformed to the corresponding homoaporphines by boron trifluoride etherate in methylene chloride in 70–90% yields.[22]

Finally, the 6-hydroxy analog **16** was oxidized with vanadium oxytrifluoride in trifluoroacetic acid to supply the homoneoproaporphine **17** in an impressive 98% yield.[22]

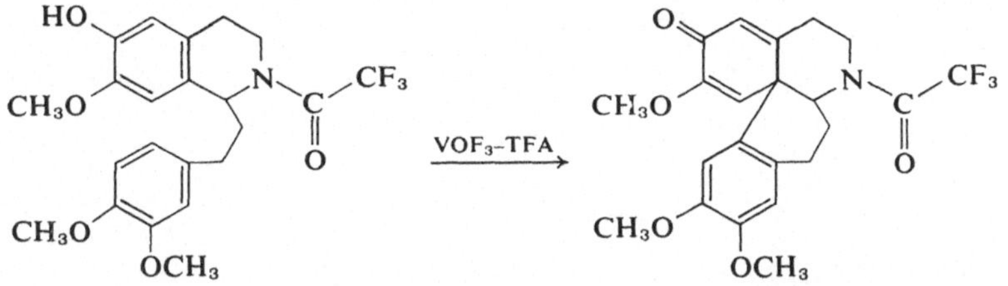

16   **17**

29.2.3. Phenolic Oxidative Coupling

Phenolic oxidative coupling on racemic homoorientaline, using cuprous chloride and oxygen in pyridine afforded kreysiginone and its diastereomer.[23]

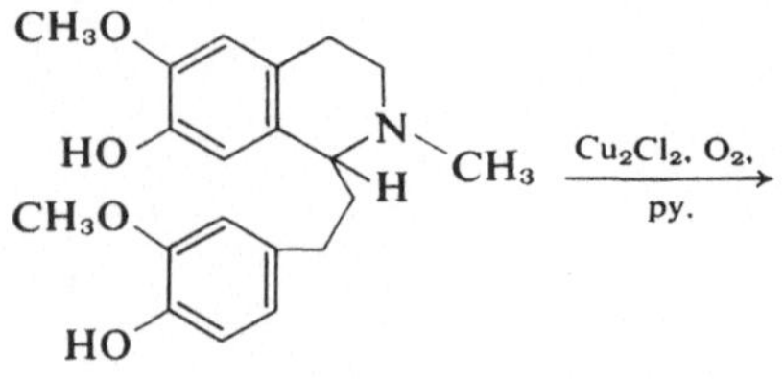

Homoorientaline

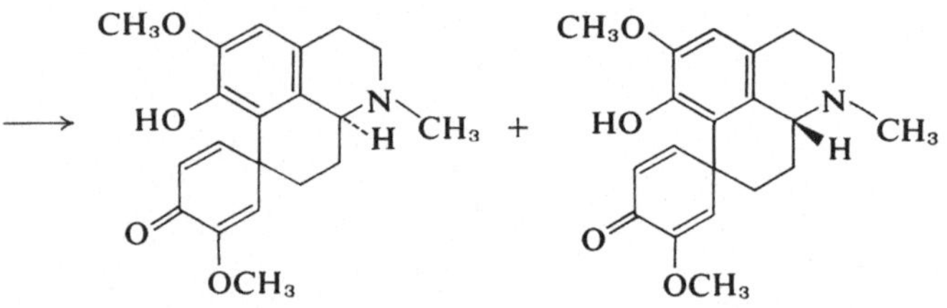

Kreysiginone (11.4%)   (26.6%)

Aqueous ferric chloride oxidation of the triphenolic phenethylisoquinoline **17a** provided a 6% yield of the homoproaporphine **17b**.[23a]

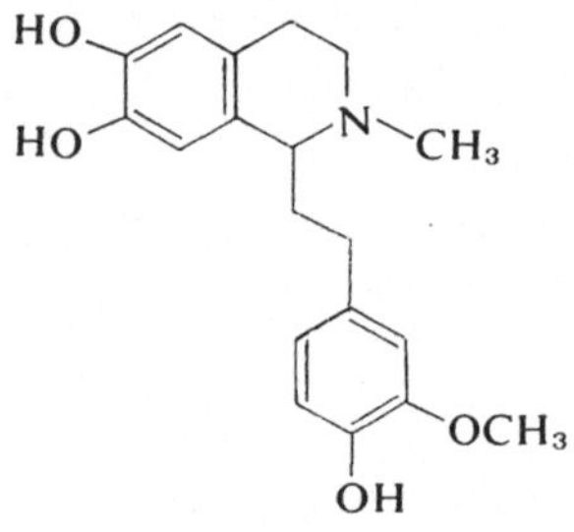

17a

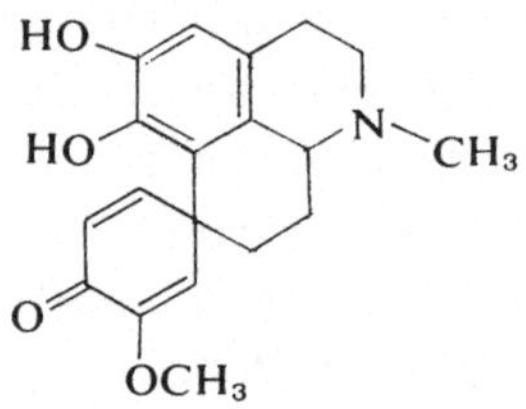

17b

29.3. Racemization of Homoproaporphines

Hydrogenation of homoproaporphines using Adams catalyst can result in racemization at C-6a.[24] Thus reduction of the homoproaporphine **18** yielded diastereomers **19** and **20**. However, when palladium was used as a catalyst, the substrate **18** provided the unepimerized ketone **21**. In a parallel experiment, compound **18** deuterated at C-6a, also furnished the nondeuterated products **19** and **20**.[25]

References and Notes

1. M. K. Yusupov, D. A. Abdullaeva, K. A. Aslanov, and A. S. Sadykov, *Dokl. Akad. Nauk SSSR*, **208**, 1123 (1973); through *Chem. Abstr.*, **79**, 79010r (1973).
2. F. Šantavý and L. Hruban, *Collect. Czech. Chem. Commun.*, **38**, 1712 (1973).
3. A. A. Trozyan, M. K. Yusupov, and K. A. Aslanov, *Khim. Prir. Soedin.*, **11**, 527 (1975); through *Chem. Abstr.*, **84**, 44477z (1976); *Chem. Natural Compounds*, 557 (1976).
3a. A. K. Kasimov, E. K. Timbekov, M. K. Yusupov, K. A. Aslanov, and A. S. Sadykov, *Izv. Akad. Nauk. Turkm. SSR, Ser. Fiz.-Tekh., Khim. Geol. Nauk*, 65 (1976); through *Chem. Abstr.*, **85**, 108840t (1976); and E. K. Timbekov, A. K. Kasimov, D. A. Abdullaeva, M. K. Yusupov, and K. A. Aslanov, *Khim. Prir. Soedin.*, **12**, 328 (1976); *Chem. Abstr.*, **86**, 121575a (1977); *Chem. Natural Compounds*, 289 (1976).
4. M. K. Yusupov, D. T. Ngo, and K. A. Aslanov, *Khim. Prir. Soedin.*, **11**, 526 (1975); through *Chem. Abstr.*, **84**, 44476y (1976); *Chem. Natural Compounds*, 555 (1976).
5. M. K. Yusupov, D. T. Ngo, K. A. Aslanov, and A. S. Sadykov, *Khim. Prir. Soedin.*, **11**, 109 (1975); through *Chem. Abstr.*, **83**, 79432n (1975).
6. M. K. Yusupov, A. A. Trozyan, K. A. Aslanov, and A. S. Sadykov, *Khim. Prir. Soedin.*, **8**, 777 (1972).
6a. A. K. Kasimov, E. K. Timbekov, M. K. Yusupov, and K. A. Aslanov, *Khim. Prir. Soedin.*, **13**, 230 (1977); through *Current Abstr. Chem.*, **66**, issue 715, item 259612 (1977).
7. F. Šantavý, P. Sedmera, G. Snatzke, and T. Reichstein, *Helv. Chim. Acta*, **54**, 1084 (1971).
8. K. Turdikulov, M. K. Yusupov, K. A. Aslanov, and A. S. Sadykov, *Khim. Prir. Soedin.*, **10**, 810 (1974).
9. K. Turdikulov, N. V. Dau, and M. K. Yusupov, *Khim. Prir. Soedin.*, **12**, 555 (1976); *Chem. Abstr.*, **86**, 121576b (1977); *Chem. Natural Compounds*, 501 (1976).

10. M. K. Yusupov, A. A. Trozyan, and K. A. Aslanov, *Khim. Prir. Soedin.*, **11**, 808 (1975); through *Chem. Abstr.*, **84**, 180441k (1976); *Chem. Natural Compounds*, 824 (1976).

11. D. A. Abdullaeva, M. K. Yusupov, A. K. Kasimov, N. V. Dau, and K. A. Aslanov, *Khim. Prir. Soedin.*, **12**, 121 (1976).

12. N. L. Mukhamedyarova, M. K. Yusupov, M. G. Levkovich, K. A. Aslanov, and A. S. Sadykov, *Khim. Prir. Soedin.*, **12**, 354 (1976); *Chem. Natural Compounds*, 308 (1976).

13. M. K. Yusupov, N. L. Mukhamedyarova, and K. A. Aslanov, *Khim. Prir. Soedin.*, **12**, 359 (1976); through *Curr. Chem. Cont. Index Chem.*, **62**, 247354 (1976); *Chem. Abstr.*, **86**, 43864c (1977).

14. A. K. Kasimov, M. K. Yusupov, E. K. Timbekov, and K. A. Aslanov, *Khim. Prir. Soedin.*, **11**, 194 (1975); through *Chem. Abstr.*, **83**, 97660q (1975); *Chem. Natural Compounds*, 202 (1976).

15. M. K. Yusupov, D. A. Abdullaeva, K. A. Aslanov, and A. S. Sadykov, *Khim. Prir. Soedin.*, **11**, 383 (1975); through *Chem. Abstr.*, **84**, 59809h (1976); *Chem. Natural Compounds*, 395 (1976); and M. K. Yusupov, D. A. Abdullaeva, F. G. Kamaev, and A. S. Sadykov, *Dokl. Akad. Nauk Uzb. SSR*, 51 (1976); through *Chem. Abstr.*, **87**, 6235y (1977).

16. M. K. Yusupov and A. S. Sadykov, *Khim. Prir. Soedin.*, **12**, 350 (1976). For the structures of the related alkaloids luteine and regeline, see N. L. Mukhamedyarova, M. K. Yusupov, M. L. Levkovich, and K. A. Aslanov, *Khim. Prir. Soedin.*, **12**, 801 (1976); and D. A. Abdullaeva, M. K. Yusupov, and K. A. Aslanov, *Khim. Prir. Soedin.*, **12**, 783 (1976), respectively; see also *Chem. Natural Compounds*, 718 (1977), and 702 (1977)

16a. M. K. Yusupov, N. L. Mukhamedyarova, A. S. Sadykov, L. Dolejš, P. Sedmera, and F. Santavý, *Collect. Czech. Chem. Commun.*, **42**, 1581 (1977). It should be mentioned in this respect that similar conclusions on the stereochemistry of the reduced proaporphines based on CD data had to be modified at a later date (see Sect. 9.2.)

17. T. Kametani, Y. Satoh, and K. Fukumoto, *J. Chem. Soc. Perkin I*, 2160 (1972).

18. T. Kametani, Y. Satoh, and K. Fukumoto, *Tetrahedron*, **29**, 2027 (1973).

19. T. Kametani, T. Nakano. C. Seino, S. Shibuya, K. Fukumoto, T. R. Govindachari, K. Nagarajan, B. R. Pai, and P. S. Subramanian, *Chem. Pharm. Bull.*, *Tokyo*, **20** 1507 (1972).

20. O. Hoshino, T. Toshioka, and B. Umezawa, *Chem. Commun.*, 740 (1972); O. Hoshino, T. Toshioka, K. Ohyama, and B. Umezawa, *Chem. Pharm. Bull.*, *Tokyo*, **22**, 1307 (1974). See also H. Hara, O. Hoshino, and B. Umezawa, *Heterocycles*, **5**, 213 (1976).

21. J. P. Marino and J. M. Samanen, *Tetrahedron Lett.*, 4553 (1973); and *J. Org. Chem.*, **41**, 179 (1976).

22. S. M. Kupchan, O. P. Dhingra, and C. K. Kim, *J. Org. Chem.*, **41**, 4049 (1976); S. M. Kupchan, O. P. Dhingra, C. K. Kim, and V. Kameswaran, *J. Org. Chem.*, **41**, 4047 (1976).

23. T. Kametani, Y. Satoh, M. Takemura, Y. Ohta, M. Ihara, and K. Fukumoto, *Heterocycles*, **5**, 175 (1976).

23a. T. Kametani, Y. Satoh, and K. Fukumoto, *Chem. Pharm. Bull.*, *Tokyo*, **25**, 922 (1977).

24. For a recent review on the epimerization and racemization of alkaloids, see T. Kametani and M. Ihara, *Heterocycles*, **5**, 649 (1976).

25. T. Kametani, K. Fukumoto, F. Satoh, and K. Kigasawa, *Heterocycles*, **3**, 921 (1975); and *J. Heterocycl. Chem.*, **13**, 29 (1976).

THE 1-PHENYLISOQUINOLINES

30.1. Introduction

Two recently isolated and optically inactive 1-phenylisoquinoline salts from *Cryptostylis erythroglossa* Hayata (Orchidaceae) are:

1,2-Dehydrocryptostyline I iodide[1]   1,2,3,4-Tetradehydrocryptostyline I iodide[1]

All known naturally occurring 1-phenylisoquinoline alkaloids have been found only in the family Orchidaceae.

30.2. Synthesis

The above two alkaloids were synthesized from the previously prepared[2] intermediate **1** in a straightforward fashion.[1]

1,2-Dehydrocryptostyline I iodide $\xleftarrow{\text{CH}_3\text{I}}$ **1** $\xrightarrow[\text{2. CH}_3\text{I}]{\text{1. Se}}$ 1,2,3,4-Tetrahydrocryptostyline I iodide

Condensation of dopamine and benzaldehyde in refluxing 1 *N* hydrochloric acid induced formation of the 1-phenyltetrahydroisoquinoline **2**; whereas for similar cyclization to the dimethoxy analog **3** from homoveratrylamine phosphoric acid was required.[3]

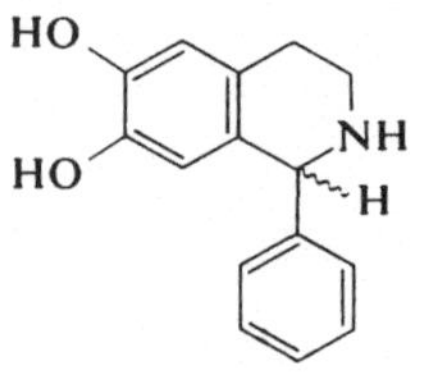

2 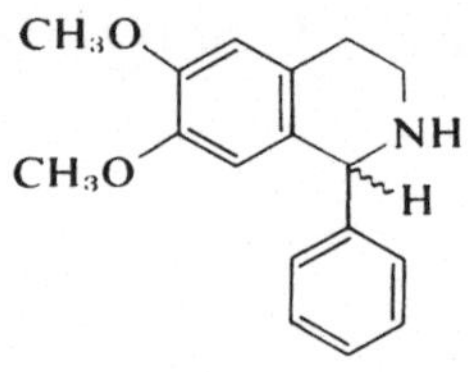   3

An interesting reaction is the Beckmann rearrangement of the β-phenyl-α,β-unsaturated ketoxime ester **4** to form the 1-phenylisoquinoline **5**. A reaction mechanism involving azacyclobutenium intermediates has been invoked[4]:

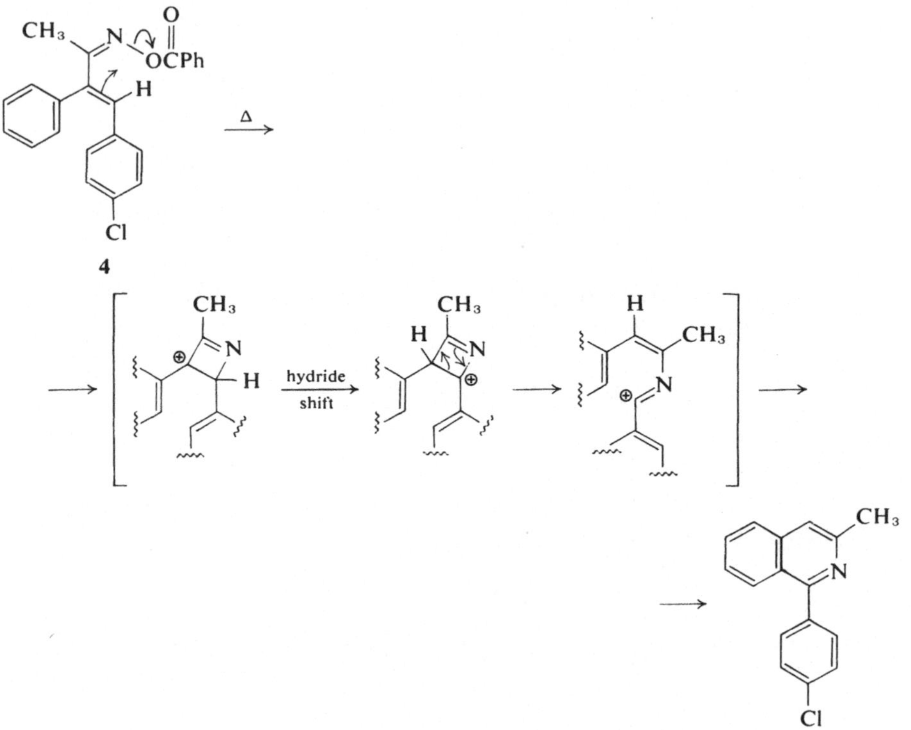

Another unusual route to 1-phenylisoquinolines is the condensation of an aromatic aldehyde such as benzaldehyde with two equivalents of an aryl-acetonitrile, e.g., phenylacetonitrile, to a 1,4-dihydro-1-phenyl-3(2H)-isoquinoline. The reaction is run in polyphosphoric acid, and proceeds through the intermediacy of a benzylidenebisphenylacetamide.[4a]

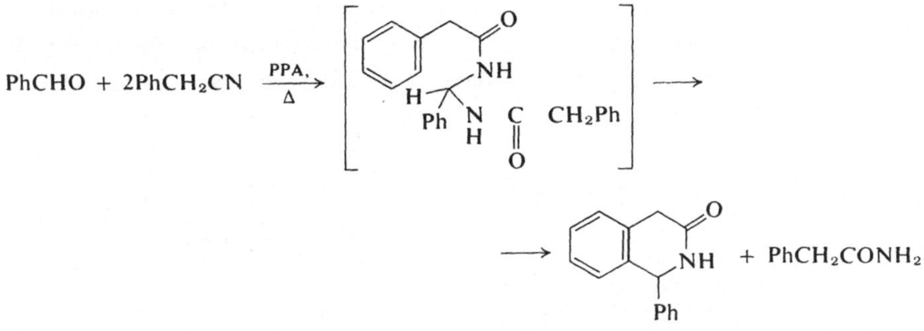

PhCHO + 2PhCH₂CN $\xrightarrow[\Delta]{\text{PPA,}}$ [...] →

→ ... NH + PhCH₂CONH₂

1,4-Dihydro-1-phenyl-
3(2H)-isoquinolinone

30.3. Stereochemistry

The full details of the X-ray crystallographic analysis of the synthetic *R*-isomer of cryptostyline II have been published along with an added proof of its stereochemistry by the aromatic chirality method.[5]

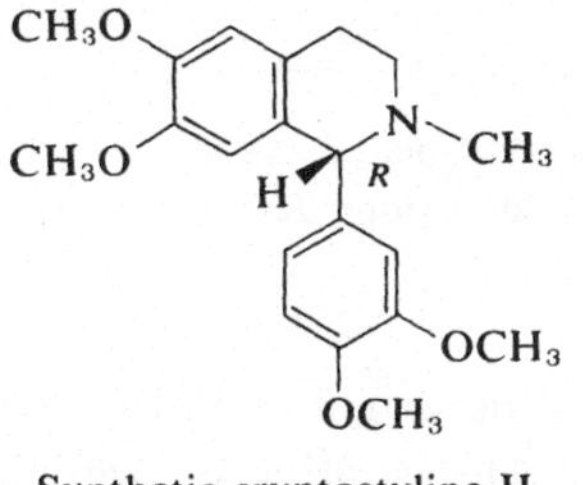

Synthetic cryptostyline II

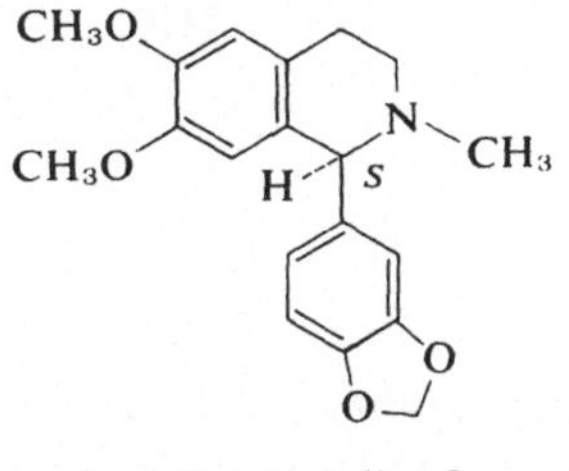

(−)-Cryptostyline I

Further proof of the *S*-configuration of the natural cryptostylines was derived from the X-ray crystallographic analysis of (−)-cryptostyline I.[6]

30.4. Biogenesis

Leander and his group have demonstrated that labeled tyrosine and dopa as well as tyramine and dopamine, are specifically incorporated in low yields into cryptostyline in *C. erythroglossa*. In addition, 3-hydroxy-4-methoxy-phenethylamine was incorporated more efficiently than the 3-methoxy-4-hydroxy analog. The latter result indicates that ring closure to the tetrahydroisoquinoline skeleton is facilitated by the *p*-hydroxy group. The location of the site of radio-activity was determined by the standard degradative methods indicated[1]:

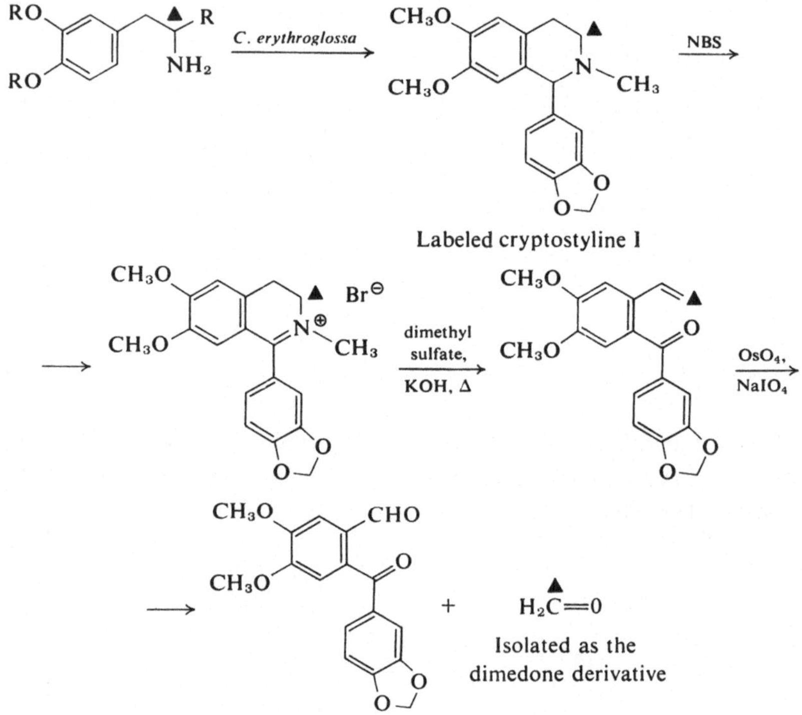

Labeled cryptostyline I

Isolated as the dimedone derivative

Dopamine has been shown to serve as the origin of both aromatic rings as well as the C-1 carbon of cryptostyline I. Degradation of isotopical crypto-styline I, obtained by feeding dopamine labeled at the β-carbon, revealed that the C-4 atom accounts for only one half of the incorporated radioactivity. The remainder was presumed to reside at C-1[7]:

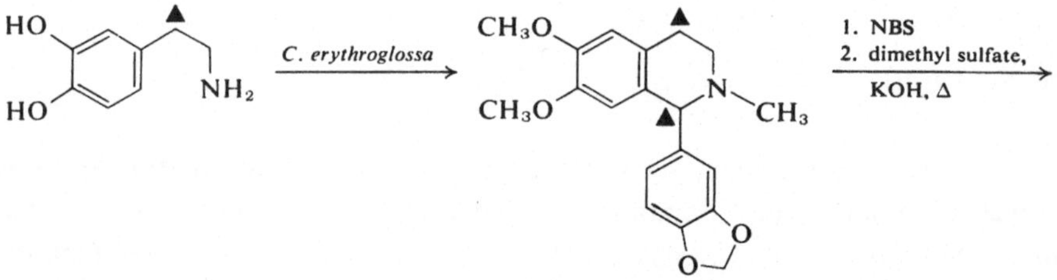

Labeled cryptostyline I

30.5. Pharmacology

A series of 1-phenyl-2-phenethyltetrahydroisoquinolines was prepared and tested in a search for antifertility agents, but the concomitant antifertility activity and estrogenicity of these derivatives could not be separated.[8]

Lactams **6** and **7** show anticonvulsant activity.[9]

6   **7**

References and Note

1. S. Agurell, I. Granelli, K. Leander, B. Luning, and J. Rosenblom, *Acta Chem. Scand.*, *Ser. B*, **28**, 239 (1974).
2. K. Leander, B. Luning, and E. Ruusa, *Acta Chem. Scand.*, **23**, 244 (1969).
3. R. Sarges, *J. Heterocycl. Chem.*, **11**, 599 (1974).
4. S. Goszczynski and M. Lozynski, *Tetrahedron Lett.*, 2355 (1972).

4a. Z. Csürös, Gy. Deák, I. Hoffmann, and A. Török-Kalmár, *Acta Chim. Acad. Sci. Hung.*, **60**, 177 (1969). For the preparation of a 1,4-dihydro-1-phenyl-3(2*H*)-isoquino-line with documented anticonvulsive activity, see Gy. Deák, K. Gáll-Istók, and L. Sterk, *Acta Chim. Acad. Sci. Hung.*, **88**, 87 (1976). See also G. Deák, K. Gáll-Istók, L. Hazai, and L. Sterk, *Synthesis*, 393 (1975). See also Ref. 9 below.

5. J. F. Blount, V. Toome, S. Teitel, and A. Brossi, *Tetrahedron*, **29**, 31 (1973).

6. K. Leander, B. Luning, and L. Westin, *Acta Chem. Scand.*, **27**, 710 (1973).

7. S. Agurell, I. Granelli, K. Leander, and J. Rosenblom, *Acta Chem. Scand., Ser. B*, **28**, 1175 (1974).

8. R. Paul, J. A. Coppola, and E. Cohen, *J. Med. Chem.*, **15**, 720 (1972).

9. G. Deák, M. Dóda, L. György, L. Hazai, and L. Sterk, *J. Med. Chem.*, **20**, 1384 (1977).

THE N-BENZYLTETRAHYDROISOQUINOLINES

No new *N*-benzyltetrahydroisoquinolines have been isolated from natural sources in recent years. However, the structure of corgoine has been confirmed by three syntheses. The first involves reduction of the *N*-benzyl quaternary salt derived from isoquinoline **1**,[1] whereas the second approach utilizes the Pictet–Spengler reaction to cyclize the substituted phenethylamine **2** (see Scheme 31.1).[2]

The third synthesis of corgoine involves heating 6-methoxy-7-hydroxy-tetrahydroisoquinoline with *p*-hydroxybenzyl alcohol to afford the alkaloid in 44% yield. The *N*-benzylation probably occurs through the intermediacy of *p*-benzoquinone methide which adds to the basic nitrogen of the isoquinoline.[3]

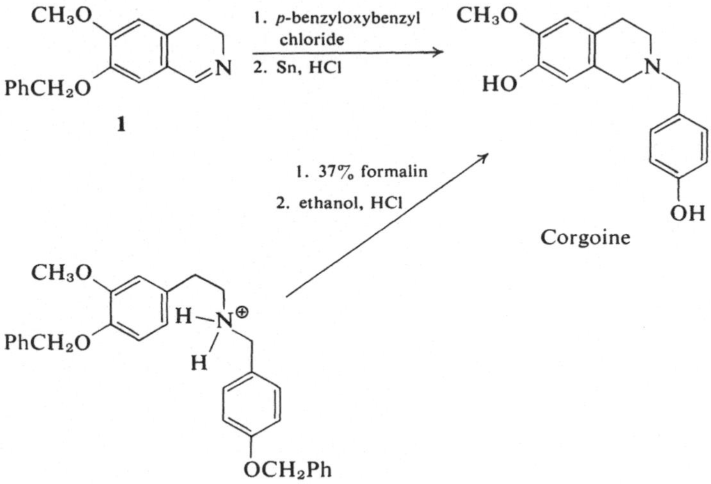

SCHEME 31.1

References

1. H. Suguna and B. R. Pai, *Indian J. Chem.*, **12**, 1141 (1974).
2. T. Kametani, K. Takahashi, C. V. Loc, and M. Hirata, *Heterocycles*, **1**, 247 (1973).
3. T. Kametani, E. Takahashi, and C. V. Loc, *Tetrahedron*, **31**, 235 (1975).

CHERYLLINE: A 4-ARYLISOQUINOLINE

Cherylline has been synthesized anew by a route which calls for a quinone methide cyclization in the final step. The methoxy bromide 2 was condensed with the benzylamine 1, and concentrated hydrochloric acid cyclized the adduct to cherylline.[1]

Cherylline (56%)

4-Phenyltetrahydroisoquinolines which do not possess oxygen activating substituents can be prepared from N-aroylephedrines such as 3. Reduction of the carbonyl group of 3 with diborane in refluxing tetrahydrofuran, followed by acid-catalyzed cyclization of the resulting N-benzylephedrine 4, supplied the required 4-phenyltetrahydroisoquinoline system.[2]

3   4 (60–90%)   (70–80%)

Substituted benzylaminoacetaldehyde diethylacetals and phenols react at room temperature in the presence of 6 N hydrochloric acid to give 4-phenyl-tetrahydroisoquinolines.[3]

References

1. T. Kametani, K. Takahashi, and C. V. Loc, *Tetrahedron*, **31**, 235 (1975).
2. D. L. Trepanier and S. Sunder, *J. Med. Chem.*, **16**, 342 (1973).
3. J. M. Bobbitt and S. Shibuya, *J. Org. Chem.*, **35**, 1181 (1970).

THE AZAFLUORANTHENES AND TROPOLOISOQUINOLINES

Azafluoranthene bases have been found to date only within the family Menisper-maceae. A recent addition to this small group of optically inactive alkaloids is norrufescine, $C_{18}H_{15}NO_4$, whose PMR spectrum is close to that of the known base rufescine. Diazomethane O-methylation of norrufescine generated rufescine. The location of the phenolic function in ring D was confirmed by a positive coupling reaction of the alkaloid with p-nitrobenzenediazonium chloride.[1]

PMR data for norrufescine, R = H
Rufescine, R = CH$_3$

Norrufescine:
λ_{max}^{EtOH} 225 sh, 248, 303, 315 sh, and 374 nm
(3.56, 3.83, 3.68, 3.36, and 2.87).
$\lambda_{max}^{EtOH-KOH}$ 230 sh, 245, 317, 382, and 495 nm
(3.64, 3.77, 3.85, 2.60, and 2.30).

Triclisine

Another new azafluoranthene is triclisine, found in *Triclisia gilletii* (Dewild) Staner.[2]

In an exciting new development, it has been shown by X-ray analysis that the orange-red imerubrine, $C_{20}H_{17}NO_5$ [λ_{max}^{EtOH} 250, 267, 295, 350, 372 sh, 394, and 450 nm (4.48, 4.52, 4.40, 4.35, 4.20, 4.11, and 3.93)], obtained from *Abuta imene* and *A. rufescens*, is the first isoquinoline alkaloid to incorporate a tropolone system.[3] A biogenetic scheme for imerubrine was proposed which proceeds from a C-7 oxygenated proaporphine. Azafluoranthenes may thus be formed in nature through the intermediacy of tropoloisoquinolines[3]:

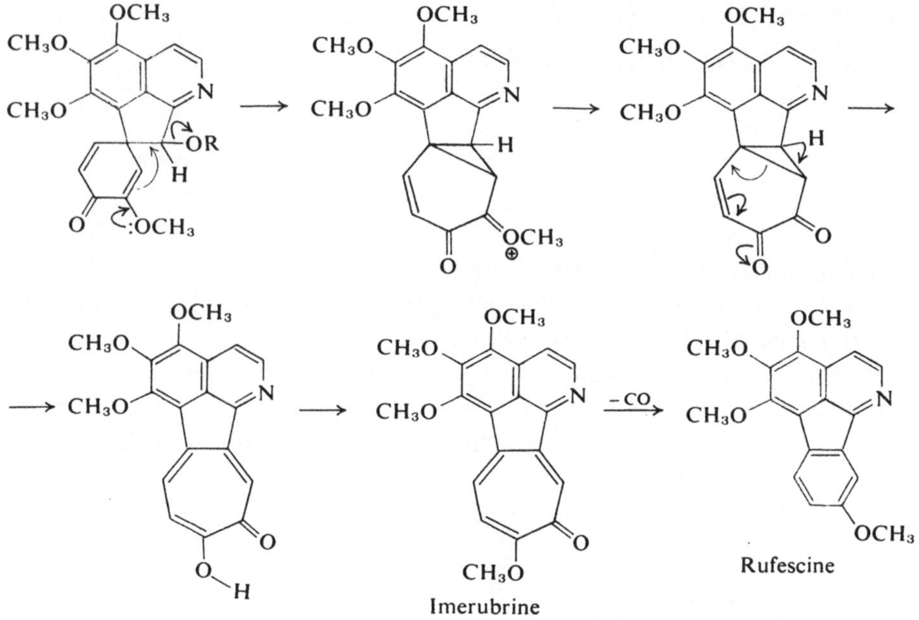

Imerubrine

Rufescine

References

1. M. P. Cava, K. T. Buck, I. Noguchi, M. Srinivasan, M. G. Rao, and A. I. DaRocha, *Tetrahedron*, **31**, 1667 (1975).
2. R. Huls, J. Gaspers, and R. Warin, *Bull. Soc.. R. Sci. Liège*, **45**, 40 (1976).
3. J. V. Silverton, C. Kabuto, K. T. Buck, and M. P. Cava, *J. Am. Chem. Soc.*, **99**, 6708 (1977).

EUPOLAURIDINE:
A 1,6-DIAZAFLUORANTHENE

Occurrence: Annonaceae and Eupomatiaceae (These two families are closely related botanically.)

Structure:

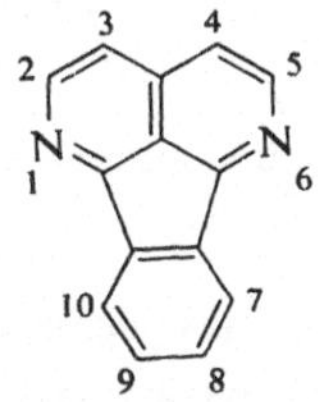

Eupolauridine

An investigation of the alkaloidal constituents of the Australian *Eupomatia laurina* R.Br. (Eupomatiaceae), yielded a number of alkaloids among which are the oxoaporphine liriodenine and the 7-hydroxynoraporphine norushinsunine. A new minor alkaloid from this source is eupolauridine, $C_{13}H_8N_2$,

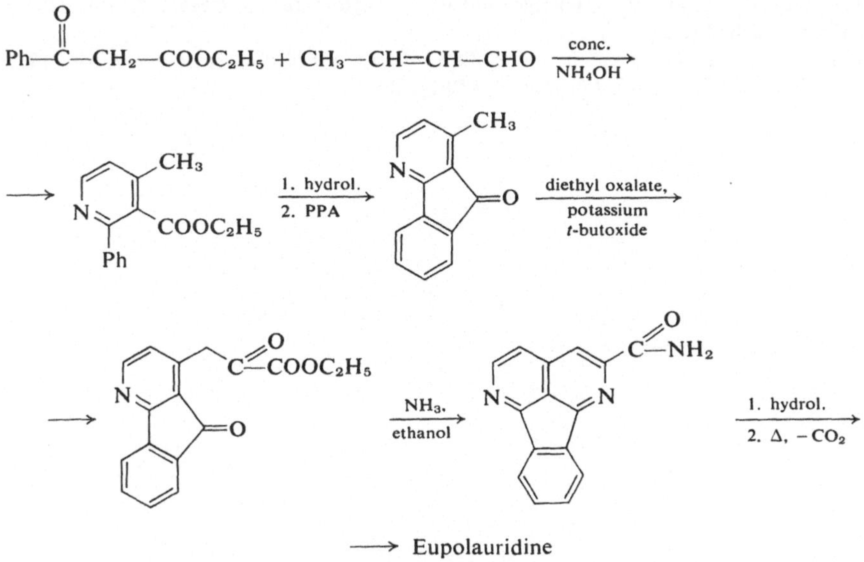

$$\text{Ph}-\overset{\overset{\displaystyle O}{\|}}{C}-CH_2-COOC_2H_5 + CH_3-CH=CH-CHO \xrightarrow[\text{NH}_4\text{OH}]{\text{conc.}}$$

⟶ Eupolauridine

SCHEME 34.1

393

λ_{max}^{EtOH} 228, 233, 278, 288, 296 sh, 335, 350, and 367 nm (4.33, 4.34, 4.23, 4.20, 3.98, 3.57, 3.81, and 3.80), whose unusual and interesting structure was proven conclusively by total synthesis (see Scheme 34.1).[1]

The same plant has also yielded the unique azaaristolactam alkaloid eupolauramine whose structure was elucidated by X-ray analysis.[2]

Eupolauramine

More recently, eupolauridine has been isolated from *Cananga odorata* (Annonaceae).[3] This alkaloid may be an aporphinoid, derived by a complex series of transformations from an aporphine or a proaporphine precursor.[1]

References

1. B. F. Bowden, K. Picker, E. Ritchie, and W. C. Taylor, *Aust. J. Chem.*, **28**, 2681 (1975); and unpublished results.
2. B. F. Bowden, H. C. Freeman, and R. D. G. Jones, *J. Chem. Soc. Perkin II*, 656 (1976).
3. M. Leboeuf and A. Cavé, *Lloydia*, **39**, 459 (1976).

AUTHOR INDEX

SUBJECT INDEX